Mass extinctions:

Processes and evidence

edited by
Stephen K. Donovan

Belhaven Press
A division of Pinter Publishers
London

© 1989, The Editor and contributors, except Chapter 3 which is copyright, the United States Department of Energy

First published in Great Britain in 1989 by
Belhaven Press (a division of Pinter Publishers),
25 Floral Street, London WC2E 9DS

Paperback edition first published in 1991

British Library Cataloguing in Publication Data

A CIP catalogue record for this book is available from the British Library

ISBN 1 85293 059 4 (HBK)
ISBN 1 85293 217 1

Typeset by Acorn Bookwork, Salisbury, Wiltshire
Printed by SRP Ltd., Exeter

CONTENTS

LIST OF ILLUSTRATIONS

LIST OF CONTRIBUTORS

Stephen K. Donovan, Department of Geology, University of the West Indies, Mona, Kingston 7, Jamaica

Antoni Hoffman, Institute of Palaeobiology, Polish Academy of Sciences, Al. Zwirki i Wigury 93, PL-02-089, Warsaw, Poland and Lamont-Doherty Geological Observatory, Columbia University, USA

Charles J. Orth, Isotope and Nuclear Chemistry Division, Los Alamos National Laboratory, Los Alamos, New Mexico 87545, USA

Martin D. Brasier, Department of Earth Sciences, University of Oxford, Parks Road, Oxford OX1 3PR, England

Stephen R. Westrop, Centre For Sedimentary Studies, Department of Geological Sciences, Brock University, St Catharines, Ontario, Canada L2S 3A1

Patrick J. Brenchley, Department of Earth Sciences, University of Liverpool, Liverpool, L69 3BX, England

George R. McGhee Jr, Department of Geological Sciences, Wright Geological Laboratory, Rutgers University, New Brunswick, New Jersey, 08903, USA

W. Desmond Maxwell, Department of Geology, The Queen's University of Belfast, Belfast, BT7 1NN, Northern Ireland

Andrew L.A. Johnson, Department of Earth Sciences, Goldsmiths' College, Rachel McMillan Building, Creek Road, London SE8 3BU, England.

Michael J. Simms, Department of Earth Sciences, University of Liverpool, Liverpool L69 3BX, England

Garland P. Upchurch Jr, University of Colorado Museum and National

Center For Atmospheric Research, P.O. Box 3000, Boulder, Colorado, 80307-3000, USA

Donald R. Prothero, Department of Geology, Occidental College, Los Angeles, California, 90041, USA

Anthony D. Barnosky, Section of Vertebrate Fossils, Carnegie Museum of Natural History, 4400 Forbes Avenue, Pittsburgh, Pennsylvania, 15213, USA

INTRODUCTION

Stephen K. Donovan

The study of any evolutionary system involves the recognition of two components—pattern and process. Before the present decade, little thought had been given to the possibility that mass extinctions might show some regular cyclic pattern, and suggestions of driving mechanisms for major extinction events varied from the plausible, such as changes in climate and/or sea level, to the wildly speculative. However, the study of mass extinctions has changed radically since Alvarez *et al.* (1980) published a paper in *Science* which provided geochemical evidence for impact by an extraterrestrial object as the driving mechanism for the Cretaceous–Tertiary (K–T) event. Subsequent studies have produced a wealth of observational and experimental data to support this hypothesis, which is now considered to be plausible by many, if not most, workers concerned with this event (Kerr, 1988; Upchurch, Chapter 10, this volume).

With the recognition of a plausible, if unexpected, driving mechanism for the K–T mass extinction, geochemical analyses were carried out at other extinction horizons to test if all such events had a common cause (see Orth, Chapter 3, this volume). This tentative speculation received support from an analysis by Raup and Sepkoski (1984), which suggested that mass extinctions in the marine realm have occurred with a periodicity of approximately 26 million years since the end of the Permian. No terrestrial driving mechanism is known that could produce such a pattern, so it was postulated that any control on the frequency of mass extinctions must be extraterrestrial. However, most other mass extinction events are not marked by an enrichment of the platinum group elements that first enabled recognition of an extraterrestrial influence at the K–T boundary (see Orth, Chapter 3, this volume) and recent statistical analyses strongly suggest that the apparent periodicity of extinctions is an artefact (Stigler and Wagner, 1987; 1988). It now remains to be explained why the K–T impact caused a mass extinction, while other major bolide impacts did not.

The true significance of the papers by Alvarez *et al.* (1980) and Raup and Sepkoski (1984) is, however, much greater than merely providing a potential process and possible pattern, respectively, for mass extinctions. They have provided the stimulus for the expansion of extinction studies from a scientific backwater to a growth area of evolutionary research. The rapidity of research since 1980 has outstripped the speed of printing and, consequently, a detailed review of major extinction phenomena does not exist. While many texts are now available on mass extinctions, these are either volumes arising from conferences, whose content is thus determined by the nature of the papers presented at the meeting, or single author volumes giving an essentially personal view. Neither of these approaches is entirely satisfactory as the basis of a primary reference work on mass extinctions.

Recognition of this deficiency has led to the present volume, which

Table 0.1. *Principal extinctions and probable causes, based on conclusions reached in Chapters 4–12 of this volume.*

Extinction event	Probable cause
Late Pleistocene	Post-glacial warming plus predation by man.
Eocene to Oligocene	Stepwise extinction associated with severe cooling, glaciation and changes of oceanographic circulation, driven by the development of the circum-Antarctic current.
End Cretaceous	Bolide impact producing catastrophic environmental disturbance.
Late Triassic	Possibly related to increased rainfall with implied regression.
End Permian	Gradual reduction in diversity produced by sustained period of refrigeration, associated with widespread regression and reduction in area of warm, shallow seas.
End Frasnian	Global cooling associated with (causing?) widespread anoxia of epeiric seas.
Late Ordovician	Controlled by the growth and decay of the Gondwanan ice sheet following a sustained period of environmental stability associated with high sea level.
Late Cambrian	Habitat reduction, probably in response to a rise in sea level, producing a reduction in number of component communities.
Late Precambrian	Complex, including widespread regression, physical stress (restricted circulation and oxygen deficiency) and biological stress (increased predation, scavenging and bioturbation).

brings together reviews of the principal extinctions identified in the fossil record. Five mass extinctions are recognised as having been the greatest crises in the history of life during the Phanerozoic (Raup and Sepkoski, 1982): end Ordovician; Frasnian–Famennian boundary; end Permian; late Triassic; and end Cretaceous. In addition, four further events are considered in this volume, each chosen for its unique contribution to our understanding of extinction phenomena. The late Precambrian event was the first extinction, following the genesis of the metazoans; the late Cambrian extinctions, best recorded by the trilobites, were geologically rapid and are known from successions that have received detailed microstratigraphic study over a broad area; the Eocene–Oligocene extinctions were the last large-scale events to have affected organic diversity in both the terrestrial and the marine environments; and the late Pleistocene extinctions of terrestrial tetrapods (essentially a taxon, rather than mass, extinction), having occurred within the range of radiocarbon dating, are known to a degree of detail impossible for earlier events.

The first three chapters of *Mass Extinctions: Processes and Evidence* provide a background to modern extinction studies, explaining the historical background to the science and outlining the criteria used in recognizing extinction phenomena. Each of the succeeding nine chapters considers a particular extinction event; taken together, they indicate an important truth in extinction studies, that each mass extinction is typified by a unique group of chemical, physical and biological environmental conditions (Table 0.1). It is therefore apparent that the driving mechanisms of mass extinction are not a particular combination of processes which have occurred sporadically (or periodically) through geological time. Rather, life on Earth is a delicate system which is constantly readjusting itself to environmental fluctuations. These fluctuations are usually small and local, but on rare occasions they reach global proportions. It is such global perturbations of the environment, which have multiple causes and effects, which are the true driving mechanisms of mass extinction. It is only by considering all of the available evidence that a valid conclusion can be reached concerning the cause of any extinction event.

This book would never have been started if Iain Stevenson of Belhaven Press had not asked me if I saw any obvious gaps in the literature of mass extinctions. My answer can be deduced from the structure of this book. Many thanks, Iain, for your help and encouragement. I also thank my fellow authors, who caught my enthusiasm for the project and have endured entreaties by letter, cable and telephone as I sought to meet our production deadline. The Department of Geology at the University of the West Indies provided invaluable logistic support, particularly paying for my endless stream of mail. I must also acknowledge the contribution of Hurricane Gilbert, which struck Jamaica while I was editing the first chapter and caused sufficient destruction that the next six chapters were edited by candlelight.

REFERENCES

Alvarez, L.W., Alvarez, W., Asaro, F. and Michel, H.V., 1980, Extraterrestrial cause for the Cretaceous–Tertiary extinction, *Science*, **208** (4448): 1095–1108.

Kerr, R.A., 1988, Huge impact is favored K–T boundary killer, *Science*, **242** (4880): 865–7.

Raup, D.M. and Sepkoski, J.J., Jr, 1982, Mass extinctions in the marine fossil record, *Science*, **215** (4539): 1501–3.

Raup, D.M. and Sepkoski, J.J., Jr, 1984, Periodicity of extinctions in the geologic past, *Proceedings of the National Academy of Science U.S.A.*, **81** (3): 801–5.

Stigler, S.M. and Wagner, M.J., 1987, A substantial bias in nonparametric tests for periodicity in geophysical data, *Science*, **238** (4829): 940–5.

Stigler, S.M. and Wagner, M.J., 1988, Response to Raup and Sepkoski, *Science*, **241** (4861): 96–9.

Chapter 1

CHANGING PALAEONTOLOGICAL VIEWS ON MASS EXTINCTION PHENOMENA

Antoni Hoffman

A HISTORICAL SKETCH

Even a quick perusal of the *Nature* and *Science* magazines over the last decade or two clearly indicates that the problem of mass extinctions—that is to say, the search for an adequate causal explanation of the disappearance of a large number of fossil groups within relatively short periods of geological time at such stratigraphic intervals as the Permian–Triassic, Cretaceous–Tertiary, Ordovician–Silurian, and Frasnian–Famennian transitions—has only recently become an important topic in the earth and life sciences. The history of research in this area is nevertheless quite long and rather complex.

The ancient Greek concept of plenitude, or fullness of the natural world, implies that no organisms that ever existed on the Earth could ultimately disappear from its surface, because their final extinction would leave an unbridgeable gap in the Great Chain of Being. Species extinction, therefore, was for long considered impossible. Eighteenth-century naturalists knew, of course, that a wide variety of fossils had no counterparts among living organisms, but this apparent anomaly was commonly explained by an as yet inadequate knowledge of life on the Earth. It appeared perfectly reasonable to assume at the time that ammonites or trilobites could still be discovered alive somewhere in the world ocean, as *Neopilina* molluscs and *Latimeria* fishes indeed have been in the twentieth century. It is only after Cuvier (1799) had first described fossil proboscideans that species extinction was established as a fact, for it was hard to believe that such large and prominent mammals could roam the Earth without being ever observed by travellers. Yet the concept of plenitude could still be reconciled with the fact of species extinction by assuming—as in fact Lamarck (1809) did—that man is the sole agent responsible for species extinction. According to Lamarck, species are essentially immortal because they are always able to adapt perfectly to any environmental configuration and man alone is capable of violating the order of the natural world. Under this assumption it was fully understandable that the species that underwent extinction were such large animals as the woolly mammoth, which could be easily considered as a valuable prey for human hunters.

With rapidly growing knowledge of the inventory of living animal species, however, and also with rapidly increasing awareness that fossils such as ammonites or trilobites, without known living counterparts, were by no means oddities but rather very common phenomena, it became undeniable in the early nineteenth century that species extinction did indeed often take place in the geological past. Within a few decades three main causal explanations for this phenomenon were put forth. Brocchi (1814) proposed that each species was created with a specified, predetermined life-span, in close analogy to individual longevity. He believed that the extinction of species simply marks the end of their lifetime and is therefore essentially independent of extrinsic, environmental factors—just as death will inevitably end the life of each person, regardless of fatal accidents or disease. Lyell (1832), in turn, regarded species extinction as a fully natural process caused primarily by the inescapable effects of continuously changing environmental conditions. In his view, each species must sooner or later encounter such a hostile configuration of environmental factors that it will become extinct, simply because its individuals will be unable to cope successfully with this new environmental context.

Both these concepts implied that species extinctions should generally occur independent of one another, without a clear-cut, orderly pattern in time. Cuvier (1825) observed, however, that fossil species in the Paris Basin disappear from the record in large clusters at certain horizons—all, or at least a great many of them, at once, as if wiped out by a single catastrophic agent at each horizon. He therefore interpreted species extinctions as due to local catastrophes which destroyed life entirely in an area, thus vacating the ecological space necessary for the area's repopulation by immigrants that survived the cataclysm elsewhere.

Cuvier's perspective on species extinction was developed into a more extreme position by Buckland (1823), who regarded the faunal breaks such as those described by Cuvier as a clear sign of world-wide catastrophes, and particularly by D'Orbigny (1852), who established a whole time series of such global holocausts as providing the main reference points for stratigraphy and the history of life. This emphasis on environmental catastrophes as the main causes of species extinction could, of course, be contested or even rejected outright, but even the most devoted followers of the Lyellian gradualism had to agree that some major events did indeed occur in life's physical environment on the Earth and must have had some impact on the fate of organic species. Darwin (1846), for example, noted that many large terrestrial mammals had lived until recently in South America and he attributed their extinction to the effects of rapid climatic changes during the ice age.

In *The Origin of Species* Darwin (1859) wrote very little of species extinction which he regarded, very much in the same vein as Lyell, as due to various environmental factors which drive species first to rarity and then to extinction. He was inclined to emphasize the role of biotic factors, foremost among them interspecific competition, much more strongly than Lyell did, but he was also explicit in his scepticism about our insight into the nature of extinction of any particular species. In spite of his gradualistic prejudice, Darwin was ready to admit that some extinctions of large taxonomic groups—for instance, the trilobites at the end of the Palaeozoic and the ammonites at the close of the Mesozoic—were 'wonderfully sudden'. Contrary to the Cuvierian tradition, however, which viewed these phenomena as an incontestable indication of mass extinction events when many organic groups had been exterminated simultaneously, Darwin suggested that their apparent rapidity might really be only artefactual, due to world-wide gaps in the fossil record which clumped together events that had actually been quite widely stretched over geological time.

This view was generally shared by the most orthodox Darwinians among palaeontologists—for example, Neumayr (1889), Andrusov (1891) and Davitashvili (1969)—who conceived of species extinction as caused primarily by interspecific competition, various physical factors, and regional environmental catastrophes, whereas they regarded the so-called mass extinctions as nothing but artefacts of the fossil record. In turn, the followers of Brocchi—for example, Beurlen (1933), Zunini (1933) and especially Schindewolf (1950)—tended to interpret the extinction of individual species in terms of their specific life cycle. Though they certainly did not deny the occurrence of regional environmental catastrophes and their implications for species extinction, they largely agreed with the Darwinians that mass extinction phenomena are wildly exaggerated by the notorious imperfection of the fossil record.

The idea of the specific life cycle was later rejected by modern evolutionary biology. The interpretation of mass extinctions—in particular the Palaeozoic–Mesozoic and Mesozoic–Cenozoic ones—as possible artefacts of gaps in the fossil record, however, has become one of the two rival perspectives on these phenomena. The other has, of course, envisaged mass extinctions as caused by real environmental catastrophes and it has been based on a more literal reading of the fossil record. The contrast between these two interpretations has largely shaped the history of debate about mass extinctions, but it has always hinged on divergent views on the precision and reliability of stratigraphic correlation among very distant areas.

For so long as the fossil record was only poorly known, Darwin's gradualist perspective appealed to the majority of geologists and palaeontologists. The subsequent progress in stratigraphy, however, has led in the twentieth century to the idea of mass extinction processes operating over the entire Earth instead of being confined to single geographic regions. Marshall (1928) and Hennig (1932), for example, regarded the Cretaceous–Tertiary boundary extinctions as a pronounced and abrupt catastrophe of global dimensions and explained them by invoking a sudden wave of cosmic radiation as the appropriately catastrophic causal factor. The stratigraphic record, however, still could be interpreted both ways and hence several workers—for example, Pavlova (1924) and Sobolev (1928)—considered the same extinctions as extended over millions of years; therefore, they referred to more mundane causal processes, such as the various geographic, climatic and biotic effects of diastrophic cycle.

This divergence in palaeontological interpretations of the completeness of the record at various stratigraphic boundaries, and consequently of the rapidity and nature of mass extinction phenomena, has persisted until today. On the one hand, several prominent geologists and palaeontologists accepted the catastrophic nature of mass extinctions; in fact, even Schindewolf (1954) and Beurlen (1956) became at some point convinced that these phenomena could not be reasonably accounted for by an accidental clustering of the ends of the life cycle in a great number of species. As a result, ever more imaginative narrative scenarios were presented which invoked a very wide variety of terrestrial and extraterrestrial causal factors as the ultimate culprits of extinction. Thus, Schindewolf (1954), Krasovskiy and Shklovskiy (1957), Liniger (1961), and Russell and Tucker (1971) proposed that mass extinctions are caused by waves of cosmic radiation produced by supernova explosions in relative proximity to our planetary system. Dyssa *et al.* (1960) claimed that intense volcanism could cause earthly radioactivity to exceed lethal levels and thus lead to extinctions. De Laubenfels (1956) invoked a bolide impact on the Earth at the Cretaceous–Tertiary boundary to explain the demise of the dinosaur. McLaren (1970) put forth the hypothesis that a whole suite of environmental consequences of a bolide impact in the ocean could best account for the pattern of Frasnian–Famennian extinctions. Hays (1971) suggested that mass extinctions could be causally related to geomagnetic field reversals. Beurlen (1956) thought that a dramatic and virtually instantaneous change in seawater chemistry had poisoned and thus exterminated the majority of marine organisms at the end of the Palaeozoic. Gartner and Keany (1978) claimed that a spillover of cold brackish water from the previously isolated Arctic Basin to the world ocean had happened at the Cretaceous–Tertiary transition and caused mass mortality among the pelagic plankton; according to McLean (1978), this event should have led to a greenhouse effect due to carbon dioxide buildup in the atmosphere, and hence to further extinctions by a dramatic climatic change. There were no limits to imagination, simply because there was no hard empirical evidence either firmly to corroborate or seriously to contradict any of these stories. Schindewolf (1954), indeed, explicitly wrote that his hypothesis was merely a 'desperate move' to explain the mystery he could not resolve in a more scientific way.

On the other hand, the advocates of more gradualistic interpretations of mass extinction phenomena could at least base their explanatory scenarios on the apparent coincidence between these large-scale biotic changes and various major geological processes. Following from such evidence, global marine regressions were often regarded as the prime causes of mass extinctions (Lichkov, 1945; Newell, 1967), even though a clear mechanism of species extinction by regression was not identified. It is only after the advent of the theory of island biogeography that Schopf (1974) could causally interpret the striking correlation he observed between the pace of Permian–Triassic extinctions in the sea and the decline in the total area of continental shelves inhabitable by shallow-water marine organisms of the time. The island biogeographic theory predicts that the smaller the available area, the smaller the number of species it can harbour; hence, the origin of the single supercontinent at the end of the Palaeozoic should lead to numerous extinctions. Given this mechanism of extinction, however, mass extinctions should be very protracted in time and they should also non-selectively affect all the organic groups present in the ocean; and Schopf was very emphatic on this point. Other authors suggested other extinction mechanisms triggered by marine regressions. For instance, Johnson (1974) suggested that the Frasnian–Famennian extinction of reef biotas had been caused by their particularly vulnerable palaeogeographic position on broad shelves where even a minor regression must have led to major environ-mental changes. Hallam (1983), in turn, invoked a wide range of environ-mental consequences of global sea-level fluctuations—from climatic to oceanographic to purely biogeographic ones.

Global marine regressions, however, are of course not the only sort of large-scale environmental phenomenon that could be correlated with mass extinctions. Many geologists and palaeontologists have always regarded climatic change, and particularly cooling or even glaciation, as the main cause of mass extinctions. The empirical evidence in support of this view is best presented by Stanley (1987; 1988). The trouble is that, given the protracted duration of mass extinction phenomena—as required by this kind of explanatory scenario—and given also the apparently great potential of various marine organisms to adapt rapidly to changes in seawater temperature, a climatic cooling hardly seems to be a sufficient cause for mass extinction; except, that is, for such unusual palaeogeographic situa-tions as the West Atlantic or the Mediterranean in the Neogene.

As clearly shown by Jablonski (1986), however, in his comparative analysis of the phenomena traditionally recognized for the main mass extinctions in the history of life on the Earth—that is, the Ordovician–Silurian, Frasnian–Famennian, Permian–Triassic, Triassic–Jurassic, and Cretaceous–Tertiary transitions—neither eustatic sea-level falls, nor global climatic changes, nor any phases of the diastrophic cycle are consistently associated with mass extinctions; and none of these explanations is adequate to explain all these five mass extinctions, let alone all the other geological time intervals that are sometimes interpreted to include events belonging to this category (the Eocene–Oligocene, Cenomanian–Turonian, earliest Toarcian, latest Precambrian, and so on). As theories aimed to explain all mass extinctions, these traditional gradualist scenarios appear, then, also unsatisfactory. This is perhaps why new empirical developments were necessary substantially to rejuvenate this field of research in the 1980s.

THE CURRENT CONTROVERSY

The finding by Alvarez *et al.* (1980) and Smit and Hertogen (1980) that geochemical anomalies, including a pronounced iridium spike, are associated with the mass extinction horizon of pelagic plankton at the Cretaceous–Tertiary boundary has for the first time provided hard evidence that could be interpreted as indicative of a causal link between biotic events and extraterrestrial influences. Iridium is a very rare element in the Earth's crust and its considerable concentration at widely distant locations in the same stratigraphic horizons could be best explained as fallout from the dust cloud raised by impact of iridium-enriched bolide on the Earth. Such an impact, in turn, would also have a wide variety of environmental consequences, beginning with an initial shock and heat wave and ending with nuclear winter-type events (Pollack *et al.*, 1983) and seawater intoxication by trace elements (Erickson and Dickson, 1987), which could easily explain mass extinction of marine as well as terrestrial organisms. The hypothesis of Cretaceous–Tertiary mass extinction by a huge bolide impact is, therefore, very appealing. Its proposal has dramatically changed the pace of research, and the tone of debate, on mass extinctions (for critical reviews of ideas and evidence see Sepkoski and Raup, 1986; Jablonski, 1986; Hoffman, 1989).

The hypothesis of an impact at the Cretaceous–Tertiary boundary is strongly supported by literally dozens of reports on geochemical (iridium and other siderophile element enrichment) and mineralogical (shocked quartz, fluffy carbon, microspherules comparable to altered impact droplets) fingerprints associated with either the marine or the palynological Cretaceous–Tertiary boundary. An iridium anomaly, associated in addition with microtektites, has also been discovered at the Eocene–Oligocene transition (Ganapathy, 1982). Therefore, when Raup and Sepkoski (1984) analysed the record of family extinctions among marine animals and found that it may indicate a 26-million-year periodicity in extinction intensity of these taxa, it was logical to assume that the uniformity of extinction mechanism, suggested by the apparent periodicity of extinction events, points to extraterrestrial impacts as the culprits of mass extinction.

A search was immediately undertaken for astronomical mechanisms that could cause such an impact periodicity, and astronomers have rapidly come up with quite an impressive array of hypotheses: the Nemesis, a twin star of our Sun, could periodically disturb the Oort cloud of comets and throw some of them toward the inner planets; the same effect could be produced by the undetected tenth planet of our solar system; oscillations of the solar system about the galactic plane could also lead to increased frequency of our planet's encounters with comets, and so on. None of these hypotheses has thus far been corroborated by evidence, and all of them have encountered more or less serious troubles, but none of them could be ultimately refuted (Sepkoski and Raup, 1986). All of them hinge, however, on the assumption of a uniformity-of-mass-extinction mechanism by extraterrestrial impacts. This assumption is in turn substantiated by the argument that if the peaks of extinction known from the fossil record are periodic, and if one or even two of these peaks are caused by impacts, then all the other peaks are most likely to be also caused by impacts.

This argument has largely shaped the controversy on mass extinctions which has continued relentlessly in the 1980s. For, on the one hand, it critically depends on the presumed validity of the hypotheses proposing the extinction periodicity and the impact causation of the Cretaceous–Tertiary (and to a lesser degree also the Eocene–Oligocene) extinctions. On the other hand, it implies that all mass extinctions in the history of life on the Earth should be associated with impact fingerprints and have a pattern compatible with a sudden and dramatic catastrophe. The latter implication has stimulated several research projects that focused on the individual biotic events which could be construed as mass extinctions, but I shall not discuss this topic here, since the evidence will be presented and evaluated in detail in the other chapters of this book. Suffice it to note here that no evidence has thus far been presented that would firmly link the pre-Cretaceous–Tertiary extinctions with bolide impacts; and the impact causation of the Cretaceous–Tertiary and Eocene–Oligocene extinctions also is highly contentious. The question of extinction periodicity, moreover, is no less controversial.

Raup and Sepkoski (1984; 1986) conducted a series of statistical analyses of the rate and intensity of extinction of marine animal families and genera during the Phanerozoic. They concluded that the observed pattern of temporal distribution of extinction peaks is best explained by assuming a significant contribution from an approximately 26-million-year periodic signal in the later Phanerozoic, and perhaps a longer-period signal in the Palaeozoic. This conclusion and my subsequent suggestion that this empirical pattern may in fact reflect nothing but a random variation in extinction intensity through time (Hoffman, 1985) have triggered a heated debate. It is important to realize that the empirical pattern of extinction peaks evidently is not strictly periodic, so that the question of the likelihood that a random process leads to a strictly periodic pattern is irrelevant. What is actually at issue in this debate is whether the empirical pattern deviates so little from periodicity that it could only be obtained with the contribution of a strong periodic signal, or whether the observed amount of regularity is so small that the pattern could also result from a random process.

The issue is far from being ultimately resolved. On the one hand, Sepkoski and Raup (1986), Sepkoski (1986) and Fox (1987) present new analyses to reinforce the hypothesis of extinction periodicity. On the other hand, there are several quantitative arguments to support my interpretation of the empirical pattern as deviating so much from strict periodicity that it may reflect randomness rather than orderliness. Thus, Kitchell and Peña (1984) demonstrated that if the magnitude of extinction peaks is taken into account alongside their spacing in time, a stochastic autocorrelation model fits the empirical pattern even better than does the periodic one. Kitchell and Estabrook (1986) show that approximately 8% of symmetrical random walks of the same length as the time series studied by Raup and Sepkoski produce patterns with their peaks spaced even more like 26-million-year periodicity than in the empirical pattern. Ross (1987) shows, moreover, that up to 35% of such random walks are statistically indistinguishable from the empirical pattern. Noma and Glass (1987) demonstrate that the historical pattern of marine animal extinction has no more periodicity than expected to occur in more than 5% of the series of numbers produced by a 32-number roulette. Lutz (1987) notes that while Raup and Sepkoski reject the hypothesis of a Poisson distribution of extinction peaks in time, their pattern can still be explained by a gamma distribution model which they cannot reject. Stigler and Wagner (1987) argue that Raup and Sepkoski's main statistical test is intrinsically biased toward 26-million-year periodicity of extinction peaks (see also Raup and Sepkoski, 1988; Stigler and Wagner, 1988). A similar opinion is also expressed by Quinn (1987), whose reanalysis of Raup and Sepkoski's data fails to discover a significant periodicity of extinction. One should also note the comments made recently by McFadden (1987) and Lutz and Watson (1988) on the pitfalls of searching, by statistical methods similar to the one employed by Raup and Sepkoski in their analysis, for periodicity of geomagnetic reversals.

These analyses strongly corroborate my view that the hypothesis of extinction periodicity is not unambiguously supported by the data and that its rival hypothesis of random distribution of mass extinctions in time is perfectly viable. Consequently, the main argument for the concept of periodic mass extinctions by large bolide impacts on the Earth is seriously undermined. For, if there is no periodicity of extinction peaks, there is also no reason to assume the uniformity of extinction mechanisms for all mass extinctions, and there is no reason to extrapolate the hypothesis of impact causation from the Cretaceous–Tertiary to the other extinction events.

In fact, the scenario of Cretaceous–Tertiary mass extinction by extra-terrestrial impact is not ultimately established either (for critical reviews see Hallam, 1987; Officer *et al.*, 1987; Hoffman, 1989). On the one hand, the trace element and microspherule distribution in the boundary clay appears to be rather incompatible with the single impact hypothesis (Hansen *et al.*, 1984; Rocchia *et al.*, 1984; 1987; Naslund *et al.*, 1986; Preisinger *et al.*, 1986; Bohor *et al.*, 1987; Crocket *et al.*, 1988). Various kinds of microspherules which were thought to represent impact droplets actually are of terrestrial origin (Hansen *et al.*, 1986; Izett, 1987). Carbon and oxygen isotope excursions fully comparable to the one at the Cretaceous–Tertiary boundary occur also well below the boundary, thus contradicting their interpretation as an indication of a single mass extinction level (Mount *et al.*, 1986; Jones *et al.*, 1987; Margolis *et al.*, 1987; Lindinger and Keller, 1987), and the palaeoceanographic record during the first few million years after the boundary disagree with the models of climatic and oceanographic events after an impact (Zachos and Arthur, 1986). All these features of the Cretaceous–Tertiary transition, however, are entirely compatible with the hypothesis of volcanic causation of the latest Cretaceous events as advocated by Hallam (1987) and Officer *et al.* (1987). In fact, Deccan flood basalts were rapidly produced within less than 1 million years in the latest Cretaceous (Duncan and Pyle, 1988; Courtillot *et al.*, 1988).

On the other hand, a recent cathodoluminescence study of shocked quartz grains at the Cretaceous–Tertiary boundary demonstrates their non-volcanic origin (Owen and Anders, 1988); and the rhodium distribution in the boundary clay is very different from that found in terrestrial rocks, but closely similar to the one recorded in meteorites (Bekov *et al.*, 1988). Moreover, tsunami deposits have been recently discovered at the Cretaceous–Tertiary boundary (Bourgeois *et al.*, 1988). These data, then, strongly suggest that an impact did indeed take place at the very end of the Cretaceous, presumably during, or very soon after, a period of extremely intense volcanic activity.

The relative contribution of the environmental consequences of impact and volcanism to the Cretaceous–Tertiary mass extinction can hardly be disentangled. The fossil record, however, documents a more complex pattern than just a severe across-the-board extinction of many organic groups at the Cretaceous–Tertiary boundary. Among the pelagic plankton the foraminifer and coccolithophorid extinctions occur at demonstrably different stratigraphic horizons (Perch-Nielsen *et al.*, 1982; Gartner and Jiang, 1984; Smit and Romein, 1985; Lindinger and Keller, 1987). Marine macroinvertebrates, in turn, were certainly undergoing significant extinctions over the last few million years of the Cretaceous (Birkelund and Hakansson, 1982; Kauffman, 1986; Ward *et al.*, 1986; Hallam, 1987; Zinsmeister *et al.*, 1987), thus ruling out their attribution solely to the terminal Cretaceous impact. This is perhaps why Alvarez *et al.* (1984) accepted the view that many organic groups, dwindling for whatever unidentified reasons toward the end of the Cretaceous, fell ultimately victim to the impact event at the Cretaceous–Tertiary boundary.

It thus appears that the Cretaceous–Tertiary mass extinction cannot be plausibly interpreted as caused solely by a bolide impact and its environmental consequences, and it is more feasible to invoke a coincidence of at least two different factors which were causally unrelated to each other (since the impact occurred at the end of a period of intense volcanism and extinctions). Hence, even if there were good reasons to assume the uniformity of extinction mechanisms for all mass extinctions, there would be no reason to accept extraterrestrial impacts as the ultimate cause for these biotic events. In the absence of strong evidence for extinction periodicity, however, the hypothesis of a uniform causation of all mass extinctions—either by impacts, as envisaged by the hypothetical astronomical mechanisms, or by episodes of extraordinary volcanism caused by mantle processes, as proposed recently by Loper *et al.* (1988)—appears absolutely unsubstantiated.

This conclusion fully pertains to the recent article by Hut *et al.* (1987) who note that the Cenomanian–Turonian, Cretaceous–Tertiary, and Eocene–Oligocene extinctions are all spread over several million years and are in fact clusters of extinction episodes, and who then propose that these mass extinction periods—and presumably also the others, for which they also hypothesize such stepwise nature—are caused by comet showers. The concept is derived from the fact that multiple impacts occurred during the Eocene–Oligocene transition and that an impact also took place during the extinctions at the Cretaceous–Tertiary boundary. The assumption of a uniformity of mass extinction mechanism then forces Hut *et al.* to envisage impact causation also for the Cenomanian–Turonian episodes of extinction and for the several latest Cretaceous episodes prior to the terminal event. There is, however, no evidence to justify this extrapolation. Moreover, both the Cenomanian–Turonian and Eocene–Oligocene data employed now by Hut *et al.* were previously explained by the same palaeontologists by reference to entirely terrestrial causal processes—oceanic anoxia and climatic change—which clearly were associated in time with the biotic events subsumed under the heading of mass extinction (Keller, 1983; Kauffman, 1984; Elder, 1987; Hansen, 1987).

It thus appears that all one can safely say at this point about the Cretaceous–Tertiary, Eocene–Oligocene, and Cenomanian–Turonian extinctions is that they occurred in clusters of extinction episodes, some of them perhaps causally related to bolide impacts on the Earth, but others apparently not. There is, moreover, strong evidence that some other extinction peaks, including the traditionally recognized Ordovician–Silurian and Frasnian–Famennian mass extinctions, are also clusters of extinction episodes rather than single catastrophes (Brenchley, 1984; House, 1985; Benton, 1986; Farsan, 1986; Hallam, 1986). This has led me to the conclusion that the major peaks of extinction may in fact be clusters of separate events more or less accidentally aggregated in time (Hoffman, 1989). They may be caused by global environmental events of various sorts—palaeoceanographic, climatic, or volcanic ones, as well as bolide impacts. Undoubtedly, extraterrestrial impacts on the Earth, enormous volcanic eruptions, oceanic anoxic events, dramatic sea-level changes and climatic fluctuations occurred repeatedly during the Phanerozoic; yet only very few of these phenomena were actually implicated in the causation of mass extinctions. Perhaps then mass extinctions are not the biotic consequences of any single phenomenon of one or another sort, but rather rare incidences of more than one major change in the physical environment accidentally clumped together within relatively short intervals of geological time, say, 2–4 million years in duration.

This hypothesis of mass extinctions as coincidences of lesser episodes caused by a variety of physical factors may superficially resemble Schindewolf's (1954) desperate move to explain the mystery that could not be resolved in a more scientific way. Occurrence of such coincidences, however, is not at all implausible or unlikely.

Consider, for example, w different kinds of physical events that occur at random, each with the probability of $1 - p$ per million years. The following formula

$$[p^{wk} + wp^{(w - 1)k}(1 - p^k)]^{n/k} = 1 - C$$

gives then the probability $1 - C$ that, during the period of n million years, none of the consecutive k-million-year intervals will contain more than one event. In other words, this formula gives an underestimate C of the lower limit on the probability that at least one coincidence (on the scale of k million years) of at least two events will occur within n million years. Assuming that, say, 5 different kinds of events may cause considerable extinctions (extraordinary volcanic eruptions, rapid climatic cooling, major transgressions and regressions, oceanic overturns, and large bolide impacts) and occur each with the average frequency of one every 50 million years, C equals 0.52 for at least one coincidence in a 2-million-year interval occurring within 100 million years and C is 0.84 for one such coincidence within 250 million years. For at least one coincidence in a 4-million-year interval, C equals under such assumptions 0.73 for 100 million years and 0.96 for 250 million years. If the average frequency of events of each kind is decreased to only 1 per 100 million years, C still reaches 0.40 for at least one coincidence in a 4-million-year interval occurring within 100 million years and 0.60 for one such coincidence within 250 million years.

Recall that C seriously underestimates the actual probability of coincidences of events, as it does not take into account the probability of coincidences that would span two consecutive time intervals. Clearly, then, coincidences (on the geological time-scale) of major environmental events accidentally aggregated within a few million years are quite likely to occur, and they may well account for at least some of the mass extinctions. This is not to say, however, that all mass extinctions are nothing but such clusters of separate extinction episodes.

For example, because of a major marine regression, the fossil record at the Permian–Triassic transition is so poor that the detailed pattern of extinction cannot be recognized on the global scale. In any event, however, no distinct episodes of extinction can be identified, and contrary to the traditional opinion that the extinctions were spread over as much as 10 million years (Schopf, 1974), more recent studies on continuous stratigraphic sections suggest a much more rapid, though by no means catastrophic, process (Sheng *et al.*, 1984). The oceanic carbon and oxygen isotope records at the Permian–Triassic transition—spanning this greatest of all mass extinctions—indicate a dramatic oceanographic event, apparently more profound than any other observed thus far in the Phanerozoic. These isotopic records have been recently reviewed by Holser and Magaritz (1987; see also Magaritz *et al.*, 1988), but Gruszczyński *et al.* (1989, Małkowski *et al*, 1989) present crucially important new data which demonstrate that, after a rapid shift toward extremely positive values, the oceanic $\delta^{13}C$ declined by more than 10 per mille, that is, down to distinctly negative values. The oceanic oxygen isotope curve essentially mimicks the carbon one.

The trigger for the latest Permian changes in $\delta^{13}C$ trend is unknown. The mass balance calculations made by Gruszczyński *et al.* (1989; Małkowski *et al*, 1989) demonstrate, however, that the late Permian rise in $\delta^{13}C$ must have been caused by a rapid removal from the ocean-atmosphere system of huge amounts of organic carbon—more than 30 times the total amount of carbon in the presently living biosphere. The subsequent drop in $\delta^{13}C$ indicates that the ocean received three times more of organic carbon than it had previously lost. The amounts of organic carbon involved in these processes, which seem to have occurred within 1–2 million years, are so huge that the processes must have encompassed and severely affected the whole Earth system. The increased burial of organic carbon, which is reflected by the carbon isotope spike, must have left much oxygen free; the reducing conditions at the sea-bottom contributed to nutrient recycling in the ocean and thus to an increase in the standing crop of the marine biosphere. However, the subsequent oxidation of organic matter, which is reflected by the carbon isotope fall, must have relied upon drawing vast amounts of oxygen from the atmosphere; it also led to nutrient sinking at the sea-bottom. The oceanic isotope records at the Permian–Triassic transition thus indicate a considerable decline in atmospheric oxygen and oceanic nutrient levels, which could well be the prime causes of mass extinction.

The isotopic records thus identify an oceanographic event as the cause of the Permian–Triassic extinctions, whereas no other causal factors have been recognized which could significantly contribute to these extraordinary biotic phenomena. Perhaps, then, this is a rare case of mass extinction phenomenon that really deserves this name—triggered indeed by a single process and, consequently, confined in duration to a reasonably coherent time interval. The other mass extinctions, however, may be caused by various independent processes. It seems though that, apart from the Cretaceous–Tertiary mass extinction, their palaeoceanography is too poorly known to rule out the possibility of their, at least partial, causation by processes similar to the one operating at the Permian–Triassic transition. Much more palaeoceanographic effort, particularly on the Palaeozoic, is needed to decipher the palaeoenvironmental context of those mass extinctions.

REFERENCES

Alvarez, L.W., Alvarez, W., Asaro, F. and Michel, H.V., 1980, Extraterrestrial cause for the Cretaceous–Tertiary extinction, *Science*, **208** (4448): 1095–1108.

Alvarez, W., Kauffman, E.G., Surlyk, F., Alvarez, L.W., Asaro, F. and Michel, H.V., 1984, Impact theory of mass extinctions and the invertebrate fossil record, *Science*, **223** (4641): 1135–41.

Andrusov, N.I., 1891, O kharakters i proiskhozhdenii sarmatskoy fauny, *Gornyi Zhurnal*, **2**: 241–80.

Bekov, G.I., Letokhov, V.S., Radaev, V.N., Badyukov, D.D. and Nazarov, M.A., 1988, Rhodium distribution at the Cretaceous/Tertiary boundary analysed by ultrasensitive laser photoionization, *Nature*, **332** (6160): 146–8.

Benton, M.J., 1986, More than one event in the late Triassic mass extinction, *Nature*, **321** (6073): 857–61.

Beurlen, K., 1933, Vom Aussterben der Tiere, *Natur und Museum*, **63** (1–3): 1–8, 55–63, 102–6.

Beurlen, K., 1956, Der Faunenschnitt an der Perm-Trias Grenze, *Zeitschrift der Deutschen Geologischen Gesellschaft*, **108** (1): 88–99.

Birkelund, T. and Hakansson, E., 1982, The terminal Cretaceous extinction in the Boreal shelf seas—a multicausal event. In L.T. Silver and P.H. Schultz (eds), *Geological implications of impacts of large asteroids and comets on the Earth, Special Paper of the Geological Society of America*, **190**: 373–84.

Bohor, B.F., Triplehorn, D.M., Nichols, D.J. and Millard, H.T., 1987, Dinosaurs, spherules, and the 'magic' layer: a new K–T boundary clay site in Wyoming, *Geology*, **15** (10): 896–9.

Bourgeois, J., Hansen, T., Wilberg, P. and Kauffman, E.G., 1988, Tsunami deposits at the Cretaceous–Tertiary boundary in east Texas, *Abstract, Third International Global Bioevents Meeting, University of Colorado at Boulder*: 10.

Brenchley, P.J., 1984, Late Ordovician extinctions and their relationship to the Gondwana glaciation. In P.J. Brenchley (ed.), *Fossils and climate*, Wiley and Sons, Chichester: 291–315.

Brocchi, G., 1814, *Conchiliologia fossile subapennina*, Milan.

Buckland, W., 1823, *Reliquiae diluvianae*, Murray, London.

Courtillot, V., Féraud, G., Maluski, H., Vandamme, G., Moreau, M.G. and Besse, J., 1988, Deccan flood basalts and the Cretaceous/Tertiary boundary, *Nature*, **333** (6176): 843–6.

Crocket, J.H., Officer, C.B., Wezel, F.C. and Johnson, G.D., 1988, Distribution of noble metals across the Cretaceous/Tertiary boundary at Gubbio, Italy: iridium variation as a constraint on the duration and nature of Cretaceous/ Tertiary boundary events, *Geology*, **16** (1): 77–80.

Cuvier, G., 1799, Mémoir sur les éspèces d'éléphants vivantes et fossiles, *Mémoires de l'Académie des Sciences de Paris*, **2**: 1–32.

Cuvier, G., 1825, *Discours sur les revolutions de la surface du globe et sur les changements qu'elles ont produites dans le règne animal*, Dufour et d'Ocagne, Paris.

Darwin, C., 1846, *Geological observations on South America*, Smith and Elder, London.

Darwin, C., 1859, *On the origin of species by means of natural selection*, Murray, London.

Davitashvili, L.S., 1969, *Prichiny vymiranya organizmov*, Nauka, Moskva.

De Laubenfels, M.W., 1956, Dinosaur extinction: one more hypothesis, *Journal of Paleontology*, **30** (1): 207–12.

D'Orbigny, A., 1852, *Cours élémentaire de paléontologie et de géologie stratigraphique*, Paris.

Duncan, R.A. and Pyle, D.G., 1988, Rapid eruption of the Deccan flood basalts at the Cretaceous/Tertiary boundary, *Nature*, **333** (6176): 841–3.

Dyssa, F.M., Nesterenko, P.T., Stovas, M.V. and Shirokov, A.Z., 1960, K voprosu o prichinakh vymiranya bolshikh grupp organizmov, *Dokladў Akademii Nauk SSSR*, **131** (1): 185–7.

Elder, W.P., 1987, The paleoecology of the Cenomanian–Turonian (Cretaceous) stage boundary extinctions at Black Mesa, Arizona, *Palaois*, **2** (1): 24–40.

Erickson, D.J. and Dickson, S.M., 1987, Global trace element biogeochemistry at the K/T boundary; oceanic and biotic response to a hypothetical meteorite impact, *Geology*, **15** (11): 1014–17.

Farsan, N.M., 1986, Faunenwandel oder Faunenkrise? Faunistische Untersuchung der Grenze Frasnium-Famennium in mittleren Südasien, *Newsletter of Stratigraphy*, **16** (3): 113–31.

Fox, W.T., 1987, Harmonic analysis of periodic extinctions, *Paleobiology*, **13** (3): 257–71.

Ganapathy, R., 1982, Evidence for a major meteorite impact on the Earth 34 million years ago: implication for Eocene extinctions, *Science*, **216** (4549): 885–6.

Gartner, S. and Jiang, M.J., 1984, The Cretaceous–Tertiary boundary in east-central Texas, *Transactions of the Gulf Coast Association of Geological Societies*, **35**: 373–80.

Gartner, S. and Keany, J., 1978, The terminal Cretaceous event: a geologic problem with an oceanographic solution, *Geology*, **6** (12): 708–12.

Gruszczyński, M., Hałas, S., Hoffman, A. and Małkowski, K., 1989, A brachiopod calcite record of the oceanic carbon and oxygen isotopic shift at the Permo/Triassic transition, *Nature*, **337** (6203): 64–8.

Hallam, A., 1983, Plate tectonics and evolution. In D.S. Bendall (ed.), *Evolution from molecules to men*, Cambridge University Press, Cambridge: 367–86.

Hallam, A., 1986, The Pliensbachian and Tithonian extinction events, *Nature*, **319** (6056): 765–8.

Hallam, A., 1987, End-Cretaceous mass extinction event: argument for terrestrial causation, *Science*, **238** (4831): 1237–42.

Hansen, H.J., Gwozdz, R., Bromley, R.G., Rasmussen, K.L., Vogensen, E.B. and Pedersen, K.R., 1986, Cretaceous–Tertiary boundary spherules from Denmark, New Zealand and Spain, *Bulletin of the Geological Society of Denmark*, **35** (1): 75–82.

Hansen, T.A., 1987, Extinction of late Eocene to Oligocene molluscs: relationship to shelf area, temperature changes, and impact events, *Palaios*, **2** (1): 69–75.

Hansen, T.A., Farrand, R.B., Montgomery, H.A., Billman, H.G. and Blechschmidt, G., 1984, Sedimentology and extinction patterns across the Cretaceous–Tertiary boundary interval in east Texas, *Cretaceous Research*, **8** (3): 229–52.

Hays, J.D., 1971, Faunal extinctions and reversals of the Earth's magnetic field, *Bulletin of the Geological Society of America*, **82** (9): 2433–47.

Hennig, E., 1932, *Wege und Wesen der Paläontologie*, Berlin.

Hoffman, A., 1985, Patterns of family extinction depend on definition and geological timescale, *Nature*, **315** (6011): 359–62.

Hoffman, A., 1989, Mass extinctions: the view of a sceptic, *Journal of the Geological Society of London*, **146** (1): 21–35.

Holser, W.T. and Magaritz, M., 1987, Events near the Permian–Triassic boundary, *Modern Geology*, **11** (2): 155–79.

House, M.R., 1985, Correlation of mid-Palaeozoic ammonoid evolutionary events with global sedimentary perturbations, *Nature*, **313** (5997): 17–22.

Hut, P., Alvarez, W., Elder, W.P., Hansen, T., Kauffman, E.G., Keller, G., Shoemaker, E.M. and Weissman, P.R., 1987, Comet showers as a cause of mass extinctions, *Nature*, **329** (6135): 118–26.

Izett, G.A., 1987, Authigenic 'spherules' in K–T boundary sediments at Caravaca, Spain, and Raton Basin, Colorado and New Mexico, may not be impact derived, *Bulletin of the Geological Society of America*, **99** (1): 78–86.

Jablonski, D., 1986, Causes and consequences of mass extinctions: a comparative approach. In D.K. Elliott (ed.), *Dynamics of extinction*, Wiley and Sons, New York: 183–229.

Johnson, J.G., 1974, Extinction of perched faunas, *Geology*, **2** (10): 479–82.

Jones, D.S., Mueller, P.A., Bryan, J.R., Dobson, J.P., Channell, J.E.T., Zachos, J.C. and Arthur, M.A., 1987, Biotic, geochemical, and Paleomagnetic changes across the Cretaceous/Tertiary boundary at Braggs, Alabama, *Geology*, **15** (4): 311–15.

Kauffman, E.G., 1984, The fabric of Cretaceous marine extinctions. In W.A. Berggren and J.A. Van Couvering (eds), *Catastrophes and earth history*, Princeton University Press, Princeton, NJ: 151–246.

Kauffman, E.G., 1986, High-resolution event stratigraphy: regional and global Cretaceous bio-events. In P.H. Walliser (ed.), *Global bio-events*, Springer, Berlin: 279–335.

Keller, G., 1983, Biochronology and paleoclimatic implications of middle Eocene to Oligocene planktic foraminiferal faunas, *Marine Micropaleontology*, **7** (4): 463–86.

Kitchell, J.A. and Estabrook, G., 1986, Was there a 26-Myr periodicity of extinctions?, *Nature*, **321** (6069): 534–5.

Kitchell, J.A. and Peña, D., 1984, Periodicity of extinction in the geological past: deterministic versus stochastic explanations, *Science*, **226** (4675): 689–92.

Krasovskiy, V.I. and Shklovskiy, I.S., 1957, Vozmozhnoye vliyanye vspyshek sverkhnovykh na evolutsyu zhizni na Zemle, *Dokladÿ Akademii Nauk SSSR*, **116** (2): 197–9.

Lamarck, J.B., 1809, *Philosophie zoologique*, Dentu, Paris.

Lichkov, B.L., 1945, Geologicheskie peryody i evolutsya zhivogo veshchestva, *Zhurnal Obshchei Biologii*, **5** (3): 157–79.

Lindinger, M. and Keller, G., 1987, Stable isotope stratigraphy across the Cretaceous/Tertiary boundary in Tunisia: evidence for a multiple extinction mechanism?, *Geological Society of America Abstracts with Programs*, **19**: 747.

Liniger, H., 1961, Über das Dinosauriersterben in der Provence, *Leben und Umwelt*, **18** (2): 27–33.

Loper, D.A., McCartney, K. and Buzyna, G., 1988, A model of correlated episodicity in magnetic-field reversals, climate, and mass extinctions, *Journal of Geology*, **96** (1): 1–15.

Lutz, T.M., 1987, Limitations to the statistical analysis of episodic and periodic models of geologic time series, *Geology*, **15** (12): 1115–17.

Lutz, T.M. and Watson, G.S., 1988, Effects of long-term variation on the frequency spectrum of the geomagnetic reversal record, *Nature*, **334** (6179): 240–2.

Lyell, C., 1832, *Principles of geology*, Murray, London.

McFadden, P.L., 1987, A periodicity of magnetic reversals?, *Nature*, **330** (6143): 26.

McLaren, D.J., 1970, Time, life, and boundaries, *Journal of Paleontology*, **44** (5): 801–15.

McLean, D.M., 1978, A terminal Mesozoic 'greenhouse': lessons from the past, *Science*, **201** (4354): 401–6.

Magaritz, M., Bär, R., Baud, A. and Holser, W.T., 1988, The carbon-isotope shift at the Permian/Triassic boundary in the southern Alps is gradual, *Nature*, **331** (6154): 337–9.

Małkowski, K., Gruszczyński, M., Hoffman, A. and Halas, S., 1989, Oceanic stable isotope composition and a scenario for the Permo-Triassic crisis. *Historical Biology*, **2**.

Margolis, S.V., Mount, J.F., Doehne, E., Showers, W. and Ward, P., 1987, The Cretaceous/Tertiary boundary carbon and oxygen isotope stratigraphy, diagenesis, and paleooceanography at Zumaya, Spain, *Paleoceanography*, **2** (4): 361–78.

Marshall, H.T., 1928, Ultraviolet and extinction, *American Naturalist*, **62** (2): 165–87.

Mount, J.F., Margolis, S.V., Showers, W., Ward, P. and Doehne, E., 1986, Carbon and oxygen isotope stratigraphy of the Upper Maastrichtian, Zumaya, Spain: a record of oceanographic and biologic changes at the end of the Cretaceous Period, *Palaios*, **1** (1): 87–91.

Naslund, H.R., Officer, C.B. and Johnson, G.D., 1986, Microspherules in Upper Cretaceous and lower Tertiary clay layers at Gubbio, Italy, *Geology*, **14** (11): 923–6.

Neumayr, M., 1889, *Stämme des Tierreiches*, Vienna.

Newell, N.D., 1967, Revolutions in the history of life, *Special Paper of the Geological Society of America*, **89**: 63–91.

Noma, E. and Glass, A.L., 1987, Mass extinction pattern: result of chance, *Geological Magazine*, **124** (4): 319–22.

Officer, C.B., Hallam, A., Drake, C.L. and Devine, J.D., 1987, Late Cretaceous and paroxysmal Cretaceous/Tertiary extinctions, *Nature*, **326** (6109): 143–9.

Owen, M.R. and Anders, M.H., 1988, Evidence from cathodoluminescence for non-volcanic origin of shocked quartz at the Cretaceous/Tertiary boundary, *Nature*, **334** (6178): 145–7.

Pavlova, M.V., 1924, *Prichiny vymiranya zhivotnykh v proshedshye geologicheskye peryody*, Nauka, Moskva.

Perch-Nielsen, K., McKenzie, J.A. and He, Q., 1982, Biostratigraphy and isotope stratigraphy and the 'catastrophic' extinction of calcareous nannoplankton at the Cretaceous/Tertiary boundary. In L.T. Silver and P.H. Schultz (eds), *Geological implications of impacts of large asteroids and comets on the Earth, Special Paper of the Geological Society of America*, **190**, 353–72.

Pollack, J.B., Toon, O.B., Ackerman, T.P., McKay, C.P. and Turco, R.P., 1983, Environmental effects of an impact-generated dust cloud: implications for the Cretaceous–Tertiary extinctions, *Science*, **219** (4582): 287–9.

Preisinger, A., Zobetz, E., Gratz, A.J., Lahodynsky, R., Becke, M., Mauritsch, H.J., Eder, G., Grass, F., Rögl, F., Sradner, H. and Surenian, R., 1986, The

Cretaceous/Tertiary boundary in the Gosau Basin, Austria, *Nature*, **322** (6082): 794–9.

Quinn, J.F., 1987, On the statistical detection of cycles in extinctions in the marine fossil record, *Paleobiology*, **13** (4): 465–78.

Raup, D.M. and Sepkoski, J.J., Jr, 1984, Periodicity of extinctions in the geologic past, *Proceedings of the National Academy of Science U.S.A.*, **81** (3): 801–5.

Raup, D.M. and Sepkoski, J.J., Jr, 1986, Periodic extinction of families and genera, *Science*, **231** (4740): 833–6.

Raup, D.M. and Sepkoski, J.J., Jr, 1988, Testing for periodicity of extinction, *Science*, **241** (4861): 94–6.

Rocchia, R., Renard, M., Boclet, D. and Bonté, P., 1984, Essai d'évaluation de la transition C-T par l'évolution de l'anomalie en iridium: implications dans la recherche de la cause de la crise biologique, *Bulletin de la Société Géologique de France*, new series, **26** (6): 1193–1202.

Rocchia, R., Boclet, D., Bonté, P., Denineau, J., Jehanno, C. and Renard, M., 1987, Comparison des distributions de l'iridium observées à la limite Cretacé-Tertiaire dans divers sites européens, *Mémoires de la Société Géologique de France*, new series, **150**: 95–103.

Ross, S.M., 1987, Are mass extinctions really periodic?, *Probability in Engineering and Information Science*, **1** (1): 61–4.

Russell, D.A. and Tucker, W., 1971, Supernovae and the extinction of the dinosaurs, *Nature*, **229** (5286): 553–4.

Schindewolf, O.H., 1950, *Grundfragen der Paläontologie*, Schweizerbart, Stuttgart.

Schindewolf, O.H., 1954, Über die möglichen Ursachen der grossen erdgeschichtlichen Faunenschnitte, *Neues Jahrbuch für Geologie und Paläontologie Monatshefte*, **1954** (10): 457–65.

Schopf, T.J.M., 1974, Permo-Triassic extinctions: relation to seafloor spreading, *Journal of Geology*, **82** (2): 129–43.

Sepkoski, J.J., Jr, 1986, Global bioevents and the question of periodicity. In O.H. Walliser (ed.), *Global bio-events*, Springer, Berlin: 47–61.

Sepkoski, J.J., Jr. and Raup, D.M., 1986, Periodicity in marine extinction events. In D.K. Elliott (ed.), *Dynamics of extinction*. Wiley and Sons, New York: 3–36.

Sheng, J.Z., Chen, C.Z., Wang, Y.G., Rui, L., Liao, Z.T., Bando, Y., Ishi, K., Nakazawa, K. and Nakamura, K., 1984, Permian–Triassic boundary in middle and eastern Tethys, *Journal of the Faculty of Science, Hokkaido University, fourth series*, **21** (1): 133–81.

Smit, J. and Hertogen, J., 1980, An extraterrestrial event at the Cretaceous–Tertiary boundary, *Nature*, **285** (5762): 198–200.

Smit, J. and Romein, A.J.T., 1985, A sequence of events across the Cretaceous–Tertiary boundary, *Earth and Planetary Science Letters*, **74** (2–3): 155–70.

Sobolev, D.N., 1928, *Zemla i zhizn. O prichinakh vymiranya organizmov*, Kiev.

Stanley, S.M., 1987, *Extinction*, Scientific American Books, San Francisco.

Stanley, S.M., 1988, Paleozoic mass extinctions: shared patterns suggest global cooling as a common cause, *American Journal of Science*, **288** (4): 334–52.

Stigler, S.M. and Wagner, M.J., 1987, A substantial bias in nonparametric tests for periodicity in geophysical data, *Science*, **238** (4829): 940–5.

Stigler, S.M. and Wagner, M.J., 1988, Testing for periodicity of extinction, *Science*, **241** (4861): 96–9.

Ward, P., Wiedmann, J. and Mount, J.F., 1986, Maastrichtian molluscan biostratigraphy and extinction patterns in a Cretaceous/Tertiary boundary section exposed at Zumaya, Spain, *Geology*, **14** (11): 899–903.

Zachos, J.C. and Arthur, M.A., 1986, Paleoceanography of the Cretaceous–Tertiary boundary event: inferences from stable isotopic and other data, *Paleoceanography*, **1** (1): 5–26.

Zinsmeister, W.J., Feldmann, R.M., Woodburne, M.O., Kooser, M.A., Askin, R.A. and Elliott, D.K., 1987, Faunal transitions across the K/T boundary in Antarctica, *Geological Society of America Abstracts with Programs*, **19** (7): 718.

Zunini, G., 1933, La morte della species, *Rivista Italiana di Paleontologia e Stratigrafia*, **39** (1): 56–102.

Chapter 2

PALAEONTOLOGICAL CRITERIA FOR THE RECOGNITION OF MASS EXTINCTION

Stephen K. Donovan

INTRODUCTION

A mass extinction is any substantial increase in the amount of extinction (i.e., lineage termination) suffered by more than one geographically widespread higher taxon during a relatively short interval of geologic time, resulting in an at least temporary decline in their standing diversity (Sepkoski, 1986, p. 278).

The palaeontologist's conception of mass extinction has undergone extreme modification following the publication of two seminal papers in the 1980s, which suggested that such events may be produced by bolide impacts (Alvarez *et al.*, 1980), or perhaps some other extraterrestrial cause, and that they have occurred periodically (Raup and Sepkoski, 1984). Prior to the present decade, mass extinctions were regarded as rather peculiar events which had happened occasionally in the geologic past and were difficult to interpret. The literature was limited and researchers worked on mass extinctions in passing, rather than in particular. Now the study of mass extinctions is a growth area of palaeontological research. Palaeontologists have linked arms with geochemists in the search for signatures of possible extraterrestrial impacts, particularly anomalously high concentrations of platinum-group elements (such as iridium) at extinction boundaries. Both mathematical and geological analyses have been published which support or criticize the geological time-scales, the taxonomy of fossil organisms and the statistical methodology used in defining periodicity in extinctions of marine organisms. The literature on mass extinctions has proliferated and extinction studies have replaced the punctuated equilibrium-phyletic gradualism debate of the 1970s as the principle area of macro-evolutionary research and discussion.

However, in this maelstrom of fact and theory, the most important and diagnostic feature of mass extinctions often seems to be forgotten. Stated simply, a mass extinction can only be said to have occurred where there has been a (geologically) rapid reduction in biological abundance and diversity on a global scale (Sepkoski's somewhat more rigid definition, above, is the best that has been published during the current debate). Other criteria, such as geochemical anomalies, must be regarded as supporting evidence, rather than conclusive proof. While extinction is a continual process, the differences between mass extinctions and the continually occurring background extinctions are, in most cases, considerable (Table 2.1). Nevertheless, it is probably incorrect to regard extinction as comprising two discrete end members. For example, the mid-Miocene extinction peak only involved five to ten families (Jablonski, 1986: 210); does this represent a high rate of background extinction or a mild 'mass' extinction? Further, it is not correct to recognize only two forms of extinction, as at least two other patterns can be recognized from the fossil record (Table 2.1). Regional 'mass' extinctions are well known, affecting organisms over a broad, but limited, area. Typical examples of regional extinctions from the fossil record are the Pliensbachian and Tithonian events in western Europe (Hallam, 1986), the decline of western Atlantic benthic molluscs in the Neogene (Stanley and Campbell, 1981; Stanley, 1986) and the separate events in the diachronous demise of many faunas of larger terrestrial vertebrates during the late Quaternary (Martin, 1967; 1986; among others; see also Barnosky, Chapter 12, this volume). It is significant to note that the last two events were limited to particular taxonomic groups and cannot, therefore, be regarded as true 'mass' extinction events, which, by definition, involve multiple higher taxa over a broad geographic area. Other extinction events were similarly limited to restricted higher taxonomic groups, such as the mid Mississippian decline of the blastoids (Ausich et al., 1988) and the various extinctions which have punctuated the geological history of the cephalopods (Sepkoski and Raup, 1986, p. 15; Teichert, 1986). These events are similar to mass extinctions in their rapidity and are broad-area to global in extent, but are limited in the variety of organisms affected. I suggest the term 'taxon extinction' (Table 2.1) as being an appropriate name for such events.

Geochemical anomalies, such as anomalous abundances of platinum-group elements, cannot be regarded as conclusive evidence of bolide-induced extinction per se, without reference to supporting palaeontological and geological evidence. Many mass extinctions do not have an associated geochemical signature (see, for example, Orth et al., 1984) and those that have an associated elemental peak do not necessarily indicate an extra-terrestrial influence. Indeed, while the bolide hypothesis must still be regarded as an unproven mechanism of mass extinction (Officer and Drake, 1983; 1985a; 1985b; Officer et al., 1987; Hallam, 1984; 1987; Bray, 1987; Izett, 1987; Crockett et al., 1988], potential terrestrial causes must be explored for any platinum-group element peak associated with an extinc-

Table 2.1. *A comparison of the essential features of mass, background, regional and taxon extinctions.*

	Background	Mass	Regional	Taxon
1. Occurrence	Continuous	Episodic	Episodic	Episodic
2. Rate	Gradual	Fast	Fast	Fast
3. Effect	Local	Global	Broad area	Global
4. Species affected	Few	Numerous	Many	Single taxon

tion event. Three examples illustrate this point particularly well (Donovan, 1987a; see also Orth, Chapter 3, this volume).

Ordovician–Silurian extinction

At two important Ordovician–Silurian boundary localities, the stratotype at Dob's Linn in southern Scotland (Wilde *et al.*, 1986) and Anticosti Island, Quebec (Orth *et al.*, 1986), high iridium abundancies have been shown to be terrestrial in origin. At the stratotype, the high iridium concentrations do not occur as a peak, but persist over a 20-million-year interval from the late Ordovician into the early Silurian. The high iridium levels are probably related to local unroofing, erosion and redeposition of upper mantle ultramafic rocks of the Ballantrae Ophiolite. On Anticosti Island, the abundance of iridium and other trace elements is proportional to the clay abundance of individual horizons. The boundary horizon layer is a calcareous clay and is therefore associated with a trace-element peak.

Frasnian–Famennian extinction

A late Devonian iridium peak in the Canning Basin of northwest Australia (Playford *et al.*, 1984; McLaren, 1985) has been mooted as evidence for a bolide impact which caused the Frasnian–Famennian mass extinction. However, subsequent geochemical analyses of the Frasnian–Famennian boundary sequences in North America and Europe have failed to reveal similar spikes in iridium abundance (McGhee *et al.*, 1984; 1986). This suggests that the Australian peak is not associated with a large body impacting with global consequences. Indeed, the Australian geochemical anomaly occurs above the level of the extinction seen in the European sequence and could not, therefore, have been associated with that event. Instead the iridium peak is found in a stromatolitic layer where the included algae are characteristically concentrators of heavy metals (see Orth, Chapter 3, this volume).

Permian–Triassic extinction

An iridium anomaly from the Permian–Triassic boundary of China (Dao-Yi *et al.*, 1985) has not been reproduced in other sections across the same interval (Clark *et al.*, 1986). Indeed, when samples from the original

Chinese sequence were reanalysed in other laboratories, the iridium peak was not reproduced (Raup, 1987, p. 3), throwing the validity of the initial report into doubt.

It is, therefore, obvious that the primary tool of mass extinction studies must be the fossil record. While geochemistry can produce supporting evidence (for example, an increase in $\delta^{13}C$ may indicate a drop in organic productivity: Hsü et al., 1985; Magaritz et al., 1986; Donovan, 1987c; Brasier, Chapter 4, this volume), it cannot be used on its own as evidence of extinction, Sepkoski (1986) rightly recognized that detailed biostratigraphy and taxonomic compilations are the true basic tools of the student of mass extinctions (or, in the broader sense, any investigation of the pattern of evolution). Both sources of data present their own particular advantages and pitfalls. Both are influenced, to varying degrees, by problems of correlation and succession; taxonomy; preservation, sampling and sedimentology; and geography. It is thus the purpose of this chapter to review the current status of the palaeontological method as used to study mass extinctions, by reference to theoretical concepts and well-documented examples. However, the precise form that a mass extinction is likely to take in the rock record must first be considered before the available data can be discussed.

THE SHAPE OF MASS EXTINCTIONS

If extinction levels are plotted per stage or series, mass extinctions appear as peaks in the graph (see Figure 2.1(a); see also Raup and Sepkoski, 1984). These peaks are therefore relatively higher than the levels of extinction which immediately precede and succeed them. While a useful working definition for identifying mass extinctions in analyses based on taxonomic compilations of this kind, it has the disadvantage that it would not recognize two events which occurred in successive stratigraphic subdivisions (Hoffman, 1985). Such closely spaced extinctions do occur, such as in the Carnian and Norian stages of the Triassic (Benton, 1986). Peaks of extinction may also be more or less pronounced. Particularly prominent mass extinctions have long been recognized at the following boundaries: Ordovician–Silurian; Frasnian–Famennian (late Devonian); Permian–Triassic; Triassic–Jurassic; and Cretaceous–Tertiary. Other extinction peaks are less intense, grading in size down to those that are indistinguishable from background extinction (see Table 2.1; see also Sepkoski and Raup, 1986; Raup and Sepkoski, 1986). Analyses such as these have formed the basis of Raup and Sepkoski's research into the apparent periodicity of mass extinctions.

Ideally, the taxonomic unit used in such analyses should be the species, but in many groups the number of undescribed fossil species is probably too numerous, combined with the incompleteness of the fossil record, to make such an approach meaningful. Similarly, the true stratigraphic range

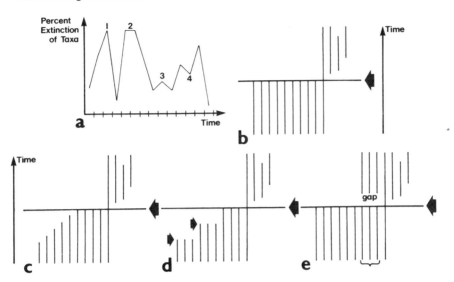

Figure 2.1(a) A hypothetical pattern of extinction, using stratigraphic subdivisions of equal length: (1) a peak in extinction, that is, a mass extinction; (2) two sequential points with high, and equal, extinctions—as neither is a peak, they would not be considered to be mass extinctions *per se*, although the level of extinction in both is equal to (1); (3) a peak, and therefore a 'mass extinction', despite being much lower than either unit in (2); (3) is so low, however, that it is probably below the level of background extinction; (4) although higher than (3), this is not a peak and therefore not a mass extinction. (b)–(e) Contrasting patterns of mass extinction. Extinction boundaries are indicated by broad arrows: (b) abrupt or catastrophic; (c) gradual; (d) stepped; (e) abrupt, showing a gap in the known stratigraphic distributions of three Lazarus species (bracketed).

of most species remains unknown. Data for marine organisms have therefore been examined at higher taxonomic levels, both genus and family. These give similar, but not identical, results (Raup and Sepkoski, 1986, Figure 1; Raup, 1987, Figure 1). While examination of some family data suggests that they are approaching completeness (Paul, 1982a; 1982b), the disadvantages of working at such a high taxonomic level are many (Teichert, 1986). As Newell (1982) pointed out, analysis of familial data smoothes out inequalities in diversity, producing generalized ranges.

Thus, analyses of this form, made at a taxonomic level higher than species and using a coarse stratigraphic scheme, can only indicate the very broad pattern of extinction. They are not sufficiently detailed to indicate whether extinction within any particular stage or series was geologically 'instantaneous' (perhaps 10 000 years or less), or whether it occurred slowly, spread over the duration of the whole stratigraphic unit. The standard method of reporting the range of a species, by assigning it to particular zones, stages, systems or series, inevitably, but misleadingly, suggests that all taxa die out at the end of stratigraphic units (Newell, 1982; Paul, 1985, Figure 5). Our stratigraphic assignments are almost invariably too coarse to interpret accurately the distribution of organisms in time (Shaw, 1971). Precise recording of the stratigraphic positions within measured sections is an essential adjunct to comparative morphology, cladistic analysis, and so on, in the determination of any pattern of evolution (see, for example, Smith and Paul, 1985).

In extinction studies, fine-scale stratigraphic determinations have given valuable insights into the likely sequence of events affecting numerous taxa, which in turn has enabled postulation of the probable process of extinction. Without such detailed knowledge of the shape of mass extinctions, discussion of process must largely be speculation. For example, consider the demise of the ammonites at the end of the Cretaceous. Hancock (1967, pp. 91–4) recognized that the number of ammonite families declined from 22 in the Cenomanian to only 11 in the Maastrichtian, with a reduction from 23 to 11 genera from the Lower to the Upper Maastrichtian (see also the very pertinent comments of Teichert, 1986). However, as the Maastrichtian had a duration of about 8 million years (Harland *et al.*, 1982), these data do no more than suggest that about three ammonite genera became extinct every million years towards the end of the Cretaceous. While Alvarez, Kauffman *et al.* (1984) have suggested that the final demise of the ammonites was at the Cretaceous–Tertiary (K–T) boundary, Ward *et al.* (1986) went further, describing the detailed distributions of both inoceramid bivalves and ammonites in a 240-metre section measured at Zumaya, Spain, which spans the interval Lower Maastrichtian to earliest Paleocene. Apart from one species, the inoceramids in this section did not survive the Lower Maastrichtian. This end Lower Maastrichtian extinction of inoceramids is well known (Kauffman, 1979). The solitary exception disappears about 10 m below the K–T boundary. The numerous ammonite species similarly do not survive to the boundary, but instead show a 'gradual' decline in diversity, the last undoubted ammonite fossil occurring 12.5 m below the K–T horizon. Other groups also disappear before the K–T boundary, while echinoids appear to range into the Palaeocene unaffected. Such a pattern does not support any theory that latest Cretaceous extinctions were simultaneous, indicating instead a slow demise, possibly driven by environmental factors, such as deterioration of climate (Stanley, 1984; 1987), rather than a sudden, catastrophic event.

A 'sudden' or a 'gradual' demise to a variety of taxa in the K–T, or any other, mass extinction event, are two of three possible models for mass extinctions that have arisen in recent years (Figure 2.1(b)–(d)). The traditional conception of a mass extinction is shown in Figure 2.1(b). Essentially this is a geologically instantaneous event which leads to the simultaneous demise of a broad range of hitherto flourishing taxa. Such a pattern appears to fit, for example, the biomere boundaries of the late Cambrian (Palmer, 1982; Westrop and Ludvigsen, 1987; Westrop, Chapter 5, this volume) and the Frasnian–Famennian event (McLaren, 1982;

1983), and has also been suggested for other horizons, such as the K–T boundary (Alvarez, Alvarez *et al.*, 1984), although the last is perhaps questionable. Detailed stratigraphic analysis suggests that the end-Cretaceous event was probably gradual or perhaps stepped (Figure 2.1(c),(d); see also Kauffman, 1979; 1984; Maurrasse *et al.*, 1985; Ward *et al.*, 1986; Hut *et al.*, 1987) for many fossil groups, although at least some taxa, such as the brachiopods, appear to have undergone an abrupt decline in diversity at the K–T boundary (Surlyk and Johansen, 1984).

Some analyses of mass extinctions, such as the terminal K–T event, have suggested that many groups were in simultaneous decline over a period of, perhaps, tens of millions of years, the final demise of many groups appearing to be due to an acceleration of the overall trend. Such a pattern is illustrated by Figure 2.1(c). An abrupt end to such a pattern of slow decline is problematic, due either to an acceleration of the existing extinction process or to the intervention of a further influence which delivers the *coup de grâce* to a number of already weakened lineages (Donovan, 1987b).

Stepped mass extinctions are hypothetical, but presumably follow a similar pattern to gradual events (Figure 2.1(d)). However, the overall pattern is of a succession of events, each of which is identified by the demise of a number of species. Such a pattern of extinctions is well known from the late Eocene to early Oligocene (Keller, 1986a; 1986b; Keller *et al.*, 1983; 1987; Hut *et al.*, 1987; Prothero, Chapter 11, this volume). It has been suggested that the driving mechanisms of such extinctions are comet showers (Hut *et al.*, 1987), but this theory is, at best, highly speculative. It has even been suggested recently that some apparently stepwise patterns of extinction can be produced by sampling error (Koch and Morgan, 1988).

It is relevant at this point to mention one other pattern of 'extinction'. It is recognized that some species appear to become extinct at event horizons, only to reappear later in the stratigraphic record (Figure 2.1(e); see also, for example, Surlyk and Johansen, 1984). Thus the demise of these at mass extinction horizons, like the first, but incorrect, report of the death of Mark Twain, is greatly exaggerated. Such species have been given the name of Lazarus taxa by Jablonski (Flessa and Jablonski, 1983; Jablonski, 1986; Raup, 1986; 1987). Lazarus species presumably survived mass extinctions in undiscovered refuges (Vermeij, 1986) and re-established their distribution when environmental conditions improved.

It is also important to recognize that our understanding of mass extinctions is based primarily on the evidence of animals. Knoll (1984) has shown that vascular land plants are not generally affected by those catastrophic events which cause mass mortalities in animal populations (but see Upchurch, Chapter 10, this volume). Tschudy and Tschudy (1986) noted four patterns of stratigraphic distribution shown by the pollen of angiosperm species that survived the K–T event. Some taxa became extinct locally, but persisted elsewhere; others evolved into new species over the boundary; a few groups, which first appeared in the late Cretaceous, persisted unaffected into the Tertiary before becoming extinct; finally, many long-ranging taxa were apparently completely unaffected by the K–T event. That environmental fluctuation did affect terrestrial vascular plants is indicated by the change in composition of palynomorph assemblages at the K–T boundary. Coincident with the boundary iridium anomaly, terrestrial plant assemblages became deficient in angiosperms and ferns replaced them as the dominant group, albeit for only a brief period (Nichols *et al.*, 1986). Angiosperms regained their position of dominance in a geologically short period of time, but this recovery took much longer than from a 'normal' environmental disturbance, such as a volcanic eruption. Mass extinctions must therefore be recognized as primarily palaeozoological events that have relatively little long-term influence on the composition of the flora, although immediately following an extinction event there may be a considerable change in the relative abundancies of plant taxa, indicating considerable sensitivity, at least in the short term.

POTENTIAL PITFALLS

Taxonomy and phylogeny

No evolutionary study is valid without accurate taxonomy or, at least, a consistent systematic approach, so that any differences of interpretation from other researchers are internally uniform. However, many modern extinction studies rely on compilations of other peoples' data (such as the *Treatise on Invertebrate Paleontology*; or Sepkoski, 1982), so there is a possibility of error. This is partially counteracted, however, by such studies being based on taxonomic divisions higher than species, particularly genera and/or families. This has the effect of counteracting any inadequacies in our knowledge of past biotas, due to imperfections in our data base or incompetent taxonomy, by 'smoothing' the data.

Nevertheless, the species is the true unit of evolution and it is necesssary to recognize how species react to extinction events. There are two obvious extremes that can be envisaged. Many species will obviously become extinct (Figure 2.2(a)), but at least some taxa will survive the event with no morphological change (Figure 2.2(b)). The latter obviously remain unaffected, but, with the mass extinction horizon exerting a subtle influence on the mind of the taxonomist, there must always be a slight chance that pre- and post-event populations will erroneously be divided into separate species. However, there is a definite probability that speciation will occur in at least some lineages, due to the sudden fluctuation in environmental conditions, resulting in a morphological jump (= 'punctuated equilibrium'; Figure 2.2(c)). Thus, sp. 1 does not become extinct in the true sense, but evolves into a new, morphologically distinct taxon, sp. 2. However, without a proper understanding of the relationships of spp. 1 and 2, it would be

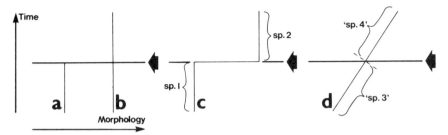

Figure 2.2 Four possible reactions of species to mass extinction. Extinction boundaries are indicated by broad arrows: (a) extinction; (b) species unaffected by extinction event, showing neither extinction nor change in morphology; (c) punctuated speciation event; spp. 1 and 2 are morphologically distinct; (d) lineage showing a gradual change in morphology. The limits of spp. 3 and 4 are arbitrary.

easy to interpret the disappearance of sp. 1 as a true extinction. Even in a lineage which shows only gradual morphologic change over the boundary (Figure 2.2(d)), a random subdivision could be made into two separate species 3 and 4, based on the expectation that sp. 3 went extinct at the event horizon. Our subdivision of such lineages into 'separate' species is normally on the basis of gaps in the fossil record, which produce apparent morphological 'jumps' (Newell, 1956). This subdivision at an event horizon could be interpreted as a 'pseudoextinction', similar to that shown by sp. 1. Sheldon (1987; 1988) has shown that at least some lineages undergo a 'random walk' morphologically, with the lack of stratigraphically continuous samples leading to erection of separate 'species' which are, in fact, only ephemeral responses to environmental fluctuation. Such morphological variations are most likely to be mistakenly recognized as new species when the environment undergoes a large-scale perturbation, such as at an extinction event.

While it is careless to oversplit species at extinction boundaries, it is equally incorrect to suggest that new higher taxa that radiate after the event are merely direct descendents from now-extinct and distantly related taxa. For example, Moore (in Rhodes, 1967) suggested that the post-Palaeozoic articulate crinoids are merely a polyphyletic group comprising survivors of the Palaeozoic inadunates, flexibles and camerates which had escaped extinction at the end of the Permian. The articulates are undoubtedly a monophyletic group (Simms, 1988) which evolved from a cladid ancestor.

Most criticisms of Raup and Sepkoski's 26-million-year extinction periodicity hypothesis have attacked the validity of the statistical approach and the choice of geologic time-scale. Patterson and Smith (1987) examined the theory from a different direction, testing the taxonomic validity of part of the original data base using cladistic methodology. They examined the evolution of fishes and echinoderms since the Upper Permian. These two groups together form about 20% of Raup and Sepkoski's taxonomic data base. Patterson and Smith's analysis produced the unexpected result that only a quarter of the families and family distributions recognized by Raup and Sepkoski were valid. The other three-quarters fell into six inappropriate groupings: monophyletic families whose last appearance was referred to the wrong stage; paraphyletic families which undergo pseudoextinction; polyphyletic families; non-monophyletic families; monotypic families; and non-marine fish families. The corrected data base of echinoderms and fishes does not show periodic peaks of extinction. Sepkoski (1987) suggested that at least some groups that are not monophyletic, such as the dinosaurs, seem to 'recognize' the K–T event (see also comments by Raup and Boyajian, 1988, pp. 112–13). However, it is the monophyletic families, which together make up the paraphyletic dinosaurs, that disappear at the end of the Cretaceous. Indeed, it has now been suggested that the 26-million-year periodicity of extinction is an artefact of the lengths of the stratigraphic units used, rather than any effect of the taxonomy used (Stigler and Wagner, 1987; 1988; Raup and Sepkoski, 1988).

Geography

Mass extinctions are, by definition, global in extent, but our knowledge of the true palaeobiogeographic distribution of many taxa is still incomplete. For obvious reasons, the fossil biotas from the more accessible parts of the globe are generally better known than those from remote and hostile regions. There are also obvious language problems. It is difficult for an English-speaking expert on any group to stay abreast of taxonomic developments in, say, Russia and China. For these reasons, our understanding of extinction events is often largely influenced by our knowledge of the fossil record of North America and Europe. For example, Eckert (1988) relied mainly upon the available North American and British literature when discussing the effect of the late Ordovician mass extinction on the crinoids (also see discussion in Donovan, 1988). This analysis was of particular interest, suggesting that crinoids suffered a peak in extinction in the Rawtheyan (late Ashgill), rather than in the mid-Hirnantian (latest Ordovician), as is more generally recognized (Brenchley and Newall, 1984; Brenchley, 1984; Brenchley, Chapter 6, this volume). Similarly, Ausich *et al.* (1988), while not describing a mass extinction *per se*, recognized a mid-Mississippian blastoid extinction event mainly on the basis of the North American and western European faunas.

These geographic limitations do not necessarily invalidate our recognition of extinction events. Raup (1982) has shown that at least a near-global environmental crisis is necessary in order to produce mass extinction, so presumably a broadly similar pattern of extinction should be apparent in any 'randomly' selected area, providing a sufficiently large palaeobiogeographic region is considered. However, it is always possible that a merely local extinction could be misinterpreted as a global event. For example, in their original announcement of an apparent periodicity of mass extinctions, Raup and Sepkoski (1984) considered the end Pliensbachian and end Tithonian to show peaks which fitted into their pattern. Conversely, Hallam (1986) examined the geographic distribution of extinction at both of these horizons and concluded that, while species-level extinctions are apparent for both events in at least some groups, faunal decline was local in scale. The late Pliensbachian–early Toarcian extinction seems to be related to a major anoxic event in western Europe, while the Tithonian extinction correlates with a major regression of the shallow epicontinental seas of Europe and Russia. In both instances, the post-extinction faunas seem to have been restocked from the Pacific region, particularly Andean South America, which suffered no major environmental upheaval. However, in subsequent reassessments of their data base, Raup and Sepkoski (1986; Sepkoski and Raup, 1986) have continued to regard the Pliensbachian and Tithonian peaks as true mass extinctions.

Preservation, sampling and sedimentation

It is widely appreciated that both the palaeontological and sedimentological records are incomplete, but does incompleteness imply inadequacy? The fossil record is our only source of information on many aspects of evolution, so the importance of adequacy should not be underestimated. Paul (1982a; 1982b; 1985) has reminded us that incomplete is not inadequate: it is therefore perfectly reasonable to interpret fossil data, rather than just record it. Further, gaps in the biostratigraphic and sedimentologic record can be recognized, or deduced mathematically, and interpreted. Yet, it is still apparent that many species appear to decline and disappear before the extinction event which presumably, although not necessarily, caused their ultimate demise (see, for example, Ward et al., 1986, discussed above). Signor and Lipps (1982) suggested that, in at least some instances, this could be due to two types of sampling effect, rather than being a true decline. First, there is the obvious problem that any species will presumably be randomly distributed in the fossil record throughout its stratigraphic range. This implies that, as we approach the true extinction level of a species, the size of any sample of the taxon must decrease, so there is a reduced probability that it will be found. It is therefore apparent that collecting must be thorough to prevent an artificial truncation of range. It is thus possible that many 'last occurrences' actually pre-date the true extinction of the taxon. Second, this implies that, if last stratigraphic occurrences are random, then even if numerous taxa became extinct at a particular event horizon, their catastrophic decline would appear to be gradual. What should appear as a distinct event will thus be 'smeared' back in time. However, this does not mean that any apparently gradual pattern of extinction before an event horizon is due to 'smearing'; it may or may not reflect the true shape of extinction. What it does suggest is that our sampling must be as comprehensive as possible to minimize any likelihood of error.

Perhaps even more important as causes of truncation of biostratigraphic ranges are undetected hiatuses in the sedimentological record, which Newell (1967) has called 'paraconformities'. These are the antithesis of the Signor–Lipps effect, producing apparently abrupt and simultaneous truncations of ranges in a diversity of taxa whose true demise need not have been catastrophic nor simultaneous. Long sequences of strata may have such 'gaps' totalling between 50% and 90% of the true duration of deposition (Newell, 1982, p. 261). It has even been suggested that the K–T boundary section at Stevns Klint in Denmark is paraconformable (Newell, 1982, Figure 5), although this differs from a recent detailed investigation by Surlyk (in Alvarez, Kauffman *et al.*, 1984, p. 1137, Figure 1). However, Newell's suggestion has received some support from a palynological study by Hultberg (1987), which suggests that deposition of the boundary Fish Clay was diachronous. This places in doubt whether extinction horizons are of utility in stratigraphic correlation, except where synchronous occurrences of multiple features can be proven (see below).

Correlation and succession

Extinction horizons are important in their own right as means of correlating between different environments. It is reasonable to assume, for example, that the end Cretaceous demise of marine organisms such as ammonites, rudists and most species of planktic foraminiferan was synchronous with the end of the dinosaurs and the temporary decline of the angiosperms' dominance on land. It seems improbable that these events could be anything but interrelated. Similarly, Benton (1986) was able to demonstrate Triassic extinctions in the late Carnian and late Norian stages which affected a broad diversity of marine and terrestrial organisms.

However, it is now less certain that the demise of the dinosaurs can be used to recognize the K–T boundary in terrestrial sequences. In North America the K–T boundary is well defined on the basis of microfloral data, so it is reasonably easy to determine whether a dinosaur bone has come from Maastrichtian or Paleocene strata. Fassett (1982; Fassett *et al.*, 1987) recognized that, in the San Juan Basin of New Mexico, the K–T boundary microflora coincided with an iridium anomaly, giving further evidence that this is the true event horizon. However, apparently unreworked dinosaur bones have been recovered from above this palynological and geochemical boundary. Similarly, Paleocene dinosaur remains are known from above this boundary in the Hell Creek Formation of Montana (Rigby *et al.*, 1987).

Conversely, in the marine environment of the Tethytan Realm, rudist bivalves have long been seen as a group which met their demise at the end of the Cretaceous when they were at the peak of their diversity (unlike the dinosaurs, which were in decline). However, the rudists are now recognized to have declined in the mid- or early upper Maastrichtian (Kauffman, 1979; 1984; Donovan, 1987b), with only a weakened stock surviving to the end of the Cretaceous. Thus, in many sections, the 'K–T boundary' has hitherto been plotted too early on the basis of rudist evidence.

SUMMARY

Mass extinctions are geologically brief periods during which there is a global reduction in diversity and biomass. The true tool for the study of mass extinctions is the fossil record, aided, but not replaced, by modern developments in geochemistry (see Orth, Chapter 3, this volume). The fossil record can be utilized either at the level of fine-scale biostratigraphic studies or in analyses of taxonomic compilations (Sepkoski, 1986, pp. 278–80). Both of these approaches give significant results, but they must be interpreted carefully to avoid the numerous potential pitfalls, some of which I have illustrated with examples from the literature.

ACKNOWLEDGMENTS

I thank Mike Benton and Simon Conway-Morris for making constructive comments on an early draft of this chapter.

REFERENCES

Alvarez, L.W., Alvarez, W., Asaro, F. and Michel, H.V., 1980, Extraterrestrial cause for the Cretaceous–Tertiary extinction, *Science*, **208** (4448): 1095–1108.

Alvarez, W., Alvarez, L.W., Asaro, F. and Michel, H.V., 1984, The end of the Cretaceous: sharp boundary or gradual transition?, *Science*, **223** (4641): 1183–6.

Alvarez, W., Kauffman, E.G., Surlyk, F., Alvarez, L.W., Asaro, F. and Michel, H.V., 1984, Impact theory of mass extinctions and the invertebrate fossil record, *Science*, **223** (4641): 1135–41.

Ausich, W.I., Meyer, D.L. and Waters, J.A., 1988, Middle Mississippian blastoid extinction event, *Science*, **240** (4853): 796–8.

Benton, M.J., 1986, More than one event in the late Triassic mass extinction, *Nature*, **321** (6073): 857–61.

Bray, A.A., 1987, Sedimentary petrology and biologic evolution—discussion, *Journal of Sedimentary Petrology*, **57** (4): 795–6.

Brenchley, P.J., 1984, Late Ordovician extinctions and their relationship to the Gondwana glaciation. In P.J. Brenchley (ed.), *Fossils and climate*, Wiley and Sons, Chichester: 291–315.

Brenchley, P.J. and Newall, G., 1984, Late Ordovician environmental changes and their effects on faunas. In D.L. Bruton (ed.), *Aspects of the Ordovician system, Palaeontological Contributions from the University of Oslo*, **295**: 65–79.

Clark, D.L., Wang, C.-Y., Orth, C.J. and Gilmore, J.S., 1986, Conodont survival and low iridium abundances across the Permian–Triassic boundary in south China, *Science*, **233** (4767): 984–6.

Crockett, J.H., Officer, C.B., Wezel, F.C. and Johnson, G.D., 1988, Distribution of noble metals across the Cretaceous/Tertiary boundary at Gubbio, Italy: iridium variation as a constraint on the duration and nature of Cretaceous/Tertiary boundary events, *Geology*, **16** (1): 77–80.

Dao-Yi, X., Shu-Lan, M., Zhi-Fang, C., Xuo-Ying, M., Yi-Ying, S., Qin-Wen, Z. and Zheng-Zhong, Y., 1985, Abundance variation of iridium and trace elements at the Permian/Triassic boundary at Shangsi in China, *Nature*, **314** (6007): 154–6.

Donovan, S.K., 1987a, Iridium anomalous no longer?, *Nature*, **326** (6111): 331–2.

Donovan, S.K., 1987b, How sudden is sudden?, *Nature*, **328** (6126): 109.

Donovan, S.K., 1987c, Confusion at the boundary, *Nature*, **329** (6137): 288.

Donovan, S.K., 1988, The British Ordovician crinoid fauna, *Lethaia*, **21** (4): 424.

Eckert, J.D., 1988, Late Ordovician extinction of North American and British crinoids, *Lethaia*, **21** (2): 147–67.

Fassett, J.E., 1982, Dinosaurs in the San Juan Basin, New Mexico, may have survived the event that resulted in creation of an iridium-enriched zone near the Cretaceous/Tertiary boundary. In L.T. Silver and P.H. Schultz (eds), *Geological implications of impacts of large asteroids and comets on the Earth, Special Paper of the Geological Society of America*, **190**: 435–47.

Fassett, J.E., Lucas, S.G. and O'Neill, F.M., 1987, Dinosaurs, pollen and spores, and the age of the Ojo Alamo Sandstone, San Juan Basin, New Mexico. In J.E. Fassett and J.K. Rigby, Jr (eds), *The Cretaceous–Tertiary boundary in the San Juan and Raton Basins, New Mexico and Colorado, Special Paper of the Geological Society of America*, **209**: 17–34.

Flessa, K.W. and Jablonski, D., 1983, Extinction is here to stay, *Paleobiology*, **9** (4): 315–21.

Hallam, A., 1984, Asteroids and extinctions—no cause for concern, *New Scientist*, **104** (1429): 30–3.

Hallam, A., 1986, The Pliensbachian and Tithonian extinction events, *Nature*, **319** (6056): 765–8.

Hallam, A., 1987, End-Cretaceous mass extinction event: argument for terrestrial causation, *Science*, **238** (4831): 1237–42.

Hancock, J.M., 1967, Some Cretaceous–Tertiary marine faunal changes. In W.B. Harland *et al.* (eds), *The fossil record, a symposium with documentation*, Geological Society, London: 91–104.

Harland, W.B., Cox, A.V., Llewellyn, P.G., Pickton, C.A.G., Smith, A.G. and Walters, R., 1982, *A geologic time scale*, Cambridge University Press, Cambridge.

Hoffman, A., 1985, Patterns of family extinction depend on definition and geological timescale, *Nature*, **315** (6011): 659–62.

Hsü, K.J., Oberhänsli, H., Gao, J.-Y., Shu, S., Haihong, C. and Krähenbuhl, U., 1985, 'Strangelove ocean' before the Cambrian explosion, *Nature*, **316** (6031): 809–11.

Hultberg, S.U., 1987, Palynological evidence for a diachronous low-salinity event in the C–T boundary clay at Stevns Klint, Denmark, *Journal of Micropalaeontology*, **6** (2): 35–40.

Hut, P., Alvarez, W., Elder, W.P., Hansen, T., Kauffman, E.G., Keller, G., Shoemaker, E.M. and Weissman, P.R., 1987. Comet showers as a cause of mass extinction, *Nature*, **329** (6135): 118–26.

Izett, G.A., 1987, Authigenic 'spherules' in K–T boundary sediments at Caravaca, Spain, and Raton Basin, Colorado and New Mexico, may not be impact derived, *Geological Society of America Bulletin*, **99** (1): 78–86.

Jablonski, D., 1986, Causes and consequences of mass extinctions: a comparative approach. In D.K. Elliott (ed.), *Dynamics of extinction*, Wiley and Sons, New York: 183–229.

Kauffman, E.G., 1979, The ecology and biogeography of the Cretaceous–Tertiary extinction event. In W.K. Christensen and T. Birkelund (eds), *Cretaceous–Tertiary boundary events II*, University of Copenhagen, Copenhagen: 29–37.

Kauffman, E.G., 1984, The fabric of Cretaceous marine extinctions. In W.A. Berggren and J.A. Van Couvering (eds), *Catastrophes and earth history*, Princeton University Press, Princeton, NJ: 151–246.

Keller, G., 1986a, Stepwise mass extinctions and impact events: late Eocene to early Oligocene, *Marine Micropaleontology*, **10** (2): 267–93.

Keller, G., 1986b, Late Eocene impact events and stepwise mass extinction. In C. Pomerol and I. Premoli-Silva (eds), *Terminal Eocene events*, Elsevier, Amsterdam: 403–12.

Keller, G., d'Hondt, S.L., Orth, C.J., Gilmore, J.S., Oliver, P.Q., Shoemaker, E.M. and Molina, E., 1987, Late Eocene impact microspherules: stratigraphy, age and geochemistry, *Meteorics*, **22** (1): 25–60.

Keller, G., d'Hondt, S.L. and Vallier, T.L., 1983, Multiple microtektite horizons in Upper Eocene marine sediments: no evidence for mass extinction, *Science*, **221** (4606): 150–2.

Knoll, A.H., 1984, Patterns of extinction in the fossil record of vascular plants. In M.H. Nitecki (ed), *Extinctions*, University of Chicago Press, Chicago: 21–68.

Koch, C.F. and Morgan, J.P., 1988, On the expected distribution of species ranges, *Paleobiology*, **14** (2): 126–38.

McGhee, G.R., Jr, Gilmore, J.S., Orth, C.J. and Olsen, E., 1984, No geochemical evidence for an asteroidal impact at late Devonian mass extinction horizon, *Nature*, **308** (5960): 629–31.

McGhee, G.R., Jr, Orth, C.J., Quintana, L.R., Gilmore, J.S. and Olsen, E.J., 1986, Late Devonian 'Kellwasser Event' mass-extinction horizon in Germany: no geochemical evidence for a large-body impact, *Geology*, **14** (9): 776–9.

McLaren, D.J., 1982, Frasnian–Famennian extinctions. In L.T. Silver and P.H. Schultz (eds), *Geological implications of impacts of large asteroids and comets on the Earth, Special Paper of the Geological Society of America*, **190**: 477–84.

McLaren, D.J., 1983, Impacts that changed the course of evolution, *New Scientist*, **100** (1385): 588–94.

McLaren, D.J., 1985, Mass extinction and iridium anomaly in the Upper Devonian of Western Australia: a commentary, *Geology*, **13** (3): 170–2.

Magaritz, M., Holser, W.T. and Kirschvink, J.L., 1986, Carbon-isotope events across the Precambrian–Cambrian boundary on the Siberian Platform, *Nature*, **320** (6059): 258–9.

Martin, P.S., 1967, Prehistoric overkill. In P.S. Martin and H.E. Wright Jr (eds), *Pleistocene extinctions, the search for a cause*, Yale University Press, New Haven, CT: 75–120.

Martin, P.S., 1986, Refuting late Pleistocene extinction models. In D.K. Elliott (ed.), *Dynamics of extinction*, Wiley and Sons, New York: 107–30.

Muarrasse, F.J.-M.R., Pierre-Louis, F. and Rigaud, J.J.-G., 1985, Upper Cretaceous to Lower Palaeocene pelagic calcareous deposits in the Southern Peninsula of Haiti: their bearing on the problem of the Cretaceous–Tertiary boundary,

Transactions of the 4th Latin-American Geological Congress, Port-of-Spain, Trinidad, 1979: 328–38.

Newell, N.D., 1956, Fossil populations, in P.C. Sylvester-Bradley (ed.), *The species concept in palaeontology, Systematics Association Publication*, **2**: 63–82.

Newell, N.D., 1967, Paraconformities. In C. Teichert and E.L. Yochelson (eds), *Essays in paleontology and stratigraphy, R.C. Moore Commemorative Volume*, Lawrence, University of Kansas Press: 349–67.

Newell, N.D., 1982, Mass extinctions—illusions or realities? In L.T. Silver and P.H. Schultz (eds), *Geological implications of impacts of large asteroids and comets on the Earth, Special Paper of the Geological Society of America*, **190**, 257–63.

Nichols, D.J., Jarzen, D.M., Orth, C.J. and Oliver, P.Q., 1986, Palynological and iridium anomalies at Cretaceous–Tertiary boundary, south-central Saskatchewan, *Science*, **231** (4739): 714–17.

Officer, C.B. and Drake, C.L., 1983, Cretaceous–Tertiary transition, *Science*, **219** (4591): 1383–90.

Officer, C.B. and Drake, C.L. 1985a, Terminal Cretaceous environmental events, *Science*, **227** (4691): 1161–7.

Officer, C.B. and Drake, C.L. 1985b, Cretaceous–Tertiary extinctions: alternative models, *Science*, **230** (4731): 1294–5.

Officer, C.B., Hallam, A., Drake, C.L. and Devine, J.D., 1987, Late Cretaceous and paroxysmal Cretaceous/Tertiary extinctions, *Nature*, **326** (6109): 143–9.

Orth, C.J., Gilmore, J.S., Quintana, L.R. and Sheehan, P.M., 1986, Terminal Ordovician extinction: geochemical analysis of the Ordovician/Silurian boundary, Anticosti Island, Quebec, *Geology*, **14** (5): 433–6.

Orth, C.J., Knight, J.D., Quintana, L.R., Gilmore, J.S. and Palmer, A.R., 1984, A search for iridium abundance anomalies at two late Cambrian biomere boundaries in western Utah, *Science*, **223** (4633): 163–5.

Palmer, A.R., 1982, Biomere boundaries: a possible test for extraterrestrial perturbation of the biosphere. In L.T. Silver and P.H. Schultz (eds), *Geological implications of impacts of large asteroids and comets on the Earth, Special Paper of the Geological Society of America*, **190**: 469–75.

Patterson, C. and Smith, A.B., 1987, Is the periodicity of extinctions a taxonomic artefact?, *Nature*, **330** (6145): 248–51.

Paul, C.R.C., 1982a, The adequacy of the fossil record. In K.A. Joysey and A.E. Friday (eds), *Problems of phylogenetic reconstruction, Systematics Association Special Volume*, **21**: 75–117.

Paul, C.R.C., 1982b, The completeness of the echinoderm fossil record. In J.M. Lawrence (ed.), *Proceedings of the 4th International Echinoderms Conference, Tampa Bay, 14–17 September, 1981*, Balkema, Amsterdam: 89.

Paul, C.R.C., 1985, The adequacy of the fossil record reconsidered, *Special Papers in Palaeontology*, **33**: 7–15.

Playford, P.E., McLaren, D.J., Orth, C.J., Gilmore, J.S. and Goodfellow, W.D., 1984, Iridium anomaly in the Upper Devonian of the Canning Basin, Western Australia, *Science*, **226** (4673): 437–9.

Raup, D.M., 1982, Biogeographic extinction: a feasibility test. In L.T. Silver and P.H. Schultz (eds), *Geological implications of impacts of large asteroids and comets on the Earth, Special Paper of the Geological Society of America*, **190**: 277–81.

Raup, D.M., 1986, Biological extinction in Earth history, *Science*, **231** (4745): 1528–33.

Raup, D.M., 1987, Mass extinctions: a commentary, *Palaeontology*, **30** (1): 1–13.

Raup, D.M. and Boyajian, G.E., 1988, Patterns of generic extinction in the fossil record, *Paleobiology*, **14** (2): 109–25.

Raup, D.M. and Sepkoski, J.J., Jr, 1984, Periodicity of extinctions in the geologic past, *Proceedings of the National Academy of Science U.S.A.*, **81** (3): 801–5.

Raup, D.M., Sepkoski, J.J., Jr, 1986, Periodic extinction of families and genera, *Science*, **231** (4740): 833–6.

Raup, D.M., Sepkoski, J.J., Jr, 1988, Testing for periodicity of extinction, *Science*, **241** (4861): 94–6.

Rhodes, F.H.T., 1967, Permo-Triassic extinction. In W.B. Harland *et al.* (eds), *The fossil record, a symposium with documentation*, Geological Society, London: 57–76.

Rigby, J.K., Jr, Newman, K.R., Smit, J., van der Kaars, S., Sloan, R.E. and Rigby, J.K., 1987, Dinosaurs from the Paleocene part of the Hell Creek Formation, McCone County, Montana, *Palaios*, **2** (3): 296–302.

Sepkoski, J.J., Jr, 1982, A compendium of fossil marine families, *Milwaukee Public Museum Contributions to Biology and Geology*, **51**: 1–125.

Sepkoski, J.J., Jr, 1986, Phanerozoic overview of mass extinction. In D.M. Raup and D. Jablonski (eds), *Patterns and processes in the history of life*, Springer-Verlag, Berlin: 277–95.

Sepkoski, J.J., Jr, 1987, Reply to Patterson and Smith, *Nature*, **330** (6145): 251–2.

Sepkoski, J.J., Jr and Raup, D.M., 1986, Periodicity in marine extinction events. In D.K. Elliott (ed.), *Dynamics of extinction*, Wiley and Sons, New York: 3–36.

Shaw, A.B., 1971, The butterfingered handmaiden, *Journal of Paleontology*, **45** (1): 1–5.

Sheldon, P.R., 1987, Parallel gradualistic evolution of Ordovician trilobites, *Nature*, **330** (6148): 561–3.

Sheldon, P.R., 1988, Making the most of the evolution diaries, *New Scientist*, **117** (1596): 52–4.

Signor, P.W., III and Lipps, J.H., 1982, Sampling bias, gradual extinction patterns and catastrophes in the fossil record. In L.T. Silver and P.H. Schultz (eds), *Geological implications of impacts of large asteroids and comets on the Earth*, *Special Paper of the Geological Society of America*, **190**: 291–6.

Simms, M.J., 1988, The phylogeny of post-Palaeozoic crinoids. In C.R.C. Paul and A.B. Smith (eds), *Echinoderm phylogeny and evolutionary biology*, Oxford University Press, Oxford: 269–84.

Smith, A.B. and Paul, C.R.C., 1985, Variation in the irregular echinoid *Discoides* during the early Cenomanian, *Special Papers in Palaeontology*, **33**: 29–37.

Stanley, S.M., 1984, Marine mass extinctions: a dominant role for temperature. In M.H. Nitecki (ed.), *Extinction*, University of Chicago Press, Chicago: 69–117.

Stanley, S.M., 1986, Anatomy of a regional mass extinction: Plio-Pleistocene decimation of the western Atlantic bivalve fauna, *Palaios*, **1** (1): 17–36.

Stanley, S.M., 1987, *Extinction*, Scientific American Books, New York.

Stanley, S.M. and Campbell, L.D., 1981, Neogene mass extinction of western Atlantic molluscs, *Nature*, **293** (5832): 457–9.

Stigler, S.M. and Wagner, M.J., 1987, A substantial bias in nonparametric tests for periodicity in geophysical data, *Science*, **238** (4829): 940–5.

Stigler, S.M. and Wagner, M.J., 1988, Response to Raup and Sepkoski, *Science*, **241** (4861): 96–9.

Surlyk, F. and Johansen, M.B., 1984, End-Cretaceous brachiopod extinctions in the Chalk of Denmark, *Science*, **223** (4641): 1174–7.

Teichert, C., 1986, Times of crisis in the evolution of the Cephalopoda, *Paläontologische Zeitschrift*, **60** (3–4): 227–43.

Tschudy, R.H. and Tschudy, B.D., 1986, Extinction and survival of plant life following the Cretaceous/Tertiary boundary event, Western Interior, North America, *Geology*, **14** (8): 667–70.

Vermeij, G.J., 1986, Survival during biotic crises: the properties and evolutionary significance of refuges. In D.K. Elliott (ed.), *Dynamics of extinction*, Wiley and Sons, New York: 231–46.

Ward, P., Wiedmann, J. and Mount, J.F., 1986, Maastrichtian molluscan biostratigraphy and extinction patterns in a Cretaceous/Tertiary boundary section exposed at Zumaya, Spain, *Geology*, **14** (11): 899–903.

Westrop, S.R. and Ludvigsen, R., 1987, Biogeographic control of trilobite mass extinctions at an Upper Cambrian 'biomere' boundary, *Paleobiology*, **13** (1): 84–99.

Wilde, P., Berry, W.B.N., Quinby-Hunt, M.S., Orth, C.J., Quintana, L.R. and Gilmore, J.S., 1986, Iridium abundance across the Ordovician–Silurian stratotype, *Science*, **233** (4761): 339–41.

Chapter 3

GEOCHEMISTRY OF THE BIO-EVENT HORIZONS*

Charles J. Orth

INTRODUCTION

This chapter has its origins in the now classic discovery, and hypothesis, reported by Alvarez *et al.* (1979; 1980). Examining rock samples taken across the Cretaceous–Tertiary (K–T) boundary at marine sedimentary sequences in Italy and Denmark, they found marked enrichment of iridium (Ir) exactly at the boundary in a thin clay parting (Figure 3.1). Relative to Solar system abundances, the Earth's crust is depleted in Ir. Alvarez and his co-workers hypothesized that the excess Ir and the clay layer came from the fallout of ejecta from impact of a large (and undifferentiated) extra-terrestrial body: an asteroid of about 10 km diameter.

In their 1980 *Science* article, Alvarez *et al.* suggested that the several months' darkness from the dust cloud would have suppressed photosynthesis, brought about the collapse of food chains and caused death by starvation: the extinction mechanism. The large-body impact aspect of their hypothesis is now supported by a large volume of hard chemical and physical evidence, and counter-arguments that suggest a volcanic source for the excess Ir at the K–T boundary are not as convincing. There are, however, some indications that deep volcanic and eruptive processes might have been initiated by the impact, possibly from a combination of excavation and fracture of the crust from the tremendous shock; the energy released by the impact of an object 10 km in diameter, of density 2 g/cm^3 and travelling at 20 km/s is about 10 000 times the energy equivalent of the

*This chapter was prepared under the auspices of the US Department of Energy, Los Alamos National Laboratory

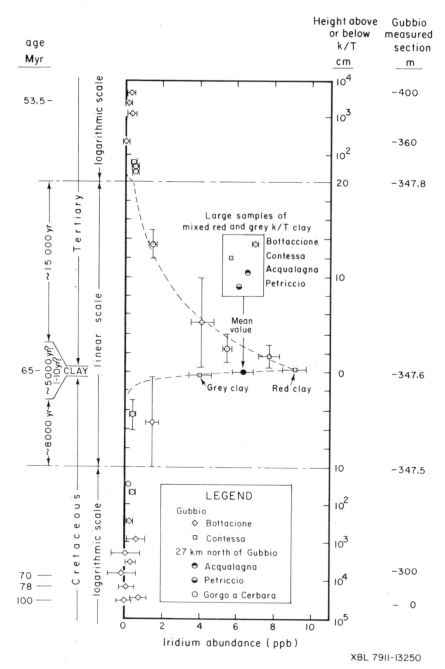

Figure 3.1. Iridium abundances per unit weight of 2M nitric acid-insoluble residues from Italian marine limestones near the Cretaceous–Tertiary boundary. Note that the vertical scale is linear in the vicinity of the K–T boundary, where details are important, but changes to logarithmic to show results from 350 m below to 50 m above the boundary (Alvarez et al., 1980).

world's nuclear arsenals. There also is evidence from soot in boundary clay of global wildfires accompanying the event (Wolbach *et al.*, 1985), which suggests a scenario of 'impact winter', nature's equivalent of 'nuclear winter'. Whether or not the impact event was responsible for the extinction, or was merely a contributing factor to a process already under way, is currently being tested and debated by palaeontologists.

Whatever the outcome of the present K–T investigations, the discovery by the Alvarez team has had a profound influence on the way that we now look at the extinctions of life in the fossil record. More than ever before, palaeontologists and biostratigraphers are examining their sections at centimetre and decimetre, rather than at metre or multi-metre, intervals, and geochemistry, the subject of this chapter, and long a tool of mineral exploration, meteoritics and oceanography, is making important contributions to our knowledge of the bio-events in the geologic record.

An immediate question that arose from the Alvarez *et al.* discovery was whether similar events were associated with one or more of the other mass extinctions. The Earth was struck numerous times during the Phanerozoic by large Solar system bodies (asteroids, comets and large meteorites), as evidenced by massive impact craters. Damage, at least to the local environment, must have been severe. Since 1981 this problem has received much attention from several research groups. The study of extinctions and their causes received new impetus when Raup and Sepkoski (1984), from their statistical analysis of disappearance rates, announced that extinctions, at least over the last 250 million years, appear to have occurred at more or less regular intervals with a period of about 26 million years. Their report led to statistical scrutiny of the ages of terrestrial impact structures by researchers (Alvarez and Muller, 1984; Rampino and Stothers, 1984), who soon reported a similar periodicity in crater age distributions. Several astrophysical mechanisms were suggested (Rampino and Stothers, 1984; Whitmire and Jackson, 1984; Davis *et al.*, 1984) that might modulate the comet flux in the inner Solar system and increase the chances for comet collisions with Earth every 26–33 million years.

Thus far, the best supporting evidence for the cyclic comet swarm hypotheses is found in the Upper Eocene record. In sedimentary sequences of approximately 36 million years ago (about 30 million years after the K–T event), two and possibly three microspherule horizons that were laid down within a 500 000-year interval have been found. A moderate Ir anomaly, about 1% the intensity of the K–T anomaly, is associated with the middle horizon, and its signatures have been tracked from the Caribbean to the equatorial Pacific and to the Indian Ocean. A recent discovery of a moderate Ir anomaly in the Middle Miocene (Asaro *et al.*, 1988) might provide another case of an impact–extinction relationship on the 26-million-year cycle pattern. Evidence of a rather large Antarctic Ocean impact in the Pliocene has been reported, but little is currently known about its effect on the environment. Several moderate Ir (platinum-group element) anomalies have been discovered at extinction boundaries that are older than the K–T event, but these appear to be the result of chemical enrichment processes in the Palaeozoic and Mesozoic oceans. The reported geochemical results on the extinction boundaries will be discussed in detail later in the chapter, after a brief discussion of the techniques and methods that are used.

METHODS

In the evolution of the study of bio-event horizons, many disciplines have joined forces in an effort to provide knowledge and answers to palaeonto-logical phenomena. In this section, the measurement techniques used in the geochemical studies of the extinction boundaries will be examined. The intention is to describe the different methods and the resultant data in a brief, pedestrian manner, so that the non-expert reader gets a grasp for what is going on, but does not get lulled into sleep by the technical details. Briefly, the types of data that the geochemist seeks to provide and interpret are: elemental abundance and ratio patterns, with emphasis on the rare platinum group; isotopic ratios, especially for carbon, oxygen and sulphur, but also for hydrogen and nitrogen; mineral composition; and the presence or absence of soot, microspherules, shocked minerals and volcanic shards. The first two types of data are considered in detail below.

Elemental abundances and ratios

Abundances for a large number of elements are generally determined by neutron activation analysis (NAA), because no chemical treatment of the samples is required, as it is in inductively coupled plasma emission spectro-scopy (ICP) and in X-ray fluorescence (XRF) measurements, where good precision is required. However, ICP and XRF also provide good sensitivity and complement the NAA measurements for elements that cannot be determined by the nuclear technique. In NAA the sample is exposed to thermal (very low-energy) neutrons in a nuclear reactor. The atomic nucleus captures a neutron, gaining the neutron mass and thereby becom-ing the next heavier isotope of the element. In general this isotope is unstable to nuclear decay and emits a beta particle and gamma rays in order to shed the excess energy, becoming a stable daughter element with atomic number (Z) increased by one—for example ^{46}Sc ($Z = 21$) decays to stable ^{46}Ti ($Z = 22$). It is the emission of mono-energetic gamma rays that makes NAA work.

In a rock sample almost every element in the periodic table is present and the resulting gamma ray spectrum, especially at early times following the neutron irradiation, is extremely complex. However, the use of high-resolution germanium detectors and proper timing of counting provides good data for about 40 elements. Timing is important because half-lives of the radionuclides range from seconds to years. The analyst arranges the counting periods to optimize certain elements. A typical gamma-ray spectrum of a shale sample is shown in Figure 3.2. Calibration is accomplished by comparing counts in the sample to be measured with those in standard rock samples obtainable from sources such as the International Atomic Energy Agency (IAEA), National Bureau of Standards and Geological Surveys, and from the irradiation of weighed amounts of pure elements or accurately known mixtures. Nickel (Ni) concentrations below about 1000 parts per million (ppm) cannot be determined with thermal neutrons because the combination of the low abundance and small cross-section of stable ^{64}Ni and the low gamma ray yield of the neutron capture product (^{65}Ni) is quite small. However, high-energy neutrons produce radioactive ^{58}Co via the neutron-in proton-out reaction on abundant ^{58}Ni, and therefore Ni determinations are made by irradiating the samples in or near the core of a reactor where neutron energies are highest.

The rare platinum (Pt)-group elements are generally the most important in discussions of the geochemsitry of the extinction boundaries. Unless some process has greatly enriched these elements (Ru, Rh, Pd, Os, Ir and Pt), they occur at parts-per-trillion (ppt) levels in Earth crustal rocks and, except for Ir, radiochemical separation procedures are required to isolate them from the other interfering radionuclides. Of the Pt group, the ideal nuclear properties of Ir make it the most sensitive for detection; at Los Alamos we can detect 0.5 ppt. This is more than adequate, even for sandstones and silicic volcanic ashes, where the Ir concentration is generally greater than 1 ppt.

Recently, the Lawrence Berkeley Laboratory team built an instrument they call an 'iridium spectrometer'. With it they can instrumentally measure iridium concentrations as low as 5–10 ppt. The instrument utilizes the fact that ^{192}Ir decays with the emission of several gamma rays and two of the strongest are emitted in prompt coincidence (one after the other). By using two germanium detectors, coupled in parallel to both fast electronics and to high-resolution circuitry set on the precise energy of the two Ir gamma rays, the interference from other radionuclides to the ^{192}Ir coincident pulses is greatly reduced. Then, by surrounding the germanium detectors (and the sample being counted) with a large scintillation counter, scattered electromagnetic radiation from high-energy coincident gamma rays emitted by abundant nuclides such as ^{46}Sc and ^{60}Co can be cancelled, further improving the signal-to-noise ratio.

The other Pt-group elements do not contain the optimal nuclear properties of Ir and radiochemical isolation is always required. However, at present, neutron activation methods are more sensitive than non-nuclear measurement techniques. In general, radiochemical separations utilize ion exchange resin columns. The sample is dissolved either by fusion with a hydroxide–peroxide flux or in strong acids. Os and Ru require special precautions in the dissolving step because they are easily oxidized to the octavalent state and escape from the solution as OsO_4 or RuO_4 gases;

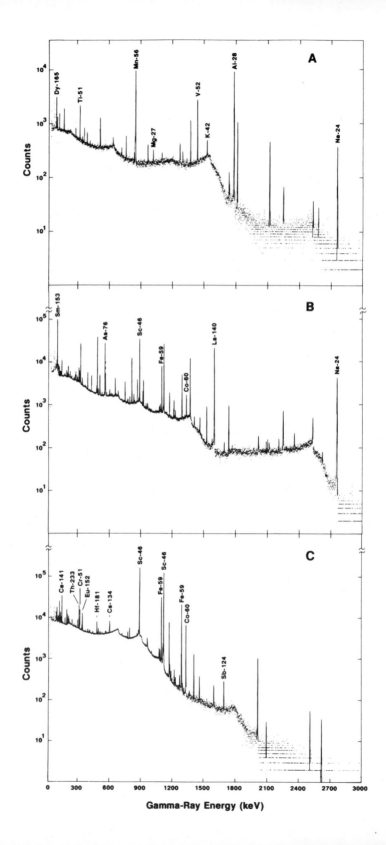

careful fusions are required. Following dissolution of the sample, an appropriate solution is poured onto the top of the column and then interfering radionuclides are selectively washed away before the Pt-group element is removed, or else the Pt-group elements are removed before the other elements. Some procedures isolate only one of the members of the Pt group, while others are capable of sequentially isolating three or more of the group with a single ion column.

The interpretation of Ir anomalies is not always straightforward and Ir results should be interpreted in light of all other available elemental abundances and ratios, the presence or absence of microspherules and shocked-mineral grains, fossil bacteria, and microfauna. In carbonate–shale sequences it is usually necessary to plot the Ir abundance pattern on a carbonate-free basis, or the Ir/Al ratio, to avoid false anomalies caused by diagenetic dissolution of carbonates.

X-ray fluorescence
The standard method for determining abundances of the common elements in rocks is by the use of X-ray fluorescence (XRF) techniques. In the XRF spectrometer continuous X-rays, enhanced by characteristic X-rays from the electron target, such as copper or tungsten, are focused (collimated) onto the rock sample. Characteristic X-rays (generally the K, L, M and N series X-rays) are produced in the atoms of the rock and detected by a high-resolution silicon crystal that is coupled to electronic amplification and counting circuitry. A small dedicated computer is generally used to analyse the complex X-ray spectra and convert the peak areas (number of X-ray events) into concentration data, much as in NAA spectra of the higher-energy gamma rays. For good rock analysis the sample should be finely pulverized and for best results the samples are generally fused with lithium borate to obtain a homogenous system.

High spatial resolution measurements
For determining abundances in micron-sized grains in rocks electron microscopy (scanning electron microscopy (SEM), for example) and proton-induced X-ray emission (PIXE) are used to produce the characteristic X-rays. In SEM a narrow beam of electrons is focused onto the target. PIXE offers greater sensitivity because there is no bremsstrahlung (continuous energy X-rays) background. However, PIXE is expensive because it requires the use of an accelerator, such as a Van de Graaff, to produce the proton projectiles.

Figure 3.2. Gamma ray spectra of marine shale taken at three intervals of time after a 45-second irradiation with reactor thermal neutrons: (a) after 20 minutes of decay; (b) after 5 days; (c) after 21 days. Element and mass number assignments are given for some of the peaks (S.R. Garcia).

Inductively coupled plasma (ICP) emission spectroscopy
This method offers good to excellent sensitivity for most elements in the periodic table, so it provides both an alternative and complement to NAA. A solution of the sample is drawn into a nebulizer and the vapour/aerosol is fed with argon gas into a radio frequency (RF) driven ion source to form the plasma and cause the emission of spectral lines. The wavelengths and brightness (intensity) of these emission lines define the element and its concentration, respectively.

Isotopic ratios

Although generally difficult to interpret, isotopic data provide another source of information about environmental conditions in the geologic record. Here, we are talking about small changes in mass in the same element. Therefore, the lighter the element, the greater the percentage change in adjacent isotopes. For example, the mass difference between the stable hydrogen isotopes, ^1H and ^2H, is 100%, compared with 6% between ^{32}S and ^{34}S (two mass units different). At uranium, the difference between adjacent masses is a mere 0.4%. Isotopic measurements are performed with mass spectrometers, where the element of interest is usually converted to gas molecules which are then ionized by electron bombardment or by heating in the presence of a high voltage potential. The high-velocity ions are focused and directed through a strong magnetic field which deflects the lighter isotopes more than the heavier ones. Controlled raising and lowering of the magnetic field 'sweeps' the different mass ions into a detector, which generates secondary electrons from collision of the ions with a dynode. The scattered electrons are multiplied by a series of dynodes and the resultant pulse is stored in a pulse-counting electronic module. The lighter elements (up to and including sulphur), with the exception of nitrogen, are usually converted to a compound gas molecule (for example, CO_2, SO_2, H_2O) and the compound mass is collected.

In geologic systems, very small differences occur in the isotopic ratios relative to terrestrial averages and, therefore, the mass distributions are compared with those for international standards. For carbon and oxygen in carbonates, Peedee Formation belemnite (PDB) is the generally accepted standard. Here, the small difference in ratios between sample and standard is measured and reported as $\delta^{13}C$ or $\delta^{18}O$ in units of parts per mille. Precision in measurement to better than 1 part in 10 000 is required.

It is generally difficult, if not impossible, to separate out a single primary source of isotope distribution. For carbon ($\delta^{13}C$) in limestone, more positive values could be the result of primary processes such as upwelling of deeper water onto the platform regions, increased photosynthesis which preferentially utilizes ^{12}C, or burial of organic carbon which is enriched in ^{12}C, and therefore removes ^{12}C from the water column. Oxygen is particularly bad, because not only are there several primary enrichment processes, but, because of its lability, it is also sensitive to diagenetic perturbations. Isotopic exchange with freshwater, for example, which is richer in ^{16}O, reduces $\delta^{18}O$.

Heavier elements like strontium (Sr) and neodymium (Nd) have also been used to provide information about the origins of sedimentary units. Isotopes of these two elements are not fractionated by biologic processes, and isotopes of Sr in seawater are a balance of heavier isotopes from old continental granites and lighter isotopes from younger basalts such as from the mid-ocean ridges (Holser *et al.*, 1986).

The osmium ratio, $^{187}Os/^{186}Os$, has been used as a diagnostic indicator of cosmic osmium in high Pt-group element layers, such as that at the K–T boundary. The argument used here is that the ratio of long-lived ^{187}Re to stable ^{186}Os in meteorites is 3.2 and in crustal rocks 400, resulting in different $^{187}Os/^{186}Os$ ratios after billions of years have passed. Meteorites now have a ratio of about unity and the Earth's crust averages (assuming no crustal cycling through the mantle) between 13 and 30. Measurements on the K–T boundary layer in Denmark and the Raton Basin by Luck and Turekian (1983) yielded ratios of 1.65 and 1.29. They concluded that the osmium could be from either a mantle or extraterrestrial source, and if extraterrestrial, and no crustal-derived contamination was present, the Denmark and Raton Basin Os (also Ir, etc.) were derived from two separate events.

The unlikely possibility that a nearby supernova burst was involved in the terminal Cretaceous event and at several other bio-event horizons has been tested by searching for the presence of plutonium-244 (Alvarez *et al.*, 1980; Orth *et al.*, 1981). This is the heaviest isotope of plutonium with a long half-life (~80 million years), but not so long that the isotope measurably exists in nature. Therefore, its presence would most likely be the result of production by a supernova burst. If we assume to first order that the ratio of heavy element formation in such a burst is similar to solar system abundances, we can estimate how much ^{244}Pu relative to Ir, after about one half-life of plutonium decay, should be found if the excess Ir at the K–T boundary were the result of such an event. There is no evidence of ^{244}Pu at the K–T or at the other boundaries examined thus far. However, the decay of ^{244}Pu renders its detection increasingly difficult with age; at the Cambro-Ordovician boundary it would have gone through six half lives (only 1.6% of what might have been there originally would have survived radioactive decay).

THE BIO-EVENT HORIZONS: GEOCHEMICAL RESULTS

In this section results of some chemical and physical measurements across the bio-event horizons are presented. It would be impossible to discuss everything that has been written on the subject and no pretence is made that this has been done. However, there should be enough information to form a basis for the interested reader to build on and perhaps contribute to our knowledge about the various extinctions and abrupt biological changes in the fossil record.

Rather than arrange the bio-events in order, this section begins with the terminal Cretaceous event, upon which the geochemical approach and methods were established, and then proceeds chronologically from the Precambrian to the latest Cenozoic.

Cretaceous–Tertiary (K–T) boundary

This discussion begins with the discovery by the Alvarez team of the Ir anomaly at the K–T boundary in a marine sedimentary sequence near Gubbio, Italy. Walter Alvarez, while at Lamont, had spent considerable time working out the palaeomagnetics of the very complete section of Upper Cretaceous and Lower Tertiary pelagic limestones in Bottaccione Gorge, which cuts through the mountains behind the town of Gubbio. Luterbacher and Premoli-Silva (1964) had discovered that there was a distinctive clay bed precisely at the K–T boundary and that all Cretaceous foraminifers, with the exception of the smallest taxa, *Globigerina eugubina* disappeared at the base of the clay. In the summer of 1977 Alvarez collected samples from these boundary beds and took them with him when he joined the faculty of the University of California at Berkeley that autumn. His father, Luis Alvarez, a Nobel laureate in physics, suggested they determine how long it took to deposit the layer of clay at the boundary by measuring its iridium content. This began a collaboration with Berkeley nuclear chemists, Frank Asaro and Helen Michel. To their great surprise they found much more Ir than could be explained by the normal rain of micrometeorites. Because of their knowledge and ingenuity, they recognized the source of the excess Ir for what more and more evidence has shown it to be: large-body impact(s) with the Earth.

The Alvarez discovery was soon followed by measurements on the other rare siderophile elements such as palladium, osmium, platinum and gold. Ganapathy (1980) reported that the ratios of these elements to each other in the Denmark boundary clay closely matched those of the chondritic or cosmic pattern. Kyte *et al.* (1980) reported similar results, but noted that these abundance patterns also resemble those observed in iron meteorites. Therefore, although the ratios are meteoritic, it is difficult to extract the class of meteorite from them.

The early reports of the K–T Ir anomaly were all from marine sequences and sceptics of the impact hypothesis argued that the enrichment of Ir was merely the result of geochemical enrichment processes in the Cretaceous ocean. If the anomaly could be found in continental sequences this argument would be weakened.

In 1980, Orth and his Los Alamos co-workers, Jere Knight and Jim Gilmore, in collaboration with Bob Tschudy and Chuck Pillmore at the

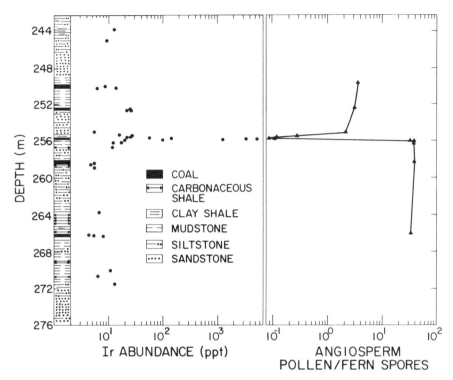

Figure 3.3. Iridium abundances and ratios of angiosperm pollen to fern spores as a function of depth and lithology in a core taken from freshwater coal swamp deposits in York Canyon, New Mexico. Note the logarithmic scale for the Ir concentrations, which are given in parts per trillion (Orth *et al.*, 1981).

United States Geological Survey (USGS) in Denver, searched for and soon found the strong Ir anomaly at the roughly-defined palynological K–T boundary in a core taken from a freshwater coal swamp sequence west of Raton, New Mexico (Figure 3.3). Subsequent higher-resolution studies by Tschudy indicated a sharp pollen break precisely at the Ir spike; several Cretaceous pollen taxa disappeared and fern spores suddenly dominated over angiosperm pollen. Over the following couple of years the Los Alamos–USGS Denver group found numerous excellent K–T boundary exposures in road cuts in the Raton Basin, and in 1983, working with a group led by Bill Clemens, found essentially identical boundary sections 1300 km to the north near Jordan, Montana.

It was in samples collected by Bruce Bohor and Don Triplehorn (USGS Denver) from their Brownie Butte, Montana, section that these investigators made their striking discovery which was to provide strong evidence for the impact hypothesis: the discovery of shocked-quartz grains in the thin Ir-rich layer (Bohor *et al.*, 1984). Shocked quartz has been found at many well-documented impact sites (Bunch, 1968; Robertson *et al.*, 1968) and at the sites of nuclear weapons tests (Short, 1968). Although Carter *et al.* (1986) have reported shocked grains in Toba ash, their paper has received much criticism from other authorities who claim the Toba features are not the same as observed at the K–T boundary and at impact sites. Further work by Bohor and his co-workers has located shocked-quartz grains in every well-preserved Ir-rich boundary clay they have examined, freshwater or marine, although in some cases the grains are extremely rare. A micrograph of one of these grains from the Brownie Butte section is shown in Figure 3.4.

Another interesting puzzle comes from comparisons of the North American sections with the more than 60 other occurrences (all in marine rock sequences) of the K–T Ir anomaly currently reported from all around the globe. In the former there is a thin (about 1–2.5 cm) kaolinitic clay (the alteration product generally found as the result of the acidic conditions of the peat bogs preceding coalification). This thin clay, which contains relatively low Ir concentrations, is overlain by the 2–5-mm thick Ir-rich layer that contains abundant shocked-quartz grains. A typical boundary

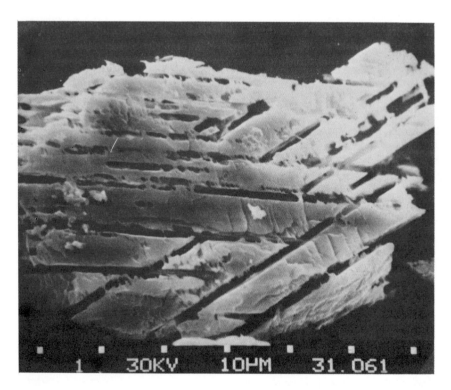

Figure 3.4. Photograph of shocked-quartz grain from a section at Brownie Butte, Montana (Bohor *et al.*, 1984).

zone in the Raton Basin is shown in Figure 3.5. Orth *et al.* (1987) have interpreted the kaolinitic clay as being derived from low-angle ejecta, suggesting that the North American sites were closer to the impact (or one of the impacts, if it was a multiple-impact event). In this scenario, the overlying Ir-rich layer with abundant shocked quartz would have resulted from bolide-rich material ejected at back angles into the stratosphere, which fell back considerably later. It is apparently the thin Ir-rich layer that is observed in well-preserved sections elsewhere around the globe.

The kaolinitic clay, to the untrained eye, resembles the tonsteins that are common in late Cretaceous and early Tertiary coal beds in the Western Interior. However, the K–T bed weathers to a greyish, and sometimes a pinkish, colour, in contrast to the usual brown or tan of the tonsteins derived from volcanic ash. Geochemical measurements, apart from Ir analyses, indicate a two- to threefold enrichment of titanium (Ti) over other tonsteins and clays, and Ti has been used as an easily determined indicator of the boundary (Gilmore *et al.*, 1984; Orth *et al.*, 1987). In general, rare-earth concentrations are lower and Cr, Sc and V higher than in other tonsteins. Hildebrand and Boynton (1987) have discussed the rare-earth pattern at the K–T boundary and suggested that the low concentrations with relative enhancement of the heavy rare earths over the light ones is typical of rather deep ocean crust (from the impact site).

Figure 3.5. Cretaceous–Tertiary boundary zone in a continental site at a roadcut in Berwind Canyon, 10 km north of Trinidad, Colorado. The arrow points to the conspicuous kaolinitic boundary clay (Orth *et al.*, 1987).

The kaolinitic clay bed is a feature thus far unique to the western North America locations, which might have been close to an impact site. A local impact structure that might have been coeval with the K–T boundary is the Manson structure in Iowa. Its diameter is about 35 km, which is too small for the proposed bolide, greater than 10 km in diameter. However, if the K–T event was associated with multiple impacts, the Manson structure could fit into the picture. Target rock in its vicinity should carry the elemental abundance patterns observed in the kaolinitic clay, that is, excess Ti and a deficiency of rare earths.

The impact scenario first hypothesized by Alvarez *et al.* (1980) is not universally accepted and massive volcanism has been suggested as an alternative for the cause of the extinction, the Pt-group metal anomalies, and the shocked-quartz and microspherules at the K–T boundary. The leading proponents for the volcanism scenario are Officer and Drake (1983; 1985) and McLean (1982; 1985). Officer and Drake argue that their analysis of the K–T boundary Ir anomaly indicates that the Ir was laid down over a time interval spanning 10 000 to 100 000 years, which is more compatible with a series of intense eruptive volcanic events occurring over a geologically short period of time. They suggest that this volcanism was related to an immense increase in mantle-plume (hot spot) activity that probably occurred simultaneously at several localities throughout the world including the Deccan traps flood basalt. Zoller *et al.* (1983) discovered high values of Ir in airborne particles from Kilauea, lending some support to the mantle-plume hypothesis. Olmez *et al.* (1986) suggested that the 10^7 km^3 of Deccan flood basalt might have emitted 3×10^{10} g of Ir, assuming it had the same Ir content as the Kilauea magma (0.28 parts per billion (ppb) and degassed the same proportion of its Ir (3 per mille). However, the Ir content of Deccan basalt ranges from 0.006 to 0.026 ppb, and therefore the global deposition would amount to from 1 to 5×10^{-10} g/cm^2 compared to the 4 to 10×10^{-8} g/cm^2 observed at the K–T boundary.

There is a global excess of chalcophiles, As, Sb and Se, at the K–T boundary and Officer and Drake (1985) interpret the chalcophile excess as a further indication that the K–T event was the result of primary volcanic processes. Orth *et al.* (1987) suggested as an alternative that deep-source volcanism might have accompanied the impact as a result of excavation and rupture of the crust.

Space limitations do not permit further discussion of the K–T boundary controversy. A thorough discussion of the impact–volcanism evidence and interpretations has been given by Alvarez (1986).

Precambrian–Cambrian boundary

Rather than an extinction of life, the Precambrian–Cambrian (PЄ–Є) boundary is roughly defined as the first appearance of hard skeletal parts in the fossil record. Although there is a world-wide distribution of stratotype candidates, only three areas are considered as realistic choices. An excellent discussion of the PЄ–Є boundary and the search for a stratotype has been given by Morris (1987). The candidate sections for the stratotype are located on the Aldan River (Ulakhan–Sulugur section) in eastern Siberia, near the town of Meishucun in Yunnan Province near Kunming, and on the Burin Peninsula in southeastern Newfoundland.

Iridium measurements on the PЄ–Є boundary were first reported by Nazarov *et al.* (1983), who discovered an Ir anomaly of 0.17 ppb, a factor of six above local background in the Ulakhan–Sulugur section. The anomaly was observed at the very base of the Cambrian in a glauconitic dolomite. Essentially identical samples were collected by J.R. Kirschvink and analysed for Ir and other elemental abundances in 1984 at the Los Alamos National Laboratory by Gilmore and Orth and at the Lawrence Berkeley Laboratory by Asaro and Michel. These laboratories both measured about 0.030 ppb in the glauconitic boundary sample. When compared to aluminum (roughly clay content) to avoid false anomalies caused by dissolution of carbonates, the Ir/Al ratios were essentially flat across the measured section, which ranged from several metres below to several metres above the fossil boundary.

Iridium anomalies have been reported at the PЄ–Є boundary in China at the Meishucun section (Hsü *et al.*, 1985) and near Zunyi in Guizhou Province. The Los Alamos team worked with X.-Y. Mao, who brought hundreds of samples from various extinction boundaries in China, including the PЄ–Є boundary. Particular attention was given to the Meishucun section and the highest Ir abundance was found to be 0.053 ppb. At Maidiping section in Sichuan Province a maximum of 0.097 ppb was measured. In the Zunyi section in Guizhou Province a very large 2.9 ppb was observed. However, chalcophiles were similarly enhanced, for example, As = 1.55%, U = 286 ppm, Mo = 6.2%, and Se = 1900 ppm. Thus, rather than an impact source for the excess Ir, it appears that it was precipitated from seawater in the sulphide-rich sediments along with the chalcophiles. In the course of our measurements at Los Alamos we have observed that Ir has both siderophile and chalcophile affinities. Thus far, I see no persuasive evidence for a large-body impact at the fuzzy PЄ–Є boundary (see also Brasier, Chapter 4, this volume).

The ratios of the stable isotopes of carbon and sulphur undergo some radical changes near the PЄ–Є boundary. Holser (1977) observed a large increase in $\delta^{34}S$ in evaporites beginning about 100 million years before the boundary and peaking at or just after the fossil boundary—the so-called Yudomski event. This increase in the ^{34}S to ^{32}S ratio might have been the result of increased burial of sulphides (lighter S), so that the ocean contained heavier S for deposition in evaporites.

Magaritz *et al.* (1986) measured a slow decrease in $\delta^{13}C$ beginning about 70 million years before the boundary, followed by an upturn that peaked at the boundary and then dropped abruptly. The decreases in $\delta^{13}C$ might have been the result of decreased ocean productivity and the increase in $\delta^{13}C$ before the boundary could have been the result of an increase in productivity.

Whatever the cause(s), it appears that large changes in C and S isotopic ratios were occurring in the world ocean.

Cambrian biomere boundaries

The name 'biomere' was proposed by Palmer (1965), who defined it as: 'a regional biostratigraphic unit bounded by abrupt non-evolutionary changes in the dominant elements of a single phylum'. For the following discussion the dominant elements are trilobites. The concept has proven useful in studies of late Cambrian faunas and biostratigraphy of North America. At several boundaries the abrupt change from a high-diversity fauna with low individual dominances to a low-diversity fauna dominated by only one or two species takes place within a few centimetres of section (Palmer, 1982). This pattern of extinction is similar to that of the calcareous plankton at the Cretaceous–Tertiary boundary and was considered to offer another possible example of an impact-related biological crisis (Palmer, 1982).

Iridium and other elemental abundances across the Marjumiid–Pterocephaliid and the Pterocephaliid–Ptychaspid biomere boundaries have been reported by Orth et al. (1984). These authors sampled marine limestone deposits in the House Range of western Utah. Although the two trilobite–brachiopod extinction boundaries could be assigned to ±4 mm of vertical section, only normal limestone amounts of Ir (2–15 ppt) were observed. Therefore, no evidence was found for large-body impacts at these two biomere boundaries. However, both of these sequences were deposited in shallow water where wave and tidal action could have dissipated local concentrations of fallout material, and the authors concluded that other sections at distant localities marked by abundant fossils and continuous deposition should be examined before a definitive statement about possible impact causes could be made.

Cambrian–Ordovician boundary

This boundary apparently does not represent a massive extinction event. However, Sepkoski (1986) reported an extinction of about 15–20% of marine genera in the Trempealeauan at the end of the Cambrian. On a local basis it coincides with the top of the Ptychaspid biomere in North America (Palmer, 1982) and also is associated with the Lower Tremadoc *Dictyonema flabelliforme* graptolitic black shales found in Scandinavia and North America that contain very high concentrations of V, Mo and U (Gee, 1980; Sunblad and Gee, 1985; Berry et al., 1986). Iridium concentrations, however, range from 0.03 to 0.05 ppb, about average for black shales (Berry et al., 1986). Comprehensive Ir measurements at this boundary are lacking and should be pursued if a suitable section, or sections, can be found, perhaps with an abundant conodont fauna.

Ordovician–Silurian boundary

According to Sepkoski (1982), the Ashgillian event at the Ordovician–Silurian (O–S) boundary could very well have been the second most severe extinction event of the Phanerozoic. He estimated that about 22% of all families became extinct during this event.

Because of the glacio-eustatic regression at the O–S boundary, stratigraphic sequences with continuous deposition across the boundary are known in few places. In 1983, O–S exposures on Anticosti Island in the Gulf of St Lawrence were strong candidates for the stratotype, because they had the most complete known depositional record with a shelly fauna and precise location of the boundary on the basis of conodont studies (McCracken and Barnes, 1981). The further report of a thin persistent clay-rich parting at the boundary provided the further incentive for a geochemical examination of the O–S zone by Orth, Gilmore *et al.* (1986). These investigators found low Ir concentrations ranging from 0.005 to 0.058 ppb in the section. Although the highest concentration was found in the boundary marl, on a carbonate-free basis or compared to Al, the Ir patterns were essentially flat across the 3–4 metre boundary sections. Carbon and oxygen isotopic measurements across the boundary indicated a rather sharp drop in their ratios at the boundary. The authors suggested that this change to lighter isotopes was probably related to a decrease in salinity in the shallowing seaway. Freshwater input from rivers would have been most influential during the shallow water conditions. A search for shocked mineral grains and microspherules was unfruitful, and the conclusion was that there was no evidence to associate a large-body impact with the O–S boundary at this locality.

In 1985, because the lack of impact signatures in the Anticosti Island sections might have been attributed to the lack of preservation in the shallow marine carbonate sequence, Wilde *et al.* (1986) examined a suite of samples they collected from the newly adopted O–S stratotype (Holland *et al.*, 1985) at Dob's Linn, Scotland. Here, the deposition was in deep water and the boundary is assigned on the basis of graptolites, specifically at the base of the *acuminatus* Zone. Persistently high Ir concentrations were observed in the section that encompassed 13 graptolite zones and a depositional interval of an estimated 20 million years. However, there is no Ir concentration spike in the interval and the Ir correlates strongly with Cr. The Cr/Ir ratio averages about 10^6, similar to the ratio in chromites. The authors concluded that the palaeogeographic and geologic reconstructions, coupled with the occurrence of ophiolites and other deep crustal rocks in the palaeo-source area, suggest that the relatively high Ir and Cr abundances observed in the boundary zone resulted from terrestrial erosion of exposed upper mantle rocks rather than from a cataclysmic extraterrestrial event. Unfortunately, the rarity of conodonts at Dob's Linn and the absence of graptolites at Anticosti Island prohibits precise time correlation between these two O–S boundary sections. More work should be done on this important boundary; it is possible that sections in south China might offer O–S sequences that contain both conodonts and graptolites.

Silurian–Devonian boundary

The Silurian–Devonian (S–D) boundary is probably unimportant in discussions of extinctions and their association with impacts. Rather than an extinction event, the S–D boundary is recognized by an increase in diversity similar to that in the Precambrian–Cambrian boundary. The stratotype (Pridolian–Lochkovium) is located at Klonk, Czechoslovakia, in a sequence of richly fossiliferous, micritic limestones and interbedded black shales that were deposited in an open-water pelagic environment at depths of tens to hundreds of metres (Chlupac and Kukal, 1986); the base of the Devonian is marked by the sudden appearance of the graptolite *Monograptus uniformis*. Thus far, no geochemical measurements across the Silurian–Devonian (S–D) boundary have been published. However, residues from a large suite of samples collected by J.L. Kirschvink for palaeomagnetic measurements have been assayed for Ir and about 40 other elements at Los Alamos by Gilmore and Orth. Very low Ir concentrations were observed, ranging from 3 to 17 ppt. It was considered that there was no evidence for a large-body impact associated with the S–D boundary exposed at the Klonk section, although further geochemical measurements might provide some knowledge about the reason for the onset of new faunal forms at the boundary.

Frasnian–Famennian (Upper Devonian) boundary

At or near the end of the Frasnian Stage of the late Devonian, at a time when warm shallow seas covered much of the world's continental areas and organic reefs were widespread, a major extinction of shallow-water metazoans occurred (McLaren, 1982). This extinction is one of the largest known (Sepkoski, 1982), but the cause has not been established. McLaren (1970) proposed that impact by a large extraterrestrial body could have been responsible, whereas others have suggested abrupt changes in temperature or sea level as the causes (McLaren, 1970; Johnson, 1974; Stanley, 1984).

In 1982, shortly after the Snowbird Conference, McGhee *et al.* (1984) began a geochemical examination of Frasnian–Famennian boundary zones at three sections in New York and one near Sinsin, Belgium. These investigators found no evidence on the basis of Ir assay for a large-body impact in the boundary interval, although even as late as 1984 the exact stratigraphic position of the boundary had not been established, being placed somewhere between the uppermost *Palmotolepis gigas* and upper *P. triangularis* conodont subzones.

In the Canning Basin of northwestern Australia there is an Upper Devonian reef complex that contains Givetian, Frasnian and Famennian reef fringed limestone platforms flanked by steeply dipping marginal-slope limestones, that interfinger with basin deposits laid down in water up to several hundred metres deep. Conodonts in the deeper slope deposits allow precise correlation with the standard European zones and subzones. Playford *et al.* (1984) reported a moderate Ir anomaly (peak concentration about 0.30 ppb) in a section on the west flank of McWhae Ridge on the southeastern end of the reef complex (Figure 3.6). The anomaly was located between the upper and middle *P. triangularis* subzones. The anomaly coincides with a prominent stromatolite bed, about 12 cm thick, that contains closely spaced microstromatolites of the extinct fossil cyanobacterium *Frutexites*. Not only Ir, but other siderophiles (not Au), some chalcophiles, rare earths and Th are enriched in the *Frutexites* horizon. There also is a drop in ^{13}C values of about 1.5 per mille extending from the base of the anomaly to the top of the section; this finding suggested a decrease in biomass for at least a million years. The authors concluded that the association of the anomaly with the extinction event may be purely coincidental, but it seemed more likely that there was some genetic relation between them, involving either an impact or an unidentified terrestrial process.

In July 1984, Playford and Orth returned to the Canning Basin and sampled other exposures. In particular, in a stream-bed exposure several kilometres from the original McWhae section, *Frutexites* stromatolites alternated with other stromatolitic beds over a vertical interval of about 0.4 m. They found that the concentrations of Ir and the other elements noted above were roughly proportional to the density of the *Frutexites* 'shrubs' or bundles of microfilaments (Figure 3.7). This observation, plus another even more dramatic case, where the *Frutexites*–Ir correlation occurred over a stratigraphic interval of over a metre, clearly demonstrated that the bacteria were merely concentrating the Ir, and so on, out of seawater. They concluded that the bacteria oxidized these elements out of seawater and those elements that formed insoluble oxidized chemical states (probably oxides) were retained in the bacterial filaments. Perhaps the local extinction was the result of upwelling of deep anoxic and sulphide-rich water into the platform and shelf communities. Extinction of grazing animals, such as snails, could conceivably have given the bacteria the opportunity to thrive. Whatever the mechanism, the work in the Canning Basin demonstrated that specialized marine bacteria can concentrate Ir and great care should be exercised in the interpretation of Ir anomalies.

The Frasnian–Famennian (F–F) extinction in Europe has been called the 'Kellwaser Event' (House, 1985), because the event preceded any historical use of the F–F stage boundary and it corresponds stratigraphically to the geographically widespread unit called the Kellwasser limestones, which generally occur as two black shale and bituminous limestone beds. A geochemical analysis of the Kellwasser Event (F–F boundary) in the

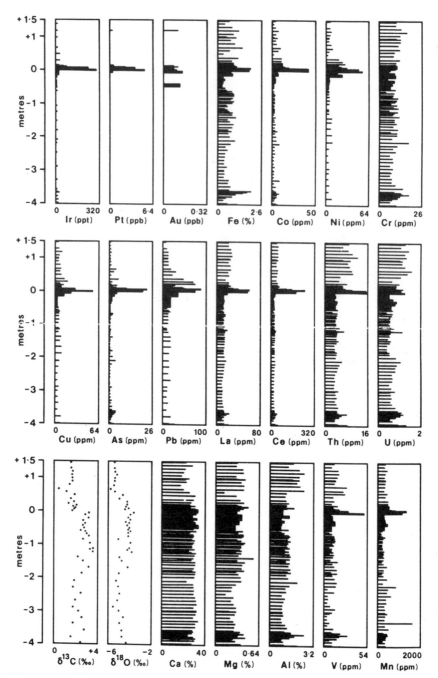

Figure 3.6. Elemental abundances for some significant elements, and isotope ratios for carbon and oxygen, across the Frasnian–Famennian boundary (0 m) at the McWhae Ridge section, Upper Devonian Virgin Hills Formation, Canning Basin, Western Australia (Playford *et al.*, 1984).

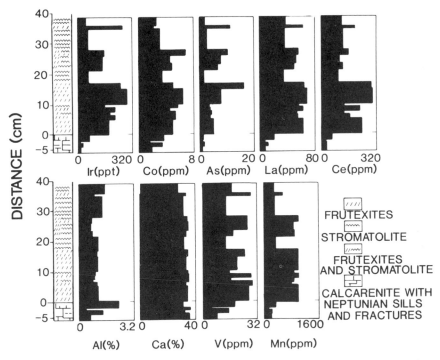

Figure 3.7. Some elemental abundance patterns across the Frasnian–Famennian boundary (0) at a section in a creek bed near McWhae Ridge, Canning Basin, Western Australia. Note that the Ir concentration is roughly proportional to the density of *Frutexites* bacterial shrubs. Other intervening stromalites do not appear to concentrate Ir, etc.

Steinbruch Schmidt, northeast of Marburg, Federal Republic of Germany, has been reported by McGhee *et al.* (1986). In a stratigraphic span of 5 metres they observed that 15 Ir peaks of 75–160 ppt rise above the local background of about 27 ppt. None of these is a distinct impact-related Ir peak, because the higher Ir values are associated with the clay partings in the limestone sequence and Ir correlates strongly with Al and other lithophiles (Sc, Cr and Th) generally associated with clay minerals. A careful search for other signatures of an impact (shocked-mineral grains and microspherules) failed to locate any of these forms. The ratio of ^{13}C to ^{12}C drops off in the upper *P. gigas* Subzone and then increases sharply in the biological crisis zone, suggesting that an abrupt bloom in phytoplankton occurred in this interval.

Thus far, three thorough examinations of the F–F extinction in widely separated localities have provided no evidence for an impact-related cause for the biological crisis: terrestrial processes such as global cooling (Stanley, 1988) provide a better alternative.

Devonian–Carboniferous (Mississippian) boundary

The Devonian–Mississippian (D–M) boundary is not recognized as a mass extinction horizon. However, the boundary is defined, on the basis of conodonts, as the first appearance of *Siphonodella sulcata*. One of the leading candidates for the stratotype is the Muhua section in southern Guizhou Province, south China, because the section contains a good succession of *S. praesulcata–S. sulcata* and Lower and Upper *duplicata* Zones. However, a 5–25-cm thick black shale intervenes just below the boundary in an otherwise unbroken limestone sequence. A moderate Ir spike of 150 ppt in the black shale has been reported by Chai *et al.* (1987). Almost all measured chalcophiles and chalcophile-like elements are also enriched in this zone; U is enriched as much as the Ir. Neither impact or volcanism can explain the anomalies, but the enchanced elements might have been precipitated from seawater in the organic- and sulphide-rich black shale unit.

The Lower Mississippian of Oklahoma has also provided some interesting geochemical results. Here the *praesulcata* Zone is absent, but this does not detract from the story because the elemental abundance anomalies occur in and above the *sulcata* Zone. Four Pt-group metal anomalies (Figure 3.8) have been reported by Orth, Quintana *et al.* (1988), two of which occur in the Welden Limestone and are thought to be the result of bacterial processes. A third anomaly occurs at the top of the Woodford Black Shale, which is overlain by a highly condensed (high density of conodonts) grey shale, and the authors suggest that the enrichments might be the result of precipitation in the black shale from its long exposure to seawater. The fourth Ir anomaly, which is the strongest (0.56 ppb), lies above the limestone unit in another shale sequence. Large enrichments of Pt (150 ppb), Os (0.15 ppb), Co (725 ppm) and Ni (1450 ppm) coincide with the Ir peak. The enrichment does not appear to be impact-related, but rather appears to arise from some terrestrial process, perhaps erosion of ultramafic rocks into the sea bottom.

Mississippian–Pennsylvanian boundary

The Mississippian–Pennsylvanian (M–P) boundary does not represent an important extinction event, but it nevertheless represents a globally documented biological change in the fossil record (Saunders and Ramsbottom, 1982). In work on a marine section near Ada, Oklahoma, Orth, Quintana *et al.* (1986) discovered a weak Ir anomaly within the conodont-defined boundary zone that was previously determined to a few decimetres by Grayson *et al.* (1985). Further conodont work by Grayson proved that the conodont change occurred precisely at the Ir anomaly. A second section in central Texas was then examined and there the Ir anomaly was much stronger (0.38 ppb). An extreme enrichment of Cr (12 000 ppm) coincided

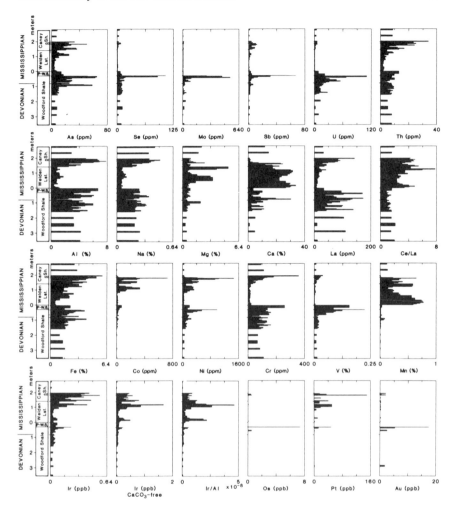

Figure 3.8. Elemental abundance patterns near the Devonian–Carboniferous boundary near Ada, Oklahoma, showing multiple iridium peaks. Note that chalcophiles are strongly enriched at the top of the Woodford Black Shale and that only the two iridium peaks in the Welden Limestone stand out when Ir is ratioed to aluminum (roughly proportional to clay mineral content). Platinum is highly concentrated in the Caney Shale (150 ppb) and Os is concentrated at the top of the Woodford (Orth, Quintana *et al.*, 1988).

with the excess Ir, as did elevated concentrations of Os (4 ppb), Pt (6 ppb), and U (380 ppm). No tangible evidence of an impact-related source for the Pt group was found and it was concluded that some terrestrial process was involved. The anomaly zone was quite phosphatic (apatite), so that upwelling might have been the cause of both the conodont crisis and the elemental enrichments.

Permian–Triassic boundary

The Upper Permian (P–Tr) extinction, according to Sepkoski (1982), was the largest in the Phanerozoic, and, in more recent articles (Raup and Sepkoski, 1984; 1986), this extinction provides a solid 245-million-year datum for their arguments of periodicity in extinctions over the last 250 million years. The earliest reported Ir measurements on the P–Tr boundary were by Asaro *et al.* (1982). They determined upper limits of less than 0.5 ppb in a mixed illite/smectite boundary clay and also in subjacent and superjacent strata from sections at Meishan near Changxing in Zhejiang Province and at Wachapo Mountain in Guizhou Province. The Baoqing Quarry of Meishan has received much attention and is under consideration for the international P–Tr boundary stratotype. A strong Ir anomaly (8 ppb) in the Meishan boundary clay (sample AG-91) was reported by Sun *et al.* (1984). They also reported 5 ppb in sample AG-92, immediately above AG-91. Prior to this report, Ir and other elemental abundance measurements were performed at Los Alamos on splits of AG-91 and 92, furnished by Y.-Y. Sun through R.R. Brooks. A mere 0.002 ppb of Ir was measured in AG-91 and 0.024 ppb in AG-92 and reported to the sample donors. Several samples of the P–Tr boundary clay from sections near Lichuan in western Hubei Province were analysed at Los Alamos in June 1984 and again the Ir concentrations were only 0.001 ppb (Clark *et al.* 1986). At the 1985 Conference on Rare Events in Geology, Sun *et al.* (1985) reported their further measurements at Meishan: 0.05–0.19 ppb of Ir in the AG-91 layer and 0.2–1.0 ppb in the AG-92 layer. At this same conference Asaro *et al.* (1985) reported that their recent measurements on the Meishan, Watchapo Mountain and Tangshan (near Nanjing) sections all resulted in Ir concentrations of less than 0.040 ppb. Therefore, there is considerable disagreement between the Chinese results for Ir and those of the Lawrence Berkeley Laboratory and the Los Alamos National Laboratory; the agreement among the three laboratories for the concentrations of the more abundant elements, however, is quite good.

An Ir anomaly also has been reported by Xu *et al.* (1985) in a P–Tr section located near Shangsi in Sichuan Province. They found 2.0 ppb in sample AG-253. One of the Chinese radiochemists (X.-Y. Mao), who worked on both the Meishan and Shangsi samples, spent ten months at Los Alamos in 1986–7. She brought several hundred samples from many boundaries, predominantly P–Tr. The Meishan and Shangsi sections were re-examined and again the Meishan AG-91 and 92 layers were observed to contain about 0.002 and 0.025 ppb, respectively. The Shangsi boundary clay and the overlying mudstone contained 0.003 and a maximum of 0.035 ppb, respectively.

The south China P–Tr boundary clay contains a rather unique trace element pattern, which is depleted in siderophiles and enriched in the lithophiles, Cs, Hf, Ta and Th, typical of continental acidic volcanic ash. In addition to the boundary clay bed, at Meishan and elsewhere when exposed, there are at least two other overlying clay beds with this same abundance pattern, presumably from the same source area or the same volcano (Clark *et al.*, 1986).

A consortium of scientists, led by Holser and Schönlaub, is currently examining a large suite of samples from a 300-metre core collected across the P–Tr boundary in the Carnic Alps in southern Austria. There are two moderate Ir peaks in the section; the lower coincides with a decrease in $\delta^{13}C$ and the upper with a pyrite-rich bed. Both peaks appear to be the result of terrestrial processes.

In south China and in Austria the P–Tr boundary is associated with a relatively abrupt decrease in carbonate $\delta^{13}C$, probably associated with the disappearance of phytoplankton (Chen, 1987; Zheng *et al.*, 1987; Holser and Schlönlaub, 1988).

Triassic–Jurassic boundary (Norian event)

Very few well-preserved Triassic–Jurassic (Tr–J) boundary sections are known. Low sea level at that time resulted in hiatuses and erosion. This boundary figures prominently in any discussions of the possible association of large-body impacts with mass extinctions because the huge Manicouagan structure (~70 km across) in Quebec has a similar age, ~210 million years. The discovery by Olsen *et al.* (1988) of abundant earliest Jurassic tetrapod fossils in the Fundy Basin, only 750 km from the Manicouagan structure, provided the incentive for geochemical studies in the freshwater Fundy sequences. Geochemical measurements are under way at the Geological Survey of Canada (Goodfellow and McLaren), Lawrence Berkeley Laboratory (Asaro and Michel) and at Los Alamos (Attrep and Orth). A potential problem with these Ir measurements is that the Manicouagan event might not have provided a signal as strong as that at the K–T boundary. Melt rock from the structure contains very low Ir concentrations, ~0.025 ppb.

Sections in Austria and Great Britain have been sampled by Hallam and his students and geochemical measurements on the samples are under way at Los Alamos and Berkeley. According to Hallam, one of the best-preserved sections might be in New York Canyon, Nevada, and work also is in progress on this section.

Middle–Upper Jurassic (Callovian–Oxfordian boundary)

Measurements performed on Callovian–Oxfordian (C–O) boundary sections in carbonate facies in Spain and southern Poland yielded anomalously high concentrations of Ir (1–2.4 ppb) and also Fe, Mn, Ni, Co, Cr, Ti, REE and Au (Brochwicz-Lewinski *et al.*, 1985). The authors observed a phase of block movements, and volcanic and hydrothermal activity, that coincide with the elemental abundance anomalies. They suggest that their observations indicate a tectonic phase, most probably triggered by the impact of one or more large cosmic bodies, with propagation of stresses along reactivated and newly formed weakness zones.

Late Cretaceous (Cenomanian–Turonian boundary)

Sepkoski (1982) and Raup and Sepkoski (1984; 1986) recognized the Cenomanian–Turonian (C–T) boundary interval as a lesser or secondary extinction event, best documented for planktonic organisms. This boundary event is included in their chain of cyclic extinctions, being one cycle (26 million years) older than the K–T event. The major extinction of typical Cenomanian taxa occurred before the stage boundary, just prior to and during the greatest sea-level rise in the Cretaceous (Kauffman, 1984). The C–T boundary is associated with a global anoxic event (the Bonarelli Anoxic Event) characterized by incursion of disaerobic to anoxic water-masses into epicontinental marine basins (Jenkyns, 1980; Arthur *et al.*, 1985).

One of the most completely preserved C–T boundary sections in North America is exposed near Pueblo, Colorado. Elemental abundance measurements on 45 metres of section, representing over 2 million years of deposition, were performed by Orth, Attrep *et al.* (1988), who reported (Figure 3.9) two weak Ir anomalies (0.11 ppb over a background of 0.017 ppb) that coincide stratigraphically with the disappearance of two Cenomanian foraminifers (*Rotalipora cushmani* and *R. greenhornensis*) and approximately with the beginning of a molluscan extinction identified by Elder (1987). Other elements also enriched with the Ir are Sc, Ti, Cr, Mn, Co, Ni, Pt and Au. This suite of elements is generally associated with mafic to ultramafic rocks, but not with undifferentiated Solar system bodies, because Sc, Ti and Mn occur in very low concentrations in meteorites. Abundances of the rare earth, alkali metal and alkaline earth elements, As, Sb and U, are either depleted in the anomaly zone or do not change compared to the rest of the section. If the enrichment mechanism involved redox processes, a change in the As, Sb and U concentrations could be expected. Furthermore, elements not generally susceptible to redox variations, Sc and Ti, are enriched. Further work on sections from New Mexico to Manitoba, Canada, and from eastern Nebraska to eastern Utah, has shown that the magnitude of the anomalies drops off rapidly to the north, being hardly detectable in Canada, and also decreases somewhat to the east and to the far west (Orth, Attrep *et al.*, 1988). The strongest Ir signals occur in southeastern Colorado, but the anomalies are still relatively strong in west Texas (Figure 3.10). Results of work on sections in Mexico, Europe and on North Atlantic DSDP cores should help determine if the excess elements came in from the proto-North Atlantic/Caribbean, were from a

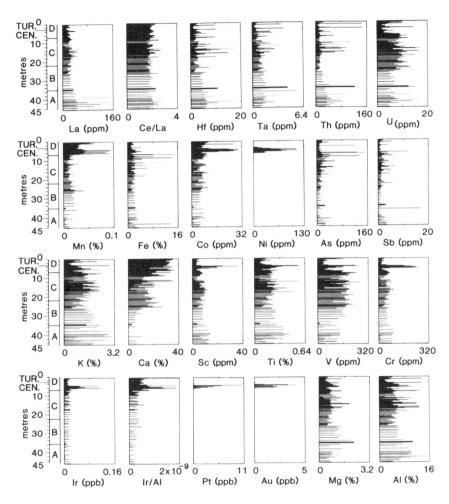

Figure 3.9. Abundance patterns for some significant elements across a marine Cenomanian–Turonian boundary section near Pueblo, Colorado. The 45 metres of section represent over 2 million years of deposition. (a) Graneros Shale; (b) Lincoln Member of Greenhorn Formation; (c) Hartland Shale Member of Greenhorn Formation; (d) Bridge Creek Limestone Member of Greenhorn Formation. Note the two iridium peaks in the lower portion of the Bridge Creek Limestone; Sc, Ti, V, Cr, Mn, Fe, Co, Ni, Pt and Au are also enriched. Peaks in the Hf, Ta and Th patterns indicate bentonitic beds from ejecta of continental (silicic) volcanic eruptions (Orth, Attrep *et al.*, 1988).

local source in southeastern Colorado, or were derived from some unknown global event. An impact source has not been ruled out, but appears unlikely. Orth and co-workers suspect the signal resulted from increased spreading centre activity in the late Cenomanian North Atlantic or Caribbean regions, perhaps associated with the deep-water opening of the Atlantic and the beginning of the North Atlantic gyre.

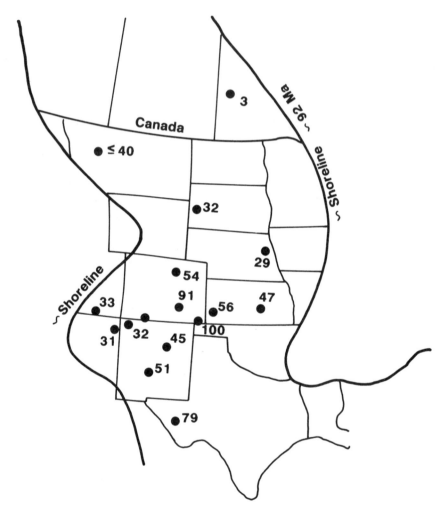

Relative Deposition of Iridium

Figure 3.10. Relative deposition of iridium in the late Cenomanian Western Interior of North America. The strongest concentrations are in southeast Colorado and the deposition decreases to the north. The shoreline of the epeiric seaway is indicated.

Impact horizons in the late Eocene

In the late Eocene, roughly 35 million years ago, about 10^{10} tons of glass, commonly referred to as the North American tektites on land and the associated microtektites in sea-bottom cores, were deposited over a wide area of southeastern North America and the Caribbean (Glass *et al.*, 1979). An Ir anomaly was discovered by Ganapathy (1982) about 30 cm below the microtektite horizon in giant piston core RC9-58 from the Venezuelan Basin. A parallel study of a core from DSDP site 149 in the eastern Caribbean (Alvarez *et al.*, 1982) also indicated an Ir anomaly. However, the peak of the microtektite abundance (and of the Ir distribution) was apparently lost in an unrecovered interval of the core. The Ir peak is associated with a crystalline microspherule layer that was deposited about 5000–20 000 years before the North American bed. The crystalline microspherules are not always present, probably as a result of corrosion of these more mafic (clinopyroxene) bodies (Glass *et al.*, 1979). The Ir-rich layer has been observed from the Caribbean, across the equatorial Pacific, to the Indian Ocean (Keller *et al.*, 1986). A third microspherule horizon has been identified (Keller *et al.*, 1986) in strata 0.5 million years older than the closely spaced pair described above. Currently, these three impact horizons offer the best supporting evidence for recent hypotheses that link impacts from periodic comet swarms with the periodic extinctions of Raup and Sepkoski (1984; 1986) (but see Prothero, Chapter 11, this volume).

The sources of the North American field, the Ir-rich horizon and the considerably older microspherule horizon are not known. The recent discovery of large spherules in DSDP 612 off the coast of New Jersey has led to speculation that a recently discovered impact structure off the coast of Nova Scotia (Jansa and Pe-Piper, 1987) might be the source locality, although it is dated at ~50 million years. A potential source for the Ir-rich bed was the Popigai event. This huge crater (100 km across) is located in Siberia and its melt rock contains high levels of Ir (~1.4 ppb).

Middle Miocene extinction

A minor extinction in the Middle Miocene stands out in the time-series analysis by Raup and Sepkoski (1986) at the generic level. Asaro *et al.* (1988) have scanned for Ir in a nearly continuous sequence of 33 metres of Middle Miocene sediments from the DSDP site 588B in the South Pacific. One Ir peak of 0.152 ppb was discovered at 11.7 million years BP over a background of 0.011 ppb. Of about 30 other elements studied, only Cr and possibly Co and Ni showed corresponding peaks. The investigators are currently searching other DSDP cores from widely separated localities. The anomaly appears to be associated with an impact and if further measurements on other sites and on the other Pt-group elements prove fruitful and positive, this discovery will provide another example of an impact-related extinction event.

Late Pliocene event

In 1981, Kyte *et al.* reported that anomalously high noble metal and meteoritic particle concentrations occur in Upper Pliocene (~2.3 million years BP) sediments from deep-sea piston core E13-3, which was recovered from the southeast Pacific about 1400 km west of Cape Horn. Their recent isotopic, mineralogic and chemical composition data from other piston cores in the general area indicated that the projectile was a low-metal mesosiderite meteorite, an asteroid and not a comet (Kyte *et al.*, 1988). It apparently fell into a deep (5 km) ocean basin and did not produce a crater on the ocean floor. They estimate the diameter of the asteroid to have been about 0.5 km. The date of this impact is about 2.3 million years BP, essentially the same as that inferred for the onset of glaciation in the Northern Hemisphere. These investigators conclude that the event probably injected at least 2 billion tonnes of water into the stratosphere, causing global high-altitude clouds which might have been sufficient to increase the Earth's albedo, resulting in a rapid decrease in the global temperature.

SUMMARY

The discovery by the Alvarez team of an abrupt and strong Ir abundance peak at the K–T boundary set the stage for a new collaboration between geochemists and palaeontologists. Working together they have provided considerable new knowledge about the terminal Cretaceous extinction event and about many of the other bio-events in the fossil record. The combination of chemostratigraphy and biostratigraphy shows great potential for exciting discoveries and contributions to geology.

In the past eight years a wealth of information has been collected on the K–T boundary, and solid chemical and physical evidence strongly supports that portion of the original Alvarez hypothesis that claimed the excess Ir was the result of the impact of an asteroid or comet. Alternative suggestions of a volcanic origin thus far rest on rather unconvincing arguments generally based on poor or erroneous data. While it is important that we as scientists keep our minds open to alternative scenarios, it is also important for those claiming alternative mechanisms to base them on good, solid, reproducible scientific evidence. Whether or not the hypothesized impact was responsible for the K–T extinction, or was merely a contributing factor to an already deteriorating environment, is currently being tested and debated by palaeontologists.

At least one section across most of the bio-event horizons in the Phanerozoic has received a combined geochemical–palaeontologic examination and thus far there is no convincing evidence of an impact–extinction relationship in those older than the terminal Cretaceous event. However, thin fallout layers are extremely sensitive to erosion and mixing processes, especially the older boundaries, and many sections from widely separated localities should be examined before any firm conclusions can be drawn. There is ample evidence of very large impacts on Earth's surface and, at the very least, there must have been tremendous violence done to the local environment. Excess Ir should be deposited from the fallout of ejecta, although not necessarily at the bio-event horizons, and these Ir peaks could provide excellent time markers in the geologic column; the trick is to find them.

Work on boundaries younger than the K–T·has turned up two or possibly three impact horizons in the late Eocene and probably one in the middle Miocene, coeval with or very close to extinction events identified by Raup and Sepkoski that fit their 26-million-year extinction periodicity derived from time-series analysis. Several Ir peaks just below the Cenomanian–Turonian boundary in North America might be impact-related, but further studies are required. An impact horizon in the southern Pacific might be associated with the onset of Pliocene glaciation in the northern hemisphere.

Further work on the consequences of massive volcanic eruptions is also important. Present knowledge indicates that the ash from explosive, silicic volcanoes contains very little Ir and magma from hot spot volcanism can contain moderate amounts, but this type of free-flowing magma does not provide the energy to disperse the ejecta over any great distance. Perhaps exceptions to these observations will be found.

Iridium anomalies also have been found to be the result of terrestrial processes, especially in the marine environment, and great care should be exercised in the interpretation of Ir peaks in the geologic column.

It is apparent that extinctions were not caused by any singular process. Although large-body collisions deposit massive amounts of energy to the impact area and have the potential of producing tremendous environmental damage, terrestrial processes can be just as devastating. Large global temperature changes (in either direction), marine regressions and loss of habitat, upwelling of anoxic, sulphide-rich water onto reef to outer-shelf communities, and massive volcanism are a few examples of potentially catastrophic terrestrial processes. Further collaborative research by geochemists and palaeontologists should provide important new information and knowledge about extinctions and their causes.

REFERENCES

Alvarez, L.W., Alvarez, W., Asaro, F. and Michel, H.V., 1980, Extraterrestrial cause for the Cretaceous–Tertiary extinction, *Science*, **208** (4448): 1095–1108.
Alvarez, W., 1986, Toward a theory of impact crises, *Eos*, **67** (35): 649, 653–5, 658.

Alvarez, W., Alvarez, L.W., Asaro, F. and Michel, H.V., 1979, Experimental evidence in support of an extraterrestrial trigger for the Cretaceous–Tertiary extinctions, *Eos*, **60** (18): 734.

Alvarez, W., Asaro, F., Michel, H.V. and Alvarez, L.W., 1982, Iridium anomaly approximately synchronous with terminal Eocene extinctions, *Science*, **216** (4548): 886–8.

Alvarez, W. and Muller, R.A., 1984, Evidence from crater ages for periodic impacts on the Earth, *Nature*, **308** (5961): 718–20.

Arthur, M.A., Schlanger, S.O. and Jenkyns, H.C., 1985, The Cenomanian–Turonian oceanic anoxic event, II. Palaeoceanographic controls on organic matter production and preservation. In J. Brooks and A. Fleet (eds), *Marine petroleum source rocks, Special Publication of the Geological Society of London*, **26**: 216–8.

Asaro, F., Alvarez, L.W., Alvarez, W. and Michel, H.V., 1982, Geochemical anomalies near the Eocene/Oligocene and Permian/Triassic boundaries. In L.T. Silver and P.H. Schultz (eds), *Geological implications of impacts of large asteroids and comets on the Earth, Special Paper of the Geological Society of America*, **190**: 517–28.

Asaro, F., Michel, H.V., Alvarez, W. and Alvarez, L.W., 1988, Impacts and multiple iridium anomalies, *Eos*, **69** (16): 301.

Asaro, F., Michel, H.V., Alvarez, W., Alvarez, L.W., Kastner, M. and Utting, J., 1985, Geochemical study of the Permian–Triassic boundary in Meishan, Tangshan and Watchapo Mountain sections in the Peoples Republic of China, *Abstracts, 27th International Geological Congress*.

Berry, W.B.N., Wilde, P., Quinby-Hunt, M.S. and Orth, C.J., 1986, Trace element signatures in *Dictyonema* Shales and their geochemical significance, *Norsk Geologisk Tidsskrift*, **66** (1): 45–51.

Bohor, B.F., Foord, E.E., Modreski, P.J. and Triplehorn, D.M., 1984, Mineralogic evidence for an impact event at the Cretaceous–Tertiary boundary, *Science*, **224** (4651): 867–9.

Brochwicz-Lewinski, W., Gasiewicz, A., Melendez, G., Sequeiros, L., Suffczynski, S., Szatkowski, K., Tarkowski, R. and Zbik, M., 1985, A possible Middle/Upper Jurassic boundary event, *Abstracts, 27th International Geological Congress*.

Bunch, T.E., 1968, Some characteristics of selected minerals from craters. In N.M. Short (ed.), *Shock metamorphism of natural materials*, Mono, Baltimore, MD: 413–32.

Carter, N.L., Officer, C.B., Chesner, C.A. and Rose, W.I., 1986, Dynamic deformation of volcanic ejecta from the Toba caldera: possible relevance to Cretaceous–Tertiary boundary phenomena, *Geology*, **14** (5): 380–3.

Chai, Z.-F., Mao, X.-Y., Ma, S.-L., Bai, S.-L., Orth, C.J., Zhou, Y.-Q. and Ma, J.-G., 1987, Geochemical anomaly of the Devonian–Carboniferous boundary at Huangmao, Guangxi, China, *Abstracts, IGCP 199, Rare Events in Geology*: 43.

Chen, Z.-S., 1987, Carbon isotope variation at the P–Tr boundary in south China, *Abstracts, IGCP 199, Rare Events in Geology*: 29.

Chlupac, I. and Kukal, Z., 1986, Reflection of possible global Devonian events in the Barrandian area. In O.H. Walliser (ed), *Global bio-events*, Springer-Verlag, Berlin: 169–79.

Clark, D.L., Wang, C.-Y., Orth, C.J. and Gilmore, J.S., 1986, Conodont survival and low iridium abundancies across the Permian–Triassic boundary in south China, *Science*, **233** (4767): 984–6.

Davis, M., Hut, P. and Muller, R.A., 1984, Extinction of species by periodic comet showers, *Nature*, **308** (5961): 715–16.

Elder, W.P., 1987, The palaeoecology of the Cenomanian–Turonian (Cretaceous) stage boundary extinctions at Black Mesa, Arizona, *Palaios*, **2** (1): 24–40.

Ganapathy, R., 1980, A major meteorite impact on Earth 65 million years ago: evidence from the Cretaceous–Tertiary boundary, *Science*, **209** (4459): 921–3.

Ganapathy, R., 1982, Evidence for a major meteorite impact on the Earth 34 million years ago: implication for Eocene extinctions, *Science*, **216** (4548): 885–6.

Gee, D.G., 1980, Basement-cover relationships in the central Scandinavian Caledonides, *Geologiska Föreningens i Stockholm Förhandlingar*, **102** (4): 455–74.

Gilmore, J.S., Knight, J.D., Orth, C.J., Pillmore, C.L. and Tschudy, R.H., 1984, Trace element patterns at a non-marine Cretaceous–Tertiary boundary, *Nature*, **307** (5948): 224–8.

Glass, B.P., Swinicki, M.B. and Zwart, P.A., 1979, Australasian, Ivory coast, and North American tektite strewnfields: size, mass and correlation with geomagnetic reversals and other Earth events, *Proceedings of the Lunar and Planetary Science Conference*, **10**: 2535–45.

Grayson, R.C., Jr, Davidson, W.T., Westergaard, E.H., Atchley, S.C., Hightower, J.H., Monaghan, P.T. and Pollard, C.T., 1985, Mississippian/Pennsylvanian (mid Carboniferous) boundary conodonts from the Rhoda Creek Formation: *Homoceras* equivalent in North America. In H.R. Lane and W. Ziegler (eds), *Toward a boundary in the middle of the Carboniferous: stratigraphy and paleontology, Courier Forschunginstitut Senckenburg*, **74**, 149–80.

Hildebrand, A.R. and Boynton, W.V., 1987, Excavation depth of the K/T impact from REE abundancies, *Geological Society of America Abstracts with Programs*, **19** (7): 702.

Holland, C.H., Ross, R.J., Jr and Cocks, L.R.M., 1985, *IUGS Circular*, **20**, *Ordovician–Silurian Working Group*.

Holser, W.T., 1977, Catastrophic chemical events in the history of the ocean, *Nature*, **267** (5610): 403–8.

Holser, W.T., Magaritz, M. and Wright, J., 1986, Chemical and isotopic variations in the world ocean during Phanerozoic time. In O. Walliser (ed.), *Global bio-events*, Springer-Verlag, Berlin: 63–91.

Holser, W.T. and Schönlaub, H.P., 1988, New insights on the Permian–Triassic boundary event from core Gartnerkofel-1 (Carnic Alps, Austria), *Abstracts, IGCP 199, Rare events in geology, Vienna, Austria, September 12–17*: 11.

House, M.R., 1985, Correlation of mid-Palaeozoic ammonoid evolutionary events with global sedimentary perturbations, *Nature*, **313** (5997): 17–22.

Hsü, K.J., Oberhänsli, H., Gao, J.-Y., Shu, S., Haihong, C. and Krähenbuhl, U., 1985, 'Strangelove ocean' before the Cambrian explosion, *Nature*, **316** (6031): 809–11.

Jansa, L. and Pe-Piper, G. 1987, Identification of an underwater extraterrestrial impact crater, *Nature*, **327** (6123): 612–14.

Jenkyns, H.C., 1980, Cretaceous anoxic events—from continents to oceans, *Journal of the Geological Society of London*, **137** (2): 171–88.

Johnson, J.G., 1974, Extinction of perched faunas, *Geology*, **2** (10): 479–82.

Kauffman, E.G., 1984, Paleobiogeography and evolutionary response dynamic in the Cretaceous western interior seaway of North America. In G.E.G. Westermann (ed.), *Jurassic–Cretaceous biochronology and paleogeography of North America, Special Paper of the Geological Association of Canada*, **27**: 273–306.

Keller, G., d'Hondt, S.L., Orth, C.J., Gilmore, J.S., Oliver, P.Q., Shoemaker, E.M. and Molina, E., 1986, Late Eocene impact microspherules; stratigraphy, age and geochemistry, *Meteoritics*, **22** (1): 25–60.

Kyte, F.T., Zhou, Z. and Wasson, J.T., 1980, Siderophile-enriched sediments from the Cretaceous–Tertiary boundary, *Nature*, **288** (5792): 651–6.

Kyte, F.T., Zhou, Z. and Wasson, J.T., 1981, High noble metal concentrations in late Pliocene sediment, *Nature*, **292** (5822): 417–20.

Kyte, F.T., Zhou, Z. and Wasson, J.T., 1988, New evidence on the size and possible effects of a late Pliocene oceanic asteroid impact, *Science*, **241** (4861): 63–4.

Luck, J.M. and Turekian, K.K., 1983, Osmium 187/osmium 186 in manganese nodules and the Cretaceous–Tertiary boundary, *Science*, **222** (4624): 613–15.

Luterbacher, H.P. and Primoli-Silva, I., 1964, Biostratigrafia del limite Cretaceo–Terziano nell' Appennino centrale, *Rivista Italiana di Paleontologia e Stratigrafia*, **70** (1): 67.

McCracken, A.D. and Barnes, C.R., 1981, Conodont biostratigraphy and paleo-ecology of the Ellis Bay Formation, Anticosti Island, Quebec, with special reference to late Ordovician–early Silurian chronostratigraphy and the systemic boundary, *Bulletin of the Geological Survey of Canada*, **329**: 51–134.

McGhee, G.R., Jr, Gilmore, J.S., Orth, C.J. and Olsen, E., 1984, No geochemical evidence for an asteroidal impact at late Devonian mass extinction horizon, *Nature*, **308** (5960): 629–31.

McGhee, G.R., Jr, Orth, C.J., Quintana, L.R., Gilmore, J.S. and Olsen, E.J., 1986, Late Devonian 'Kellwasser Event' mass extinction horizon in Germany: no geochemical evidence for a large-body impact, *Geology*, **14** (9): 776–9.

McLaren, D.J., 1970, Time, life and boundaries, *Journal of Paleontology*, **44** (5): 801–15.

McLaren, D.J., 1982, Frasnian–Famennian extinctions. In L.T. Silver and P.H. Schultz (eds), *Geological implications of impacts of large asteroids and comets on the Earth, Special Paper of the Geological Society of America*, **190**: 477–84.

McLean, D.M., 1982, Deccan volcanism: the Cretaceous–Tertiary marine boundary timing event, *Geological Society of America Abstracts with Programs*, **14** (7): 562.

McLean, D.M., 1985, Deccan Traps mantle degassing in the terminal Cretaceous marine extinctions, *Cretaceous Research*, **6** (3): 235–59.

Magaritz, M., Holser, W.T. and Kirschvink, J.L., 1986, Carbon-isotope events across the Precambrian–Cambrian boundary on the Siberian Platform, *Nature*, **320** (6059): 258–9.

Morris, S.C., 1987, The search for the Precambrian–Cambrian boundary, *American Scientist*, **75** (2): 157–67.

Nazarov, M.A., Barsukova, L.D., Kolesov, G.M. and Alekseev, A.S., 1983, Iridium abundancies in the Precambrian–Cambrian boundary deposits and sedimentary rocks of Russian Platform, *Abstract, Lunar and Planetary Science Conference*, **14**: 546–7.

Officer, C.B. and Drake, C.L., 1983, The Cretaceous–Tertiary transition, *Science*, **219** (4591): 1383–90.

Officer, C.B. and Drake, C.L., 1985, Terminal Cretaceous environmental effects, *Science*, **227** (4691): 1161–7.

Olmez, I., Finnegan, D.L. and Zoller, W.H., 1986, Iridium emissions from Kilauea volcano, *Journal of Geophysical Research*. **91** (B1): 653–63.

Olsen, P.E., Shubin, N.H. and Anders, M.H., 1988, New early Jurassic tetrapod assemblages constrain Jurassic–Triassic tetrapod extinction event, *Science*, **237** (4818): 1025–9.

Orth, C.J., Attrep, A., Jr, Mao, X.-Y., Kauffman, E.G., Diner, R. and Elder, W.P., 1988, Iridium abundance maxima in the upper Cenomanian extinction interval, *Geophysical Research Letters*, **15** (4): 346–9.

Orth, C.J., Gilmore, J.S. and Knight, J.D., 1987, Iridium anomaly at the Cretaceous–Tertiary boundary in the Raton Basin. In S.G. Lucas and A.P. Hunt (eds), *New Mexico Geological Society Guidebook, 38th Field Conference, northeastern New Mexico*, University of New Mexico, Albuquerque: 265–70.

Orth, C.J., Gilmore, J.S., Knight, J.D., Pillmore, C.L., Tschudy, R.H. and Fassett, J.E., 1981, An iridium abundance anomaly at the palynological Cretaceous–Tertiary boundary in northeastern New Mexico, *Science*, **214** (4527): 1341–3.

Orth, C.J., Gilmore, J.S., Quintana, L.R. and Sheehan, P.M., 1986, Terminal Ordovician extinction: geochemical analysis of the Ordovician/Silurian boundary, Anticosti Island, Quebec, *Geology*, **14** (5): 433–6.

Orth, C.J., Knight, J.D., Quintana L.R., Gilmore, J.S. and Palmer, A.R., 1984, A search for iridium abundance anomalies at two late Cambrian biomere boundaries in western Utah, *Science*, **223** (4633): 163–5.

Orth, C.J., Quintana, L.R., Gilmore, J.S., Barrick, J.E., Haywa, J.N. and Spesshardt, S.A., 1988, Pt-group metal anomalies in the Lower Mississippian of Oklahoma, *Geology*, **16** (7): 637–40.

Orth, C.J., Quintana, L.R., Gilmore, J.S., Grayson, R.C., Jr and Westergaard, E.H., 1986, Trace-element anomalies at the Mississippian–Pennsylvanian boundary in Oklahoma and Texas, *Geology*, **14** (12): 986–90.

Palmer, A.R., 1965, Biomere—a new kind of biostratigraphic unit, *Journal of Paleontology*, **39** (1): 149–53.

Palmer, A.R., 1982, Biomere boundaries: a possible test for extraterrestrial perturbation of the biosphere. In L.T. Silver and P.H. Schultz (eds), *Geological implications of impacts of large asteroids and comets on the Earth, Special Paper of the Geological Society of America*, **190**: 469–75.

Playford, P.E., McLaren, D.J., Orth, C.J., Gilmore, J.S. and Goodfellow, W.T., 1984, Iridium anomaly in the Upper Devonian of the Canning Basin, Western Australia, *Science*, **226** (4673): 437–9.

Rampino, M.R. and Stothers, R.B., 1984, Terrestrial mass extinctions, cometary impacts and the Sun's motion perpendicular to the galactic plane, *Nature*, **308** (5961): 709–12.

Raup, D.M. and Sepkoski, J.J., Jr, 1984, Periodicity of extinctions in the geologic past, *Proceedings of the National Academy of Science U.S.A.*, **81** (3): 801–5.

Raup, D.M. and Sepkoski, J.J., Jr, 1986, Periodic extinctions of families and genera, *Science*, **231** (4740): 833–6.

Robertson, P.B., Dence, M.R. and Vos, M.A., 1968, Deformation in rock-forming minerals from Canadian craters. In N.M. Short (ed.), *Shock metamorphism of natural materials*, Mono, Baltimore, MD: 433–52.

Saunders, W.B. and Ramsbottom, W.H.C., 1982, Mid-Carboniferous, biostratigraphy and boundary choices. In W.H.C. Ramsbottom *et al.* (eds), *Biostratigraphic data for a mid-Carboniferous boundary, Subcommission on Carboniferous stratigraphy, Publication* **8**, Leeds, England: 1–5.

Sepkoski, J.J., Jr, 1982, Mass extinctions in the Phanerozoic: a review. In L.T.

Silver and P.H. Schultz (eds), *Geological implications of impacts of large asteroids and comets on the Earth, Special Paper of the Geological Society of America*, **190**: 283–9.

Sepkoski, J.J., Jr, 1986, Phanerozoic overview of mass extinctions. In D.M. Raup and D. Jablonski (eds), *Patterns and processes in the history of life*, Springer-Verlag, Berlin: 277–95.

Short, N.M., 1968, Nuclear explosion-induced microdeformation of rocks: an aid to the identification of meteorite impact structures. In N.M. Short (ed.), *Shock metamorphism of natural materials*, Mono, Baltimore, MD: 185–210.

Stanley, S.M., 1984, Marine mass extinctions: a dominant role for temperature. In M.H. Nitecki (ed.), *Extinctions*, University of Chicago Press, Chicago: 69–117.

Stanley, S.M., 1988, Paleozoic mass extinctions: shared patterns suggest global cooling as a common cause, *American Journal of Science*, **288** (4): 334–52.

Sun, Y.-Y., Chai, Z.-F., Ma, S.-L., Mao, X.-Y., Xu, D.-Y., Zhang, Q.-W., Yang, Z.-Z., Sheng, J.-Z., Chen, C.-Z., Rui, L., Liang, X.-L., Zhao, J.-M. and He, J.-W., 1984, The discovery of iridium anomaly in the Permian–Triassic boundary clay in Changxing, Zhijing, China, and its significance. In G. Tu (ed.), *Developments in geoscience*, Science Press, Beijing: 235–45.

Sun, Y.-Y., Chai, Z.-F., Ma, S.-L., Mao, X.-Y., Xu, D.-Y., Zhang, Q.-W., Yang, Z.-Z., Sheng, J.-Z., Chen, C.-Z., Rui, L., Liang, X.-L., Zhao, J.-M. and He, J.-W., 1985, Discovery of anomalies of platinum group elements at the Permian–Triassic boundary in Changxing, Zhejiang, China and their significance, *Abstracts, 27th International Geological Congress*, **8**: 309.

Sunblad, K. and Gee, D.G., 1985, Occurrence of an uraniferous-vanadiniferous graphitic phyllite in the Koli Nappes of the Stekenjokk, central Swedish Caledonides, *Geologiska Föreningens i Stockholm Förhandlingar*, **106** (3): 269–74.

Whitmire, D.P. and Jackson, A.A., 1984, Are periodic mass extinctions driven by a distant solar companion?, *Nature*, **308** (5961): 713–14.

Wilde, P., Berry, W.B.N., Quinby-Hunt, M.S., Orth, C.J., Quintana, L.R. and Gilmore, J.S., 1986, Iridium abundancies across the Ordovician–Silurian stratotype, *Science*, **233** (4761): 339–41.

Wolbach, W.S., Lewis, R.S. and Anders, E., 1985, Cretaceous extinctions: evidence for wildfires and search for meteoritic material, *Science*, **230** (4722): 167–70.

Xu, D.-Y., Ma, S.-L., Chai, Z.-F., Mao, X.-Y., Sun, Y.-Y., Zhang, Q.-W. and Yang, Z.-Z., 1985, Abundance variation of iridium and trace elements at the Permian/Triassic boundary at Shangsi in China, *Nature*, **314** (6007): 154–6.

Zheng, Y., Xu, D.-Y., Zhang, Q.-W., Sun, Y.-Y. and Ye, L.-F., 1987, Some stable isotope anomalous events across the Permian/Triassic boundary, *Abstracts, IGCP 199, Rare Events in Geology*: 34.

Zoller, W.H., Parrington, J.R. and Phelan Kotra, J.M., 1983, Iridium enrichment in airborne particles from Kilauea volcano, *Science*, **222** (4628): 1118–21.

Chapter 4

ON MASS EXTINCTION AND FAUNAL TURNOVER NEAR THE END OF THE PRECAMBRIAN

Martin D. Brasier

INTRODUCTION

The fossil record indicates that the appearance of metazoan animals proceeded through three main phases. At first, in the late Precambrian interval prior to the Varangerian glaciation, megascopic animals were lacking; this was an interval of unicellular procaryotes and eucaryotes, and some multicellular algae. A major radiation of soft-bodied metazoans in Redkinian times (the 'Ediacara fauna') then followed the Varangerian glaciation (Figure 4.1; see also Sokolov and Fedonkin, 1984; Glaessner, 1984). Finally, the skeletal remains of invertebrate groups appeared close to the base of the Cambrian in Nemakit-Daldynian and Tommotian, or Meishucunian, times (Brasier, 1979; Rozanov and Sokolov, 1984; Luo *et al.*, 1984; Cowie and Brasier, 1988).

The main elements of the Ediacara fauna have been interpreted by Glaessner (1984) as the impressions of cnidarian polyps and medusoids, annelid worms and a few arthropods. At the other extreme, Seilacher (1984) has argued that this fauna represents a heterogenous group of trace fossils and unidentified organisms, perhaps of phyla unique to their times. The opinion of Sokolov and Fedonkin (1984) lies somewhere in between these views.

Whatever the interpretation, a large proportion of Ediacaran assemblages comprise discoidal impressions of 'medusoids' superficially resembling jellyfish remains. More distinctive are the 'quilted' remains of petalona-means such as *Charnia* sp. and *Pteridinium* sp., and segmented remains of worm-like *Dickinsonia* sp. and *Spriggina* sp. Assemblages of this type, with *Charnia* sp. as a characteristic element, occupy a distinctive position in the stratigraphic successions of five continents (Figure 4.1).

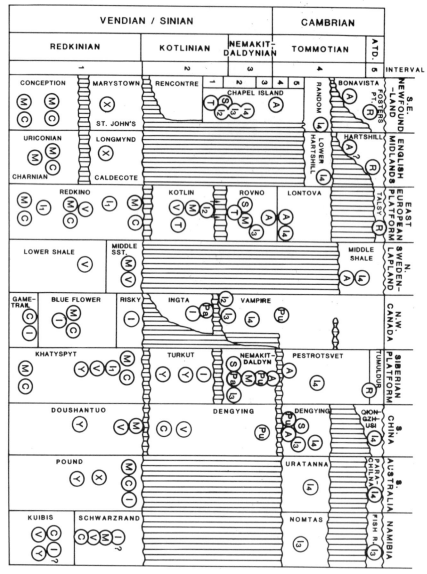

Figure 4.1. Suggested correlation of some important fossiliferous late Precambrian–Cambrian successions (from sources in text). Key to symbols: A, small shelly fossils of the *Aldanella attleborensis* assemblage (*sensu* Brasier, in Cowie and Brasier, 1989) of Tommotian/Meishucunian type; C, undifferentiated *Charnia* spp. assemblages of the Ediacara fauna; M, undifferentiated medusoid assemblages of the Ediacara fauna; I, ichnofossil assemblages of undifferentiated Precambrian type (*sensu* Narbonne et al., 1987); I$_1$, ichnofossil assemblages of Precambrian type (zone I of Crimes, 1987); I$_2$, ichnofossil assemblages with *Harlaniella podolica*, of Kotlinian type (*sensu* Narbonne et al., 1987); I$_3$, ichnofossil assemblages of zone II/*Phycodes pedum* type (Crimes, 1987; Narbonne et al.,

Petalonameans have not been reported with trace fossil assemblages of *Phycodes pedum* or younger type (for example, Crimes, 1987; Narbonne *et al.* 1987), nor have they been reported with skeletal fossils of Nemakit-Daldynian to Tommotian type. Did mass extinction bring about the disappearance of the Ediacara fauna at some time before the start of the Nemakit-Daldynian?

It might be wondered whether the very poor fossil record of soft-bodied faunas permits one to discuss extinction. This would be true of, say, the Burgess Shale fauna, but the Ediacaran assemblage was quite different: reports are known from many parts of the world (Glaessner, 1984; Donovan, 1987), often from shallow-water sediments of a type well represented throughout the succeeding fossil record. Why then did this assemblage disappear?

Several lines of evidence have been put forward that imply extinction around the end of the Precambrian: uniqueness of the Ediacara soft-bodied fauna (for example, Seilacher, 1984); extinction of ichnofossils *Harlaniella podolica* and *Palaeopascichnus delicatus* (for example, Narbonne *et al.*, 1987); perturbations in stable carbon isotopes and trace-element anomalies (for example, Hsü *et al.*, 1985; Xu *et al.*, 1985; Zhang *et al.*, 1987). Before going on to look at these various aspects, it is necessary to begin with a review of emerging evidence for faunal change through late Precambrian to earliest Cambrian times.

FAUNAL SUCCESSION IN THE LATEST PRECAMBRIAN TO EARLIEST CAMBRIAN

The successions discussed below are related diagrammatically in Figure 4.1.

Newfoundland and England (Avalon Platform)

A very comprehensive succession of events has now been reported from the Avalon Platform, southeastern Newfoundland, by Narbonne *et al.*

1987) I_4, ichnofossil assemblages of zone III/*Rusophycus avalonensis* type (Crimes, 1987; Narbonne *et al.*, 1987); Pa, small shelly fossils of the *Protohertzina anabarica* assemblage (*sensu* Brasier, in Cowie and Brasier, 1989) of Nemakit-Daldynian (or earlier) type; Pu, small shelly fossils of the *Purella antiqua* assemblage (*sensu* Brasier, 1988) of late Nemakit-Daldynian to early Tommotian type; R, small shelly fossils of the *Rhombocorniculum insolutum* assemblage (*sensu* Brasier, in Cowie and Brasier, 1989) of late Tommotian–early Atdabanian type; S, *Sabellitides cambrienesis* and related organic tubes; T, algal ribbons of *Tyrasotaenia* sp.; V, algal ribbons of *Vendotaenia* sp.; X, problematical *Arumberia* and *Aspidella* assemblages of the Ediacara fauna; Y, Yudomian stromatolite and microphytolite assemblages.

(1987). Here, the succession is many kilometres thick, though an unconformity separates Conception Group strata, with the *Charnia masoni*,
other petalonameans and medusoids, from a later cycle in the St John's
Group (with problematical *Aspidella* sp. and *Arumberia* sp.) and Marys-
town Group. There is also a major break between these often volcaniclastic
strata and the succeeding Rencontre Formation–Chapel Island Formation–
Random Formation cycle of the Burin Peninsula. Here, the *Harlaniella
podolica* ichnofauna (plus *Palaeopaschichnus* sp. and algal ribbons of
Tyrasotaenia sp.) is joined at a higher level by organic tubes of *Sabellidites
cambriensis*. The *H. podolica* ichnofauna is then abruptly replaced by a
more diverse and complex *Phycodes pedum* ichnofauna low in Member 2,
though chondrophores survived into higher levels. The first skeletal assemblages occur in Member 4 of the Chapel Island Member, with tubes of
'*Ladatheca*' sp. and *Platysolenites antiquissimus* and the gastropod *Aldanella
attleborensis*; these may mark the base of the Tommotian Stage.

A very comparable succession is known from inliers and subsurface
strata in the English Midlands, broadly substantiating the outlines of this
sequence (Cowie and Brasier, 1989). But here, a longer break may
separate deposits of Charnian and Tommotian age, much as is seen in the
Avalon and Bonavista Peninsulas of Newfoundland.

Eastern Europe and Scandinavia

A relatively complete sequence of events has also been traced on the East
European or Baltic Platform, where Sokolov and Fedonkin (1984; 1985;
1986) have found elements of the classic Ediacara fauna in the Redkinian
Stage. According to these authors, this assemblage suffered a diminution
in size and diversity during the ensuing Kotlin Stage, when only small
medusoidal impressions were left. This is true also of the trace fossils,
characterized by *Harlaniella podolica*. Organic ribbons of *Vendotaenia* sp.
are also characteristic. A few medusoidal taxa (*Kullingia* sp. and *Nemiana*
sp.) survived into the Rovno Formation (Sokolov and Fedonkin, 1984;
1986), which bears tubes of *Sabellidites cambriensis* and elements of the
Phycodes pedum ichnofossil assemblage. The first skeletal fossils (*Platysolenites* sp. and *Aldanella* spp.) appear higher in the Rovno Formation, or
near the base of the Lontova Formation. It has been suggested that these
mark the base of the Tommotian Stage (Urbanek and Rozanov, 1983).

A comparable sequence of events can be traced in the successions of
Lappland and Finnmark (e.g. Banks, 1970; Foyn and Glaessner, 1979;
Vidal, 1981). The Lower Sandstone, Lower Shale and Middle Sandstone
(with *Kullingia* sp., vendotaeniids) may be of Redkinian age. The Middle
Shale with *Platysolenites antiquissimus* and *Aldanella kunda* is probably of
Tommotian to Atdabanian age.

North America

The most important successions in North America occur in the Wernecke and Mackenzie Mountains of northwestern Canada (Narbonne and Hoffman, 1987; Aitken, 1988), where the Avalonian 'story' is repeated yet again. For simplicity, the Mackenzie Mountain nomenclature is followed here (Aitken, 1988). The petalonamean *Pteridinium* sp. occurs with the first traces (*Gordia* sp., *Torrowangea* sp.) in the Blueflower Formation of the Mackenzies, while the medusoids *Cyclomedusa* sp. and *Beltanelliformis* sp. also occur, with similar traces, at this level in the Wernecke Mountains. The overlying Risky Formation is succeeded by a hiatus widespread across the Canadian Cordillera. Where preserved, the succeeding Ingta Formation lacks the Ediacara fauna, but bears larger trace fossils, taken to be of Kotlinian age. A limestone, with protoconodont *Protohertzina anabarica*, caps this unit. The overlying Vampire Formation successively bears *Harlaniella podolica* and *Phycodes pedum* assemblages of Kotlinian and Nemakit-Daldynian aspect, while a Nemakit-Daldynian, or Meishucunian, type of skeletal assemblage occurs higher, above a disconformity surface (Nowlan *et al.*, 1985).

All the evidence from Avalonia, Baltica and NW Canada clearly points to the disappearance of the *Charnia* fauna prior to the appearance the *Phycodes pedum* assemblage, that heralds the diversification of more penetrative burrows and arthropod limbs. This interregnum is marked by medusoid assemblages of restricted diversity, by shallow horizontal traces of *Harlaniella podolica* and *Palaeopaschichnus* sp., and by levels with abundant organic remains of *Vendotaenia* sp. and *Tyrasotaenia* sp.

Siberia

On the Siberian Platform, Ediacaran assemblages with *Charnia* sp., petalonameans, medusoids and problematica are known from about six localities from the middle Yudoma dolomites of northern Siberia (Sokolov and Fedonkin, 1984; Khomentovsky, 1986). Higher strata lack the Ediacara fauna, though the medusoid *Cyclomedusa* ex gr. *davidi* is reported to have survived into Nemakit-Daldynian strata, alongside the appearance of the *Phycodes pedum* ichnofossil assemblage, sabelliditid tubes and the *Anabarites trisulcatus–Protohertzina anabarica* skeletal fauna (see, for example, Khomentovsky, 1986; Fedonkin, 1988).

Although *Harlaniella podolica* is not present, the interval between the *Charnia* fauna and the Nemakit-Daldynian skeletal fauna may compare with presumed Kotlinian strata in the Avalon, Baltic and NW Canada regions. Yudomian stromatolite assemblages appear to be consistent with this (Sokolov and Fedonkin, 1984).

China

Above the Nantuo tillite on the Yangtze Platform, the Doushantuo Formation contains Yudomian stromatolite assemblages and *Vendotaenia* sp.
(Xing *et al.*, 1985). A specimen of *Charnia* sp. recovered from the Shibantan
Member of the overlying Dengying Formation, Yangtze Gorges (Xing *et
al.*, 1985) may be Redkinian, but its high position suggests it could also be
Kotlinian. Higher strata contain siliceous annulated tubes of *Sinotubulites*
sp. (Chen *et al.*, 1981), while the first skeletal assemblages (Zone I) appear
to be late Nemakit-Daldynian to early Tommotian on the basis of fauna and
stable carbon isotopes (Brasier and Singh, 1987; Brasier and Magaritz,
1989).

On the North China Platform, the Liaoning medusoid assemblage of
northeastern China overlies sabelliditid worm remains. This fauna could be
of Kotlinian age (see, for example, Sokolov and Fedonkin, 1984).

Australia

The Pound Quartzite of South Australia contains the 'classic' Ediacaran
fauna (Glaessner, 1984). This unit is succeeded disconformably by sandstones of the Uratanna Formation, with arthropod resting traces (*Rusophycus* sp.). The Parachilna Formation follows disconformably, bearing a
Phycodes pedum ichnofossil assemblage (Daily, 1972; 1973). The first
skeletal assemblages (archaeocyathans, small shelly fossils, trilobites),
from the Ajax and Pararra Limestones, suggest a much younger, mid- to
late Atdabanian, age. Therefore, when compared with the foregoing,
much of the sequence appears to be missing.

Africa

The evidence from the Kuibis and Schwarzrand Formations, Nama Group
of Namibia in southern Africa, is more paradoxical. Medusoids and largely
endemic petalonameans occur here with diminutive traces of (?)*Skolithos*
sp., *Bergaueria* sp., *Diplocraterion* sp. and other traces normally restricted
to assemblages of post-Ediacaran type (Crimes and Germs, 1982; Crimes,
1987; Narbonne *et al.*, 1987). Reefs of calcareous *Cloudina* sp. tubes also
occur in the Kuibis Formation (Germs, 1983) giving the whole assemblage
a suspiciously young flavour.

Is the Nama assemblage really of Ediacarian age? Several lines of
evidence suggest it is. *Vendotaenia* and palynomorphs together suggest
comparison with the contemporaneous Redkinian to Kotlinian Stage of the
USSR (Germs *et al.*, 1986), and the ichnofossil *Phycodes pedum* only
appears above the Ediacaran assemblage, in the Fish River Formation

(Crimes and Germs, 1982). The peculiar character of the Nama assemblage could be explained by its isolated position on the southern flank of Gondwana.

EXTINCTION?

The best successions of Canada and the Soviet Union suggest that the petalonamean fauna disappeared from the fossil record at some time before the appearance of the earliest skeletal and diverse trace fossil asssemblages of Nemakit-Daldynian type. A restricted assemblage of medusoids locally seems to have carried on into Nemakit-Daldynian strata. From these various lines of evidence, a sequence of biotic disappearances (extinctions?) can be put forward for consideration:

(a) Widespread disappearance of the *Charnia* fauna, with petalonameans, 'annelids', 'protoarthropods', and so on, at about the end of the Redkinian Stage.
(b) Widespread disappearance of the late Precambrian ichnofossils *Harlaniella* sp. and *Palaeopaschichnus* sp., and of the Proterozoic macro-scopic algae *Vendotaenia* sp. and *Chuaria* sp., at about the end of the Kotlinian Stage.
(c) Disappearance of the *Sabellidites* and *Tyrasotaenia* biota at some time from the Tommotian(?) onwards.
(d) Progressive dwindling of preserved medusoids, and medusoid diversity, from Redkinian to Tommotian times.

Was each of these various 'disappearances' due to factors bringing about true extinction? Some possibilities are considered below.

Evolutionary 'pseudoextinction'

The risk of overlooking 'pseudoextinction' among paraphyletic groups has recently been highlighted by Patterson and Smith (1987). Cambrian skeletal elements that superficially seem traceable back to Ediacaran precursors include trilobites (cf. *Spriggina* sp.), triradiate tubes of *Anabarites* sp. (cf. *Skinnera* sp., *Albumares* sp., *Tribrachidium* sp.; see, for example, Sokolov and Fedonkin, 1984) and quadriradiate tubes of eoconulariids (cf. *Cono-medusites* sp.; see, for example, Glaessner, 1971). However, while some elements of the Ediacara fauna may have been primitive ancestors of Cambrian and younger descendents, this is uncertain. The disappearance of so many taxa can hardly be dismissed as 'pseudoextinction'. Ancestors of the great bulk of early skeletal remains (annelid tubes, molluscs, brachiopods, sponges) are not recognized in the Ediacara fauna, while the petalonamenan fauna, in particular, seems to have been unique to this period.

Non-preservation

The widespread occurence of the Ediacara fauna is very remarkable when contrasted with later lagerstätten deposits. Even the mode of preservation (three dimensional moulds in shallow-water sandstones and deeper-water tuffs) is unusual. Could the disappearance of this extraordinary Ediacara fauna have been due to changes in the conditions of preservation? Some credence might seem to be given to this view from reports of *Dickinsonia*-like remains in the trilobite-bearing Lower Cambrian of Kazakhstan (Borovikov, 1976; cautiously confirmed by Glaessner, 1984, p. 144). What, however, were those changing taphonomic controls? Factors commonly implicated include rising levels of atmospheric oxygen, increased scavenging and increased bioturbation, all leading to the destruction of dead remains.

It is conceivable that atmospheric oxygen levels were rising during the late Precambrian, in tandem with the increasing burial of carbonaceous matter (see, for example, Knoll *et al.*, 1986). That might seem to answer our question and put an end to the matter. There is, however, reason to suspect that oxygen levels in the water column were also declining during the late Precambrian (see below). If so, such conditions might be expected to have favoured the preservation of soft-bodied faunas (cf. the Burgess Shale or the Posidonia Shales). Indeed, part of the evidence for this anoxia is the widespread preservation of sapropelic organic matter, organic-walled vendotaeniids and sabelliditid worm tubes in the Kotlinian and Nemakit-Daldynian. We may therefore turn the question around: why was the *Charnia* fauna not preserved in deeper-water facies close to the oxygen minimum zone of Kotlinian to Cambrian times? Is it because its elements were already extinct?

An increase in scavenging is difficult to detect, but the increase in bioturbation from Precambrian to Cambrian times is conspicuous and well documented. Was it, as Gehling (1987) has suggested, that 'more efficient bioturbation in the Cambrian . . . put an end to the Ediacaran style of preservation'? Here again, we come up against a partial mismatch of the evidence, for while some traces may be attributable to scavengers in the Redkinian, the explosive radiation of burrowers and trail producers (including scavengers and grazers), and of more pervasive bioturbation, did not follow until Nemakit-Daldynian times. This was long after the major decline of the classic *Charnia* fauna.

Another aspect deserves mention: that evolution of bacterial fermenters or fungal saprophytes may have altered the preservation potential of the Ediacara fauna (see, Sokolov, 1976; Sokolov and Fedonkin, 1984). For example, the first (?) actinomycetes are preserved on vendotaeniid thalli and sapropelic films in the Vendian.

Is it possible that some taxa escaped preservation merely because of a drastic reduction in size? Microscopic phosphatic moulds from the Middle

Cambrian of Australia bear surfaces superficially resembling *Dickinsonia* sp. (J.H. Shergold, pers. comm., 1988), whereas the latter is known to have reached lengths of 1.2 m in Ediacaran times (Runnegar, 1982).

The list of taphonomic questions rolls on, and many of them seem currently unresolvable. Further work on taphonomy, particularly taphonomic changes across the late Precambrian–early Cambrian interval, is urgently needed. However, it seems reasonable to suggest that the disappearance of the *Charnia* fauna near the end of Redkinian times was not a taphonomic effect alone, because similar soft-body preservation continued into the Kotlinian and beyond. Another factor may therefore have been involved. The 'taphonomic filter' seems much more likely to have operated at the time of increasing bioturbation in the Nemakit-Daldynian, or with the increased oceanic circulation of Tommotian times.

Predation and scavenging

Biological changes at the Precambrian–Cambrian boundary have often been ascribed to the effects of increased predation or grazing pressures (see, for example, Hutchinson, 1961; Stanley, 1976; Brasier, 1979). We are therefore bound to ask whether Precambrian floras and faunas succumbed to new predatory pressures, *à la* Dodo.

Perhaps the most direct evidence for such pressures comes from the protoconodont *Protohertzina anabarica*. This may have been the jaw apparatus of chaetognath worms (Bengtson, 1983), a group of predators known for their voracious eating habits. Such elements appear widely among the earliest skeletal assemblages of Nemakit-Daldynian to Tommotian type (Brasier and Singh, 1987; Cowie and Brasier, 1989). A few rare occurrences in NW Canada and Iran may even lie within the Kotlinian interval.

The morphology of some trilobites also indicates a predatory way of life (see, for example, Whittington, 1980). Could such arthropods be implicated in the demise of the *Charnia* fauna? It is tempting to infer this by means of trace fossils. Crimes (1987) mentions possible *Monomorphichnus* sp. at the level of Ediacaran assemblages in Australia and (in Sepkoski, 1983) also from the Kotlinian of the East European Platform. This line of reasoning is weak, however, because these traces may have been made by either non-predatory trilobites or other arthropod groups. Even if it were so, there is no direct evidence for an explosion of predators and scavengers in Kotlinian times that could have brought about mass extinction of the *Charnia* fauna. This line of argument is better applied to the subsequent disappearances, such as those of *Harlaniella* sp., *Palaeopaschichnus* sp., *Vendotaenia* sp. and *Chuaria* sp.

Sea level, regression and tectonism

The terminal Precambrian period was essentially one of rising sea levels from a low stand in the Varangerian glaciation (see, for example, Khomentovsky, 1986; Mens and Pirrus, 1986). This was a very major eustatic rise and some aspects of the Cambrian evolutionary explosion may be attributable to it (Brasier, 1979; 1982). Within such a scenario, two kinds of situation are known to coincide with mass extinctions: anoxic transgressions and cooling regressions (Brasier, 1987). Here we shall examine the evidence for an abrupt fall in sea level coincident with the decline of the *Charnia* fauna.

Aitken (1988) implies there were four 'grand cycles' from inshore clastic to offshore carbonate in northwestern Canada, between the second glaciation of the Keele Formation to the first appearance of the trilobites. The Ediacara fauna here occurs through the second grand cycle, but is absent from the first and third, and later, cycles. Indeed, the third cycle, or Ingta Formation and equivalents, seems to be represented by a disconformity or hiatus across the Canadian Cordillera (cf. Fritz and Crimes, 1983).

A major regression therefore seems to be indicated, but can it be traced elsewhere? On the eastern side of the Laurentian craton, traces of *Neonereites* sp. can be recognized in the upper part of the first cycle and medusoids occur in the thick second grand cycle above the Port Askaig tillite of the Scottish Hebrides. This second cycle (the Jura Quartzite) marks an accelerated rate of deposition in shallow water, abruptly terminated by thick pelagic shales and turbidite conglomerates (see, for example, Harris *et al.*, 1978). The latter change is perhaps related to a phase in the separation of the Baltic from the Laurentian craton (see, for example, Anderton, 1985). From Figure 4.1, it is apparent that this break can be aligned so as to correspond with a post-Ediacaran hiatus in South (and also central) Australia and Namibia; and with a new cycle marked by barren and stromatolitic dolomites on the Siberian Platform. The similar Dengying cycle in South China bears evaporites which may compare with the enormously thick Hormuz Salt Formation of Oman and Iran. On the East European Platform, the Kotlinian is widely disconformable over the Redkinian and differs considerably from it in style: it contains no volcanic rocks and few palaeontological and mineralogical indications of marine conditions (Mens and Pirrus, 1986). Most importantly, such a break also coincides with the transformation of Avalonia from volcanic arc into horsts and grabens; that is, coinciding with the final phases of the Cadomian orogeny of Armorica. Molassic red beds are found at about this level in Newfoundland, Wales, England and Armorica.

Marginal marine to non-marine conditions, or non-deposition, therefore seem to have been widespread in Kotlinian times. Is there evidence for glaciation at this level? Guan *et al.* (1986) have independently argued that the Luoquan glaciation of the North China Platform lies at a level between the Doushantuo and Dengying cycles of South China; that is, higher than the Nantuo tillites of Varangerian age. Although sabelliditid worm tubes occur in underlying beds, there is still insufficient stratigraphic control on this interpretation (but see Harland, in Cowie and Brasier 1989).

Evidence for a glacial regression awaits further analysis, but there are clear indications for regression and major plate tectonic changes that directly affected parts of the Laurentian and Baltic cratons (with rifting), as well as those of Avalonia and Armorica (with accretion) in this interval.

Anoxia

As mentioned above, anoxic conditions may be widespread during the initial stages of transgressions and some extinctions (for example, those at the Frasnian–Famennian, Pliensbachian–Toarcian and Cenomanian–Turonian boundaries) may be attributable to their effects (Brasier, 1987). Since Recent cnidarian medusae are known to be sensitive to oxygen levels in the water column, serving as indices of pollution and stagnation (Benovic *et al.*, 1987), the late Precambrian biota might also have been vulnerable.

Several lines of evidence summarized by Brasier (1989) suggest there may have been widespread anoxia during Kotlinian to Nemakit-Daldynian times. A global maximum in $\delta^{13}C$ of carbonates was possibly due to increased carbon burial (Aharon *et al.*, 1987; Brasier and Magaritz, 1989). A broadly contemporaneous maximum in the $\delta^{34}S$ of sulphates may have been due to increased pyrite formation, associated with carbon burial elsewhere (Holser, 1977; Holser *et al.*, 1986). Next, there is the evidence for an episode of widespread carbonaceous preservation of vendotaeniid and chuariamorphid algae, or sabelliditid worms, known from north-western Canada, eastern Canada, the Baltic/East European Platform, southern Spain, Iran, the Siberian Platform and the Yangtze Platform (Brasier 1989). Finally a major build-up of phosphorus in a stratified reservoir is also inferred at this time (Cook and Shergold, 1984).

Could an 'Oceanic Anoxic Event', associated with a major regressive-transgressive pulse, have eliminated elements of the Ediacara fauna? It seems compatible with some observed or inferred (and admittedly general-ized) contrasts between Ediacaran soft-bodied and Cambrian skeletal faunas; larger/smaller; lesser/greater relative surface area; abranchiate/branchiate; passive diffusive respiration/active pumping respiration; no biomineral skeleton/biomineral skeleton for toxic metal secretion. How-ever, much more information is needed in order to substantiate the existence and timing of such an event.

Bolides

The spectre of a bolide impact and associated mass extinction at the Precambrian–Cambrian boundary has been raised by Hsü *et al.* (1985).

These authors found an iridium anomaly and pronounced light $\delta^{13}C$ spike in South China. This level lies above a metalliferous hardground in black shales at the base of the Qionghzusi Formation (Figure 4.1), well into the skeletal explosion of invertebrates. Hence, the signatures are probably related to local condensation and complex chemistries developed about the pycnocline, and cannot yet be related to anything more than the local disappearance of the first skeletal assemblages (for example, *Protohertzina anabarica*, *Anabarites trisulcatus*; their descendants reappeared in South Australia).

Minimum values in stable carbon isotope values are also widely known from underlying dolomites and phosphorites of Meishucunian/Tommotian age. There is good reason to attribute these to rejuvenated circulation of stagnant water masses (Aharon *et al.*, 1987; Brasier, 1989, rather than to mass extinction (Xu *et al.*, 1985; Zhang *et al.*, 1987).

Direct evidence for a major meteorite impact is now at hand, however. Intriguingly, this occurred during deposition of the Bunyeroo Formation (Gostin, 1987) shortly *before* the appearance of the Ediacara fauna in South Australia.

CONCLUSION

The causes of particular extinctions cannot be stated with confidence. With the soft-bodied Ediacara fauna, its widespread preservation is perhaps as much of a puzzle as its disappearance, so that even the existence of 'mass extinction' events is in some doubt.

For the present, attention is drawn to evidence for a succession of profound palaeoenvironmental changes, broadly coincident with phases in the decline and disappearance of the late Precambrian biota.

Habitat area

Widespread (?global) regression and emergence of continental shelves was coincident with plate tectonic changes and (?)the Luoquan glaciation in Kotlinian times.

Ediacaran faunas are reduced to a few medusoid taxa during this interval. If correctly correlated (and that is questionable), *Charnia* sp. from South China was a late survivor.

Physical stress

The glacial evidence is uncertain, but widespread restricted circulation and oxygen deficiency are inferred in the lower water-column, perhaps coincident with climatic and sea-level change, in Kotlinian and Nemakit-Daldynian times. Other physically stressed environments (non-marine, brackish, hypersaline) were also widespread during this interval.

Organic remains of macroscopic algae (*Vendotaenia* sp., *Tyrasotaenia* sp., chuariamorphida) and worm tubes (*Sabellidites* sp.) were widely preserved at this time. Medusoids are scarce and trace fossils tend to be small, few and of low diversity.

Biological stress

Scavenging, predation and pervasive bioturbation increased rapidly in Nemakit-Daldynian times.

Small horizontal traces of *Harlaniella podolica* and *Palaeopaschichnus delicatus* were rapidly replaced by the larger and more penetrative *Phycodes pedum* ichnofossil assemblage at this time. Biomineralized jaws of *Protohertzina anabarica* and skeletal fossil assemblages became more widespread. Macroscopic algal remains disappeared through this interval, while a few rare medusoids were all that remained of the Ediacara fauna.

REFERENCES

Aharon, P., Schidlowski, M. and Singh, I.B., 1987, Chronostratigraphic markers in the end-Precambrian carbon isotope record of the Lesser Himalaya, *Nature*, **327** (6124): 699–702.

Aitken, J.D., 1988, First appearance of trace fossils in Mackenzie Mountains, northwest Canada, in relation to the highest glacial deposits and the lowest small shelly fossils. In E. Landing and G.M. Narbonne (eds), *Trace fossils, small shelly fossils and the Precambrian–Cambrian boundary, Bulletin of the New York State Museum*, **463**.

Anderton, R., 1985, Sedimentation and tectonics in the Scottish Dalradian, *Scottish Journal of Geology*, **21** (4): 407–36.

Banks, N.L., 1970, Trace fossils from the late Precambrian and Lower Cambrian of Finnmark, Norway. In T.P. Crimes and J.C. Harper (eds), *Trace fossils, Geological Journal Special Issue*, **3**, Seel House Press, Liverpool: 19–34.

Bengston, S., 1983, The early history of the Conodonta, *Fossils and Strata*, **15**: 5–19.

Benovic, A., Justic, D. and Bender, A., 1987, Enigmatic changes in the hydromedusan fauna of the northern Adriatic Sea, *Nature*, **326** (6113): 597–600.

Borovikov, L.I., 1976, The first find of fossil remains of *Dickinsonia* in the Lower Cambrian deposits of the U.S.S.R., *Dokladÿ Akademii Nauk SSSR*, **231**: 1182–4 (in Russian).

Brasier, M.D., 1979, The Cambrian radiation event. In M.R. House (ed.), *The origin of major invertebrate groups, Systematics Association Special Volume*, **12**: 103–59.

Brasier, M.D., 1982, Sea-level changes, facies changes and the late Precambrian–early Cambrian evolutionary explosion, *Precambrian Research*, **17** (1): 105–23.

Brasier, M.D., 1988, Foraminiferid extinction and ecological collapse during global biological events. In G.P. Larwood (ed.), *Extinction and survival in the fossil record, Systematics Association Special Volume* **34**: 37–64.

Brasier, M.D. 1989, Ocean-atmosphere chemistry and evolution across the Precambrian–Cambrian boundary. In J.H. Lipps and P.W. Signor, III (eds), *Origins and early evolutionary history of the Metazoa*, Plenum, New York. (In press).

Brasier, M.D. and Magaritz, M., 1989, Towards an integrated carbon isotope–small shelly fossil stratigraphy for the Precambrian–Cambrian boundary, *Abstracts, 28th International Geological Congress, Washington D.C., July.*

Brasier, M.D. and Singh, P., 1987, Microfossils and Precambrian–Cambrian boundary stratigraphy at Maldeota, Lesser Himalaya, *Geological Magazine,* **124** (4): 323–45.

Chen, M., Chen Y. and Qian, Y., 1981, Some tubular fossils from Sinian–Lower Cambrian boundary sequences, Yangtze Gorges, *Bulletin of the Tianjin Institute of Geology and Mineral Resources, Chinese Academy of Geological Sciences,* **3**: 117–24 (in Chinese).

Cook, P.J. and Shergold, J.H., 1984, Phosphorus, phosphorites and skeletal evolution at the Precambrian–Cambrian boundary, *Nature,* **308** (5956): 231–6.

Cowie, J.W. and Brasier, M.D., 1989, *The Precambrian–Cambrian boundary,* Oxford University Press, Oxford.

Crimes, T.P., 1987, Trace fossils and correlation of late Precambrian and early Cambrian strata, *Geological Magazine,* **124** (2): 97–119.

Crimes, T.P. and Germs, G.J.B., 1982, Trace fossils from the Nama Group (Precambrian–Cambrian) of southwest Africa (Namibia), *Journal of Paleontology,* **56** (4): 890–907.

Daily, B., 1972, The base of the Cambrian and the first Cambrian faunas, *University of Adelaide Centre for Precambrian Research, Special Paper,* **1**: 13–42.

Daily, B., 1973, Discovery and significance of basal Cambrian Uratanna Formation, Mt Scott Range, Flinders Ranges, South Australia, *Search,* **4** (6): 202–5.

Donovan, S.K., 1987, The fit of the continents in the late Precambrian, *Nature,* **327** (6118): 139–41.

Fedonkin, M.A., 1988, Paleoichnology of the Precambrian–Cambrian transition in the Russian Platform and Siberia. In E. Landing and G.M. Narbonne (eds), *Trace fossils, small shelly fossils and the Precambrian–Cambrian boundary, Bulletin of the New York State Museum,* **463**.

Foyn, S. and Glaessner, M.F., 1979, *Platysolenites,* other animal fossils and the Precambrian–Cambrian transition in Norway, *Norsk Geologisk Tidsskrift,* **59** (1): 25–46.

Fritz, W.H. and Crimes, T.P., 1983, Lithology, trace fossils and correlation of Precambrian–Cambrian boundary beds, Cassiar Mountains, north-central British Columbia, Canada, *Geological Survey of Canada Paper,* **83–13**: 1–24.

Gehling, J.G., 1987, Two assemblages of fossils from the Ediacara Member, of the Pound Subgroup, South Australia, *Abstracts of the International Symposium on the Terminal Precambrian and Cambrian Geology, Chinese Academy of Geological Sciences, Yichang.*

Germs, G.J.B., 1983, Implications of a sedimentary facies and depositional environmental analysis of the Nama Group, Namibia (South West Africa), *Special Publication of the Geological Society of South Africa,* **11**: 89–114.

Germs, G.J.B., Knoll, A.H. and Vidal, G., 1986, Latest Proterozoic microfossils from the Nama Group, Namibia (South West Africa), *Precambrian Research,* **32** (1): 45–62.

Glaessner, M.F., 1971, The genus *Conomedusites* Glaessner & Wade and the diversification of the Cnidaria, *Paläontologische Zeitschrift,* **43** (1): 7–17.

Glaessner, M.F., 1984, *The dawn of animal life,* Cambridge University Press, Cambridge.

Gostin, V.A., 1987, Giant meteorite impact in Australia during the late Precambrian, *Abstracts of the International Symposium on Terminal Precambrian and Cambrian Geology, Chinese Academy of Geological Sciences, Yichang*: 29.

Guan, B., Wu, R., Hambrey, M.J. and Geng, W., 1986, Glacial sediments and erosional pavements near the Cambrian–Precambrian boundary in western Henan Province, China, *Journal of the Geological Society of London*, **143** (2): 311–23.

Harris, A.L., Baldwin, C.T., Bradbury, H.J., Johnson, H.D. and Smith, R.A., 1978, Ensialic basin sedimentation: the Dalradian supergroup. In B.E. Leake and D. Bowes (eds), *Crustal evolution in NW Britain and adjacent areas*, Seel House Press, Liverpool: 115–38.

Holser, W.T., 1977, Catastrophic chemical events in the history of the ocean, *Nature*, **267** (5610): 403–8.

Holser, W.T., Magaritz, M. and Wright, J., 1986, Chemical and isotopic variations in the world ocean during Phanerozoic time. In O.H. Walliser (ed.), *Global bioevents*, Springer-Verlag, Berlin: 63–74.

Hsü, K.J., Oberhänsli, H., Gao, J.-Y., Shu, S., Haihong, C. and Krähenbuhl, U., 1985, 'Strangelove ocean' before the Cambrian explosion, *Nature*, **316** (6031): 809–11.

Hutchinson, G.E., 1961, The biologist poses some problems. In M. Sears (ed.), *Oceanography, American Association for the Advancement of Science Publication*, **67**: 85–94.

Khomentovsky, V.V., 1986, The Vendian System of Siberia and a standard stratigraphic scale, *Geological Magazine*, **123** (4): 334–48.

Knoll, A.H., Hayes, J.M., Kaufman, A.J., Swett, K. and Lambert, I., 1986, Secular variation in carbon isotope ratios from Upper Proterozoic successions of Svalbard and east Greenland, *Nature*, **321** (6073): 832–8.

Luo, H., Jiang, Z., Wu, X., Song, X. and Ouyang, L., 1984, *Sinian–Cambrian boundary stratotype section at Meishucun, Jinning, Yunnan, China*, People's Publishing House, Yunnan.

Mens, K. and Pirrus, E., 1986, Stratigraphical characteristics and development of Vendian–Cambrian boundary beds on the East European Platform, *Geological Magazine*, **123** (4): 357–60.

Narbonne, G.M. and Hoffman, H.J., 1987, Ediacaran biota of the Wernecke Mountains, Yukon, Canada, *Palaeontology*, **30** (4): 647–76.

Narbonne, G.M., Myrow, P.M. and Landing, E., 1987, A candidate stratotype for the Precambrian–Cambrian boundary, Fortune Head, Burin Peninsula, southeastern Newfoundland, *Canadian Journal of Earth Sciences*, **24** (7): 1277–93.

Nowlan, G.S., Narbonne, G.M. and Fritz, W.H., 1985, Small shelly fossils and trace fossils near the Precambrian–Cambrian boundary in Yukon territory, Canada, *Lethaia*, **18** (3): 233–56.

Patterson, C. and Smith, A.B., 1987, Is the periodicity of extinctions a taxonomic artefact?, *Nature*, **330** (6145): 248–51.

Rozanov, A.Yu. and Sokolov, B.S. (eds), 1984, *Stage subdivision of the Lower Cambrian. Stratigraphy*, Nauka, Moscow.

Runnegar, B., 1982, Oxygen requirements, biology and phylogenetic significance of the late Precambrian worm *Dickinsonia* and the evolution of the burrowing habit, *Alcheringa*, **6** (2): 223–39.

Seilacher, A., 1984, Late Precambrian and early Cambrian Metazoa: preservational or real extinctions? In H.D. Holland and A.F. Trendall (eds), *Patterns of change in Earth evolution*, Springer-Verlag, Berlin: 159–70.

Sepkoski, J.J., Jr, 1983, Precambrian–Cambrian boundary: the spike is driven and the monolith crumbles, *Paleobiology*, **9** (2): 199–206.

Sokolov, B.S., 1976, Precambrian Metazoa and the Vendian–Cambrian boundary, *Paleontologicheskiĭ Zhurnal*, **1** (1): 13–18 (in Russian).

Sokolov, B.S. and Fedonkin, M.A., 1984, The Vendian as the terminal system of the Precambrian, *Episodes*, **7** (1): 12–19.

Sokolov, B.S. and Fedonkin, M.A., 1985, *Vendian System: Volume 2*, Nauka, Moscow (in Russian).

Sokolov, B.S. and Fedonkin, M.A., 1986, Global biological events in the late Precambrian. In O.H. Walliser (ed.), *Global bio-events*, Springer-Verlag, Berlin: 105–8.

Stanley, S.M., 1976, Fossil data and the Precambrian–Cambrian evolutionary transition, *American Journal of Science*, **276** (1): 56–76.

Urbanek, A. and Rozanov, A.Yu., 1983, *Upper Precambrian and Cambrian palaeontology of the East-European Platform*, Wydawnictwa Geologiczne, Warsaw.

Vidal, G., 1981, Aspects of problematic acid-resistant organic-walled microfossils (acritarchs) in the Upper Proterozoic of the north Atlantic region, *Precambrian Research*, **15** (1): 9–23.

Whittington, H.B., 1980, Exoskeleton, moult stage, appendage morphology, and habits of the Middle Cambrian trilobite *Olenoides serratus*, *Palaeontology*, **23** (1): 171–204.

Xing, Y. *et al.*, 1985, *Late Precambrian palaeontology of China, Geological Memoirs, series 2*, **2**, Geological Publishing House, Beijing (in Chinese).

Xu, D.-Y., Zhang, Q.-W., Sun, Y.-Y. and Yan, Z., 1985, Three main mass extinctions—significant indicators of major natural divisions of geological history in the Phanerozoic, *Modern Geology*, **9** (1): 1–11.

Zhang, Q.-W., Xu, D.-Y., Sun, Y., Yang, Z.-Z. and Chai, Z.-F., 1987, The rare event at the Precambrian–Cambrian boundary and the stratigraphic position of this boundary, *Modern Geology*, **11** (1): 69–77.

Chapter 5

TRILOBITE MASS EXTINCTION NEAR THE CAMBRIAN– ORDOVICIAN BOUNDARY IN NORTH AMERICA

Stephen R. Westrop

INTRODUCTION

Cambrian trilobites display a volatile evolutionary history that rivals the well-known 'boom and bust' pattern recorded by Mesozoic ammonoids (Stanley, 1979). The rapid evolutionary rate is reflected in the major role that trilobites have played in the development of Cambrian biostratigraphic zonations. This biostratigraphic importance has led to the accumulation of a considerable body of data on the distribution of trilobites in time and space which is, in turn, ideally suited for a variety of macro evolutionary studies.

Three mass extinctions have been identified in the Upper Cambrian sequence of trilobite faunas in North America (Stitt, 1971a; 1977; Palmer, 1979) and at least one may be present in the Lower Cambrian, at the top of the *Olenellus* Zone (Palmer, 1982). Palmer (1965) used these extinctions to define suprazonal biostratigraphic units called 'biomeres'. More recently, Ludvigsen and Westrop (1985) argued that biomeres are, in fact, nothing more than stages and have proposed that the three Upper Cambrian biomeres provide the framework for a revised stadial nomenclature for North America.

Although of smaller scale than the major Phanerozoic events (Raup and Sepkoski, 1982), the Cambrian trilobite mass extinctions offer a unique setting for the investigation of forcing mechanisms and for an evaluation of the patterns of clade sorting. They may be regarded as a replicated set of 'natural experiments' performed on the same group of organisms, so that considerable potential exists for the identification of factors common to all of the extinctions. Moreover, unlike the major Phanerozoic events, comparisons between extinctions are unencumbered by major differences in the taxonomic composition of the faunas. Upper Cambrian shelf trilobite faunas of North America display a high degree of endemism (see, for example, Taylor, 1977) and so may be treated in isolation. By reducing the scale of analysis from global to provincial, it is possible to examine the palaeoenvironmental and palaeoecologic context of the extinctions in some detail. The roles of factors such as the geographic and environmental distribution of the trilobites, and of palaeogeographic changes recorded by lithofacies shifts, may be brought into sharp focus. This chapter presents the results of an analysis of the faunal changes across the upper boundary of the Sunwaptan Stage [= 'Ptychaspid Biomere' (Westrop and Ludvigsen, 1987, Figure 1)], the youngest of the three Upper Cambrian extinctions. This stratigraphic level lies within the Cambrian–Ordovician boundary interval (the exact position of the Cambrian–Ordovician boundary is currently under consideration by a committee of the International Commission on Stratigraphy).

DATA

The study is based on 39 large samples (containing a total of more than 8000 trilobites; see Ludvigsen and Westrop, 1983, for details) from shallow marine shelf carbonates and upper slope lime mudstones (the Middle Carbonate and Outer Detrital belts of Palmer, 1971). Faunas from near-shore siliciclastic depositional systems (the Inner Detrital Belt of Palmer, 1971) are poorly known (see, for example, Raasch, 1951) and have not been included in the analysis. Sample sites (Figure 5.1(a); see also Ludvigsen and Westrop, 1983) are spread widely over North America and include the District of Mackenzie (Ludvigsen, 1982), southern Alberta (Westrop, 1986), southern Oklahoma (Stitt, 1971b; 1977), New York State (Taylor and Halley, 1974), Quebec (Rasetti, 1944; 1945), Vermont (Raymond, 1924) and western Newfoundland (Ludvigsen and Westrop, in Ludvigsen *et al.*, in press). Patterns of faunal change were examined within each of four biostratigraphic intervals—in ascending order, the Apopsis, Depressa, Typicalis and Brevispicata intervals (Figure 5.1b); see also Westrop and Ludvigsen, 1987). The extinctions are concentrated in a 'critical period' that comprises the Apopsis and Depressa intervals (Westrop and Ludvigsen, 1987).

PATTERNS OF EXTINCTION

Magnitude of the extinction

Previous evaluations of the severity of the extinction suggested that it was of truly catastrophic proportions and Stitt (1975; 1977) went so far as to

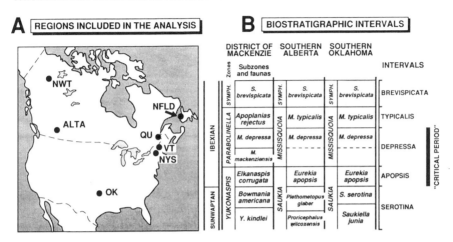

Figure 5.1. (a) Geographic regions included in the analysis (modified from Ludvig-sen and Westrop, 1983). Key: ALTA = southern Alberta; NFLD = western Newfoundland; NWT = District of Mackenzie, Northwest Territories; NYS = New York State; OK = southern Oklahoma; QU = Quebec; VT = Vermont. (b) Biostratigraphic intervals used in the study and their correlation with zonations established for the District of Mackenzie (Ludvigsen, 1982), southern Alberta (Westrop, 1986) and southern Oklahoma (Stitt, 1971b; 1977). The extinctions are concentrated in a 'critical period' that comprises the Apopsis and Depressa intervals (modified from Westrop and Ludvigsen, 1987).

claim that all shelf polymerid trilobites were eliminated. However, this early work was restricted to a single biofacies type (*Euptychaspis-Eurekia* Biofacies of Ludvigsen and Westrop, 1983) in a few stratigraphic sections in southern Oklahoma and central Texas. More recent studies (Ludvigsen and Westrop, 1983; Westrop and Ludvigsen, 1987), dealing with several biofacies and a greatly expanded geographic region, have shown that the actual magnitude was much lower—a little less than half of North American trilobite families (10 out of 24; 42%) are lost during the 'critical period' of extinctions (Figure 5.2).

Opinion has also varied on the actual duration of the extinction interval, at least partly because of differences in sampling technique. In his study of the lower and upper boundaries of the Marjuman Stage, Palmer (1982; 1984) examined a limited vertical interval on a microstratigraphic scale and concluded that the extinctions were very rapid, perhaps comparable to the rate postulated for the Cretaceous–Tertiary interval by some workers (see, for example, Alvarez *et al.*, 1980). By contrast, Westrop and Ludvigsen (1987, Figure 8; see also Ludvigsen, 1982; Ludvigsen and Westrop, 1983) investigated taxonomic turnover patterns across the top of the Sunwaptan Stage through a more extended stratigraphic sequence and concluded that the extinctions took place over a more protracted interval (through up to

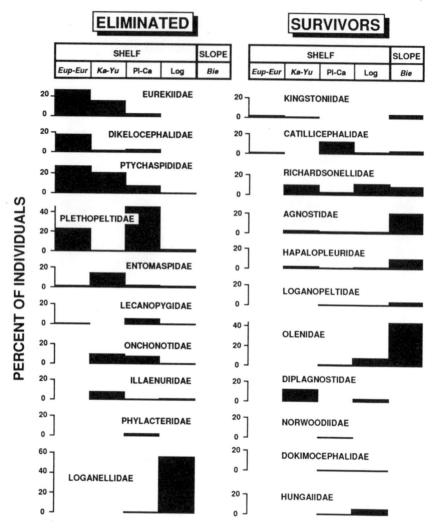

Figure 5.2. Distribution and relative abundances of families across a suite of biofacies during the Serotina interval (that is, immediately preceding the extinction interval). Biofacies (Ludvigsen and Westrop, 1983) are: *Eup-Eur, Euptychaspis-Eurekia*; *Ka-Yu, Kathleenella-Yukonaspis*; *Pl*-Ca, *Plethopeltis*-catillicephalid; Log, loganellid; *Bie, Bienvillia*. Families that are eliminated during the 'critical period' of extinctions are confined to shelf biofacies, whereas most of those that survive the 'critical period' extend from the shelf into the slope (*Bienvillia* biofacies). The figure is redrawn from Westrop and Ludvigsen (1987), with changes based on the following taxonomic revisions which are discussed in detail by Ludvigsen and Westrop, in Ludvigsen *et al.*, (in press). The Dikelocephalidae includes the paraphyletic family Saukiidae. The Plethopeltidae are confined to the shelf—material originally (Ludvigsen and Westrop, 1983, Figure 2) identified as *Plethopeltis* from the *Bienvillia*

26 m in some stratigraphic sections) and were associated with a vertical sequence of biofacies replacements. It does, however, appear from Palmer's work that taxonomic turnover at individual biofacies replacements within the extinction interval may be abrupt, suggesting that the overall pattern of diversity decline may prove to be stepwise through several metres of section.

Selectivity of extinction

The pattern of survival of taxa during an episode of mass extinction is of considerable interest for two reasons. First, analysis of the ecology of those taxa that survive might throw some light upon the extinction mechanism (see Flessa *et al.*, 1986, for discussion). Second, recent work by Jablonski (1986) on selective survival of mollusc clades during the Cretaceous–Tertiary event suggests that mass extinctions may operate by rules that differ from those that govern taxonomic turnover during normal 'background' times.

The families present in the data set used by Ludvigsen and Westrop (1983, Figure 2; revisions to the taxonomy were made based on more recent work by Ludvigsen and Westrop, in Ludvigsen *et al.*, in press) in a study of trilobite biofacies of the Cambrian–Ordovician boundary interval were divided into two categories: those that were known to have been eliminated during and those that were known to have survived the 'critical period' of extinctions. The abundance and distribution of each family were then plotted along a shelf-to-slope sequence of biofacies defined for the biostratigraphic interval prior to the 'critical period' (Figure 5.2; see also

---

biofacies (collection SR 1) in fact represent the kingstoniid trilobite *Acheilus*. The Onchonotidae is a new family established by Ludvigsen and Westrop (in Ludvigsen *et al.*, in press) for *Onchonotus* and *Yukonaspis* (the latter was erroneously assigned to the Eurekiidae by Ludvigsen, 1982). Restudy of the types of *Onchonotus foveolatus* Rasetti from the Lower Ordovician of Quebec shows that it is in fact a member of *Bellaspidella* Rasetti, a genus which is assigned by Ludvigsen and Westrop (in Ludvigsen *et al.*, in press) to the family Dokimocephalidae; this change means that the Onchonotidae do not survive the 'critical period'. The Hungaiidae is regarded as the senior synonym of the Dikelocephalinidae. The Phylacteridae is a new family that includes *Phylacterus* and *Westonaspis*. Three families that were not encountered in the data set also survive the 'critical period'. The Leiostegiidae and Eulomiidae are known from shelf margin settings (Ludvigsen and Westrop, in Ludvigsen *et al.*, in press; Palmer, 1968). Representatives of the Missisquoiidae are rare in Upper Sunwaptan shelf margin-derived boulders from the Cow Head Group (Ludvigsen and Westrop, in Ludvigsen *et al.*, in press) and have also been discovered in additional material from the upper slope *Bienvillia* biofacies (collection SR 1 of Ludvigsen and Westrop, 1983). When these taxa are included, 42% (10 out of 24) of families that occur in shelf biofacies are eliminated during the 'critical period'.

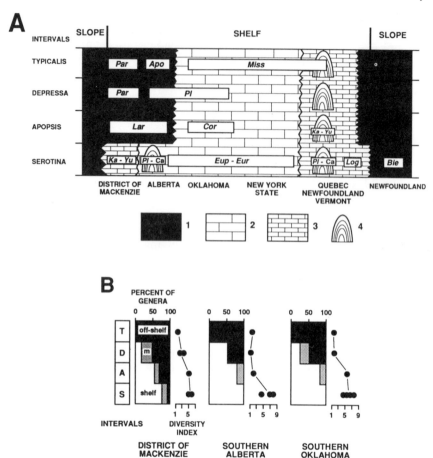

Figure 5.3. (a) Biofacies and lithiofacies distribution patterns in the Sunwaptan Stage–Ibexian Series boundary interval along a generalized transect (not to scale) between the District of Mackenzie and Newfoundland. Biofacies (Ludvigsen and Westrop, 1983) are: *Kathleenella-Yukonaspis* (*Ka-Yu*), *Plethopeltis*-catillicephalid (*Pl*-Ca), *Euptychaspis-Eurekia* (*Eup-Eur*), loganellid (Log), *Bienvillia* (*Bie*), *Larifugula* (*Lar*), *Corbinia* (*Cor*), *Parabolinella* (*Par*), *Plethopeltis* (*Pl*), *Apoplanias* (Apo) and *Missisquoia* (*Miss*). Lithofacies are: (1) outer detrital belt, dark coloured lime mudstones in Newfoundland and the District of Mackenzie or subtidal shales and storm deposits in southern Alberta; (2) light-coloured grainstones and packstones in shallow carbonate bank settings; (3) light-coloured wackestones in subtidal ramp and shelf-margin settings; and (4) algal build-ups in outer shelf and shelf-margin settings. A pronounced lithofacies and biofacies shift took place at the stage boundary (base of Apopsis interval) in the District of Mackenzie and Alberta. In the Serotina Interval, the four carbonate belt biofacies have lithofacies-specific distributions whereas the distributions of some younger biofacies cut across lithofacies boundaries. Only a single biofacies (*Miss*) is recognized in the carbonate belt during the Typicalis Interval. Figure modified from Westrop and Ludvigsen (1987).

Westrop and Ludvigsen, 1987). The results indicate that survival of clades was not random—trilobite families that extended from the shelf into the slope fared significantly better than those that were confined to shelf biofacies (G-test, $p < 0.005$; see also Westrop and Ludvigsen, 1987). Most of the families present in the slope also occur outside of North America and comparison between pandemic and endemic taxa showed that pandemic families suffered significantly lower extinction than endemic families (G-test, $p < 0.005$; this modifies the result of an earlier analysis (Westrop and Ludvigsen, 1987), made prior to taxonomic revision of the data set, in which statistical significance could not be demonstrated for the relationship between pandemicity and survival). Jablonski (1986) has demonstrated a similar relationship between geographic range and survival of North American mollusc clades during the Cretaceous–Tertiary extinctions—pandemic clades fared significantly better than endemic clades.

Biofacies changes during the extinctions

A profound reorganization of the biofacies structure of the North American shelf occurred during the extinction interval (Ludvigsen and Westrop, 1983) and appears to have been initiated by onlap in parts of the outer shelf, possibly in response to a sea-level rise (Westrop and Ludvigsen, 1987; Ludvigsen *et al.*, 1988). Cluster analysis (Ludvigsen and Westrop, 1983) outlined four shelf carbonate biofacies prior to the 'critical period' (Serotina interval) and a fifth occurred in upper slope settings (Figure 5.3(a)). Biofacies have relatively narrow environmental ranges and are typically lithofacies-specific in distribution. At the base of the 'critical period' (Apopsis interval), onlap is recorded in the outer shelf by elimination of the carbonate bank in Alberta (Westrop, 1986) and the District of Mackenzie (Ludvigsen, 1982). This facies shift is reflected by large-scale cratonward migration of trilobites from shelf margin and off-shelf sites. Immigration continued throughout the 'critical period' and was thus concurrent with the diversity decline (Figure 5.3(b); see also Ludvigsen and Westrop, 1983; Westrop and Ludvigsen, 1987). In interior sites such as Oklahoma, Apopsis and Depressa interval biofacies include immigrants

---

(b) Changes in species diversity and biogeographic affinites of genera through the Sunwaptan Stage–Ibexian Series boundary interval in the District of Mackenzie, southern Alberta and southern Oklahoma. For each region, changes in biofacies in the percentage of genera characteristic of shelf, shelf-margin and off-shelf sites are shown in the left column. Genera were allocated to each category on the basis of known or inferred distribution prior to the stage boundary. The right column shows changes in species diversity, expressed as an index (number of species/log number of individuals in collection). The decline in species diversity is not confined to a single stratigraphic horizon and is concurrent with major immigration of taxa from shelf margin and off-shelf sites. Figure modified from Westrop and Ludvigsen (1987).

from the shelf margin (for example, *Apatokephaloides* and *Ptychopleurites*), with immigrants from off-shelf sites (for example, *Apoplanias*) appearing later (Typicalis interval). In outer shelf regions, such as the District of Mackenzie, immigrants from off-shelf sites (for example, *Bienvillia* and *Parabolinites*) appear immediately at the base of the Apopsis interval. This extensive immigration disrupted the earlier, well-differentiated biofacies distribution pattern and the number of biofacies recognized is progressively reduced through merger. By the Typicalis interval, only a single biofacies is recognized in shelf carbonate environments and, in contrast to biofacies of the Serotina interval, its distribution cuts sharply across lithofacies boundaries (Figure 5.3(a)).

INTERPRETATION

Extinction model

One of the more striking features of the extinction interval is the progressive reduction in the level of biofacies differentiation. This is a dynamic process which reflects expansion of environmental ranges of taxa previously confined to shelf margin and off-shelf sites. The overall reduction in the level of spatial differentiation of the faunas invites comparison with the model proposed by Schopf (1979) for the Late Permian global extinctions of shallow marine invertebrates. Schopf (see also Flessa and Imbrie, 1973; Valentine *et al.*, 1978; Wise and Schopf, 1981) used a biogeographic approach that emphasized the influence of provincialism on global species richness. As noted by Jablonski *et al.* (1985), provinces include numerous endemic species, so that changes in the number of faunal provinces must also lead to fluctuations in total global diversity. During the Permian, final asssembly of the 'supercontinent' of Pangaea reduced the number of provinces (perhaps by as much as 42%; Schopf, 1979) and this led to the extinctions. Application of Schopf's approach to a lower hierarchical level suggests that diversity within a single faunal province, such as the North American shelf, can be reduced significantly by a decrease in the number of component communities or biofacies (Westrop and Ludvigsen, 1987; see also Valentine, 1973). This is most likely to be achieved by elimination or restriction of habitats. From this perspective, the Sunwaptan extinctions can be viewed as the consequence of a breakdown in biogeographic and ecologic structure that was initiated by sea-level rise and onlap in the outer part of the shelf. Some biofacies were eliminated, whereas others were merged by migration from shelf margin and off-shelf sites. Associated changes, including increased species packing in habitats and a reduction in the geographic ranges of some taxa, would have elevated extinction rates and led to a decline in species richness. As was noted by Westrop and Ludvigsen (1987), slope environments are likely to expand during times of sea-level rise and are able to act as refuges and as sources of immigrants onto the shelf. This provides a simple explanation for the observed pattern of family sorting during the extinction interval—those families with representatives in off-shelf refuges would have had a higher probability of survival than those confined to the shelf.

Progressive expansion of biofacies ranges during the extinction interval reflects an increase in the proportion of wide-ranging eurytopic taxa present. This shift in composition might have led to a reduction in speciation rates (low speciation rates among wide-ranging eurytopes have been discussed by several authors, including Hansen, 1980; Jablonski, 1982) which would have placed additional downward pressure on diversity (Westrop and Ludvigsen, 1987).

In the interpretation outlined above, the extinctions are viewed as an indirect result of a physical environmental change that was limited to parts of the outer shelf. Other workers have proposed that a shelf-wide environmental change, such as a catastrophic lowering of water temperature (Stitt, 1975; 1977) or oxygen levels (Palmer, 1984) led directly to the extinctions. However, independent sedimentologic evidence for a major physical environmental perturbation over the entire shelf is lacking. In several regions, including southern Oklahoma (Stitt, 1971b; 1975), central Texas (Miller *et al.*, 1982) and western Utah (Hintze *et al.*, 1988), shallow, warm water carbonate sedimentation continued through the extinction interval without significant change in environment.

Some workers (for example, Stitt, 1975; see also Brady and Rowell, 1976) have suggested that a modest temperature change might be capable of producing an extinction without altering the prevailing sedimentary regime. However, species present in the extinction interval tend to be characterized by short stratigraphic ranges (Stitt, 1971a; 1975; Westrop and Ludvigsen, 1987). Moreover, this pattern is evident in representatives of shelf families that were eliminated in the 'critical period' (for example, *Corbinia*, *Eurekia* and *Plethopeltis*) as well as those that belong to surviving families that migrated from shelf margin and slope environments (for example, *Parabolinella*, *Missisquoia* and *Apatokephaloides*). It is difficult to use a temperature decline to explain concurrent rapid turnover in both warm water clades *and* surviving clades that appear to have been eurythermal (see also Westrop and Ludvigsen, 1987).

Macroevolutionary implications

The presence of non-random sorting of families during the 'critical period' of extinctions raises questions of causality. Commonly, explanations of selective survival are couched in terms of the adaptations of individual organisms and Stitt's (1975) temperature model for Cambrian trilobite mass extinctions provides an example. By contrast, recent hierarchical approaches to evolutionary theory (see, for example, Vrba and Eldredge, 1984; Sober, 1984; Eldredge, 1985; Vrba and Gould, 1986) have emphasized that sorting can be based on properties that are emergent above the individual level; geographic distribution is a leading candidate for an emergent property of this type (Jablonski *et al.*, 1985). From this perspective, selective survival is a consequence of biodistributional properties, rather than the result of some kind of adaptive superiority.

In the interpretation adopted here, family survival is influenced greatly by the presence of representatives in refuges outside of the shelf. In other words, it proposes that sorting of families was the result of differences in an emergent property, geographic distribution. Alternative hypotheses based on differences in physiologic aptations, such as temperature tolerances, involve upward causation from the individual level. Upward causation of clade sorting should be evident from patterns of turnover at lower taxonomic levels. That is, selective survival of organisms due to advantageous phenotypic properties will lead to lower extinction rates for the species to which they belong (see also Gilinsky, 1986, pp. 251–2). This will, in turn, lead to lower extinction rates for the clades that contain those species. Thus, hypotheses concerning the focal level of clade sorting can be evaluated from species stratigraphic range data during the extinction— upward causation from the individual level should be reflected in differences in species turnover rate between eliminated and surviving clades. That is, species from surviving clades should display longer stratigraphic ranges (see Westrop, 1989, for further discussion).

Ranges of species recorded from the shelf during the 'critical period' of extinctions were compiled to the nearest subzone and converted into metres using stratigraphic thicknesses of each subzone in the most complete, monofacial (mainly shallow-water carbonates) sequence available, Chandler Creek, Oklahoma (Stitt, 1977; 1983). Only portions of the species ranges above the 'critical period' were considered. That is, this study considers only the histories of species after the beginning of the extinction interval. The data set includes both immigrants from off-shelf sites and holdovers from shelf faunas. There is no significant difference between stratigraphic ranges of species from families which were eliminated during the 'critical period' and those from families which survived (Fgirue 5.4(a); Mann–Whitney U-test, $p > 0.1$). This result corroborates the observation, made in several earlier biostratigraphic studies in widely separated parts of North America (for example, southern Oklahoma (Stitt, 1971a; 1975; 1977); central Texas (Longacre, 1970); western Utah (Miller *et al.*, 1982); District of Mackenzie (Ludvigsen, 1982); southern Alberta (Westrop, 1986)), that ranges of all species present on the shelf are very short in stage boundary intervals. Because sorting of families during the 'critical period' cannot be predicted from patterns of species turnover, explanations involving upward causation from the individual level are inappropriate. Rather, the role of geographic and environmental range indicates that the sorting process is focused above the individual level, with the presence of representatives in off-shelf sites significantly increasing the probability of family survival (see Westrop, 1989, for further discussion).

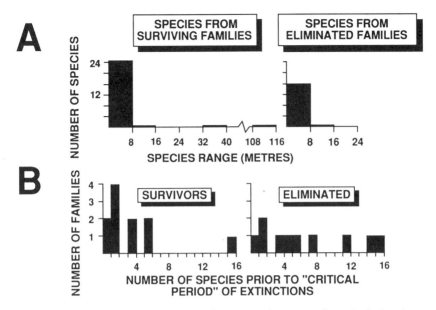

Figure 5.4 (a) Frequency distributions of stratigraphic ranges of species belonging to families eliminated during, and to families surviving, the 'critical period' of extinctions. In all cases, ranges were measured upwards from the base of the 'critical period' of extinction. The distributions are not significantly different (Mann–Whitney *U*-test, *p* > 0.1).
(b) Frequency distributions of family species richness during the Serotina Interval (that is, prior to the 'critical period' of extinctions). Surviving families do not differ significantly in species richness from those that were eliminated (Mann-Whitney *U*-test, *p* > 0.1), so that large clade size did not increase probability of survival. (Both figures redrawn from Westrop (1989).

It remains to be seen whether the sorting pattern described here represents a case of true clade selection, because selection demands that traits be both emergent and heritable (Vrba, 1984; Vrba and Gould, 1986). It is uncertain whether geographic or environmental distribution acts as a heritable property at the clade level (but see Jablonski, 1987, for an example at the species level). As noted by Vrba and Gould (1986), sorting based on non-heritable properties would fall under Cracraft's (1982; 1985) 'extrinsic control' hypothesis. In any event, the results of this analysis underscore the importance of a hierarchical approach to macroevolutionary studies.

High species richness may contribute greatly to clade survival during 'background' times between mass extinctions (Flessa and Jablonski, 1985), but evidence presented by Jablonski (1986) for Upper Cretaceous molluscs indicates that this trait may not influence the outcome of mass extinction. Comparison of family sizes prior to the extinctions at the end of the Sunwaptan (Fig. 5.4) provides further support for Jablonski's contention that 'background' and mass extinction may operate by different rules. Surviving families are not significantly different in species richness from those that were eliminated (Mann–Whitney U-test, $p > 0.1$). Geographic and environmental deployment of component species, rather than their actual numbers, played the most important role in family survival.

CONCLUSIONS

The trilobite extinctions at the top of the Sunwaptan Stage, which eliminated about half of the North American shelf trilobite families, occurred through an interval of up to 26 m of strata in some localities. Rapid turnover of taxa was concurrent with a major biogeographic and ecologic reorganization characterized by extensive cratonward migration of genera previously confined to shelf margin and off-shelf sites. These changes were initiated by onlap in the outer part of the shelf, probably in response to a sea-level rise. However, there is no evidence for a major, shelf-wide physical environmental change, such as a decline in water temperature. The overall result was a progressive reduction in levels of biofacies differentiation through the extinction interval, including a shift from lithofacies-related occurrences to biofacies distributions that cut sharply across lithofacies boundaries. The pattern of change is consistent with a simple biogeographic model in which diversity within a faunal province is influenced profoundly by a change in the number of component communities or biofacies.

Survival of families is not random, with those that extend from shelf to upper slope environments (most of which were also pandemic in distribution) faring significantly better than those confined to shelf habitats (most of which were also endemic to North America). However, a comparison of species present on the shelf during the 'critical period' of extinctions (Apopsis and Depressa intervals) showed that those that belong to surviving families do not differ in stratigraphic ranges from those from families that were eliminated. Because the pattern of family sorting cannot be predicted from turnover among their component species, explanations involving upward causation from the individual level (that is, those couched in terms of aptations of individual organisms) are inappropriate. Selective survival is more likely to be based on differences in a property that is emergent above the individual level, geographic and environmental distribution. That is, those families with representatives in upper slope refuges had a much higher probability of survival. Thus, a full understanding of causality of the extinction patterns can be gained only from a hierarchical approach to macroevolutionary analysis. Finally, species richness, a factor that increases clade survival during periods between mass extinctions, did not influence trilobite family survival during the extinction discussed here. This observation provides additional support for Jablonski's (1986) suggestion that 'background' and mass extinctions are governed by different rules.

ACKNOWLEDGMENTS

I thank Rolf Ludvigsen for his advice and encouragement during the development of this research, much of which was an outgrowth of our joint work on Upper Cambrian trilobite systematics and biostratigraphy. Funding was provided by Natural Sciences and Engineering Research Council of Canada Operating Grant 41197.

REFERENCES

Alvarez, L.W., Alvarez, W., Asaro, F. and Michel, H.V., 1980, Extraterrestrial cause for the Cretaceous–Tertiary extinctions, *Science*, **208** (4448): 1095–1108.

Brady, M.J. and Rowell, A.J., 1976, An Upper Cambrian subtidal blanket carbonate, eastern Great Basin, *Brigham Young University Geology Studies*, **23**: 153–63.

Cracraft, J., 1982, A non-equilibrium theory for the rate control of speciation and extinction and the origin of macroevolutionary patterns, *Systematic Zoology*, **31** (4): 348–65.

Cracraft, J., 1985, Biological diversification and its causes, *Annals of Missouri Botanical Gardens*, **72** (4): 794–822.

Eldredge, N., 1985, *Unfinished synthesis. Biological hierarchies and modern evolutionary thought*, Oxford University Press, New York.

Flessa, K.W., Erben, H.K., Hallam, A., Hsü, K.J., Hüssner, H.M., Jablonski, D., Raup, D.M., Sepkoski, J.J., Jr, Soulé, M.E., Sousa, W., Stinnesbeck, W. and Vermeij, G.J., 1986, Causes and consequences of extinction—group report. In D.M. Raup and D. Jablonski, (eds), *Patterns and processes in the history of life*, Springer-Verlag, Berlin; 235–57.

Flessa, K.W. and Imbrie, J., 1973, Evolutionary pulsations: evidence from Phanerozoic diversity patterns. In D.H. Tarling and S.K. Runcorn (eds), *Implications of continental drift for the earth sciences*, **1**, Academic Press, London: 247–85.

Flessa, K.W. and Jablonski, D., 1985, Declining Phanerozoic background extinction rates: effect of taxonomic structure, *Nature*, **313** (5999): 216–18.

Gilinsky, N.L., 1986, Species selection as a causal process, *Evolutionary Biology*, **20**: 249–73.

Hansen, T., 1980, Influence of larval dispersal and geographic distribution on species longevity in neogastropods. *Paleobiology*, **6** (2): 193–207.

Hintze, L.F., Taylor, M.E. and Miller, J.F., 1988, Upper Cambrian–lower Ordovician Notch Peak Formation in western Utah, *United States Geological Survey Professional Paper*, **1393**.

Jablonski, D., 1982, Evolutionary rates and modes in Late Cretaceous gastropods: role of larval ecology, *Proceedings of the Third North American Paleontological Convention*, **1**: 257–62.

Jablonski, D., 1986, Background and mass extinctions: the alternation of macroevolutionary regimes, *Science*, **231** (4734): 129–33.

Jablonski, D., 1987, Heritability at the species level: analysis of geographic ranges of Cretaceous molluscs, *Science*, **238** (4825): 360–3.

Jablonski, D., Flessa, K.W. and Valentine, J.W., 1985, Biogeography and paleobiology, *Paleobiology*, **11** (1): 75–90.

Longacre, S.A., 1970, Trilobites of the Upper Cambrian Ptychaspid Biomere, Wilberns Formation, central Texas. *Paleontological Society Memoir*, **4**.

Ludvigsen, R., 1982, Upper Cambrian and lower Ordovician trilobite biostratigraphy of the Rabbitkettle Formation, western District of Mackenzie, *Life Sciences Contributions, Royal Ontario Museum*, **134**.

Ludvigsen, R., Pratt, B.R. and Westrop, S.R., 1988, The myth of an eustatic sea level drop near the base of the Ibexian Series, *New York State Museum Bulletin*, **462**: 65–70.

Ludvigsen, R. and Westrop, S.R., 1983, Trilobite biofacies of the Cambrian–Ordovician boundary interval in northern North America, *Alcheringa*, **7** (2): 301–19.

Ludvigsen, R. and Westrop, S.R., 1985, Three new Upper Cambrian stages for North America, *Geology*, **13** (2): 139–43.

Ludvigsen, R., Westrop, S.R. and Kindle, C.H., in press, Sunwaptan (Upper Cambrian) trilobites of the Cow Head Group, western Newfoundland, *Palaeontographica Canadiana*.

Miller, J.F., Taylor, M.E., Stitt, J.H., Ethington, R.L. and Taylor, J.F., 1982, Potential Cambrian–Ordovician boundary stratotype sections in the western United States. In M.G. Bassett and W.T. Dean (eds), *The Cambrian–Ordovician boundary: sections, fossil distributions and correlations, National Museum of Wales Geological Series*, **3**: 155–80.

Palmer, A.R., 1965, Biomere, a new kind of biostratigraphic unit, *Journal of Paleontology*, **39** (1): 149–53.

Palmer, A.R., 1968, Cambrian trilobites of east-central Alaska, *United States Geological Survey Professional Paper*, **559-B**.

Palmer, A.R., 1971, The Cambrian of the Great Basin and adjacent areas. western United States. In C.H. Holland (ed), *Lower Palaeozoic rocks of the world. 1. Cambrian of the New World*. Wiley and Sons, London: 1–78.

Palmer, A.R., 1979, Biomere boundaries re-examined, *Alcheringa*, **3** (1): 33–41.

Palmer, A.R., 1982, Biomere boundaries: a possible test for extraterrestrial perturbation. In L.T. Silver and P.H. Schultz (eds), *Geological implications of impacts of large asteroids and comets on the Earth, Geological Society of America Special Paper*, **190**: 469–76.

Palmer, A.R., 1984, The biomere problem: evolution of an idea, *Journal of Paleontology*, **58** (3): 599–611.

Raasch, G.O., 1951, Revision of Croixan dikelocephalids, *Transactions of the Illinois Academy of Science*, **44**: 137–51.

Rasetti, F., 1944, Upper Cambrian trilobites from the Lévis Conglomerate, *Journal of Paleontology*, **18** (3): 229–58.

Rasetti, F., 1945, New Upper Cambrian trilobites from the Lévis Conglomerate, *Journal of Paleontology*, **19** (5): 462–78.

Raup, D.M. and Sepkoski, J.J., Jr, 1982, Mass extinctions in the marine fossil record, *Science*, **215** (4539): 1501–3.

Raymond, P.E., 1924, New Upper Cambrian and lower Ordovician trilobites from Vermont, *Proceedings of the Boston Society of Natural History*, **37** (4): 389–446.

Schopf, T.J.M., 1979, The role of biogeographic provinces in regulating marine faunal diversity through geologic time. In J. Gray and A.J. Boucot (eds), *Historical biogeography, plate tectonics and the changing environment*, Oregon State University Press, Corvalis: 449–57.

Sober, E., 1984, *The nature of selection*, MIT Press, Cambridge, MA.

Stanley, S.M., 1979, *Macroevolution: pattern and process*, Freeman, San Francisco.

Stitt, J.H., 1971a, Repeating evolutionary pattern in late Cambrian trilobite biomeres, *Journal of Paleontology*, **45** (2): 178–81.

Stitt, J.H., 1971b, Late Cambrian and earliest Ordovician trilobites, Timbered Hills and lower Arbuckle Groups, western Arbuckle Mountains, Murray County, Oklahoma. *Oklahoma Geological Survey Bulletin*, **110**.

Stitt, J.H., 1975, Adaptive radiation, trilobite paleoecology and extinction, Ptychaspid Biomere, late Cambrian of Oklahoma, *Fossils and Strata*, **4**: 381–90.

Stitt, J.H., 1977, Late Cambrian and earliest Ordovician trilobites, Wichita Mountains area, Oklahoma, *Oklahoma Geological Survey Bulletin*, **124**.

Stitt, J.H., 1983, Trilobites, biostratigraphy and lithostratigraphy of the Mackenzie Hill Limestone (lower Ordovician), Wichita and Arbuckle Mountains, Oklahoma. *Oklahoma Geological Survey Bulletin*, **134**.

Taylor, M.E., 1977, Late Cambrian of western North America: trilobite biofacies, environmental significance and biostratigraphic implications. In E.G. Kauffman, and J.E. Hazel (eds), *Concepts and methods of biostratigraphy*, Dowden, Hutchison and Ross, Stroudsburg, PA: 245–97.

Taylor, M.E. and Halley, R.B., 1974, Systematics, environment and biogeography of some late Cambrian and early Ordovician trilobites from eastern New York State. *United States Geological Survey Professional Paper*, **834**.

Valentine, J.W., 1973, *Evolutionary paleoecology of the marine biosphere*. Prentice Hall, Englewood Cliffs, N.J.

Valentine, J.W., Foin, F.C. and Peart, D., 1978, A provincial model of Phanerozoic marine diversity, *Paleobiology*, **4** (1): 55–66.

Vrba, E.S., 1984, What is species selection?, *Systematic Zoology*, **33** (3): 318–28.

Vrba, E.S. and Eldredge, N., 1984, Individuals, hierarchies and processes: towards a more complete evolutionary theory. *Paleobiology*, **10** (2): 146–71.

Vrba, E.S. and Gould, S.J., 1986, The hierarchical expansion of sorting and selection: sorting and selection cannot be equated, *Paleobiology*, **12** (2): 217–28.

Westrop, S.R., 1986, Trilobites of the Upper Cambrian Sunwaptan Stage, southern Canadian Rocky Mountains, Alberta, *Palaeontographica Canadiana*, **3**.

Westrop, S.R., 1989, Macroevolutionary implications of mass extinction—evidence from an Upper Cambrian stage boundary, *Paleobiology*, **15** (1): 46–52.

Westrop, S.R. and Ludvigsen, R., 1987, Biogeographic control of trilobite mass extinction at an Upper Cambrian 'biomere' boundary, *Paleobiology*, **13** (1): 84–99.

Wise, K.P. and Schopf, T.J.M., 1981, Was marine faunal diversity in the Pleistocene affected by changes in sea level?, *Paleobiology*, **7** (4): 394–9.

Chapter 6

THE LATE ORDOVICIAN EXTINCTION

Patrick J. Brenchley

INTRODUCTION

In a period of about a million years near the end of the Ordovician, many species, genera and families belonging to complex benthic and pelagic communities became extinct. In a statistical sense this was a mass extinction, but ecologically it was less severe than the end Permian and end Cretaceous events; there was no major loss of higher taxa, nor was the ecological structure of Lower Palaeozoic communities permanently disrupted. Nevertheless, the abrupt extinction of more than 50% of species in some groups left many niches to be filled in the early Silurian.

The Ordovician Period started with one of the great biotic radiations in Phanerozoic history, when the Cambrian fauna was succeeded, and largely replaced, by the skeletonized benthic suspension-feeding communities of the Ordovician, characterized by brachiopods, pelmatozoans, bryozoans and corals. At the same time the trilobites diversified, as did the plankton, represented by graptolites and conodonts. Progressively through the Ordovician the benthic communities diversified and established themselves in deeper water across the full width of the Ordovician shelves (Sepkoski and Sheehan, 1983), to establish the community patterns that persisted throughout the rest of the Lower Palaeozoic. By Caradoc times the faunal diversity was particularly high because much of the ecological diversification had been achieved; the sea level stood particularly high, producing very extensive shallow shelf seas across the interior of continents, and the continents were widely dispersed, giving rise to biogeographic differentiation and a high proportion of endemic species.

During the Caradoc and early Ashgill faunal diversity fell (Figure

6.2(a)–(c)), as provinciality diminished by convergence of continental plates. During this period the eastern Avalonian plate, including Britain and other parts of northwest Europe, approached Baltica, eliminating the Tornquist Ocean. By late Caradoc the faunas of the two areas were very similar (Cocks and Fortey, 1982). Similarly, the eastern Avalonian plate was approaching Laurentia during late Ordovician times (Soper and Hutton, 1984; Cocks and Fortey, 1982), eliminating the Iapetus Ocean. By the early Silurian both brachiopods and trilobites were similar on both sides of the Iapetus Suture at the generic, and in some instances at the species, level (McKerrow and Cocks, 1976).

The changing diversity of brachiopods illustrates well the effect of this mid-Ordovician extinction caused by the coalescing of previously distinct biogeographic provinces. From a mid-Caradoc peak in diversity of more than 200 genera, there was a decline to a late Caradoc level of about 150 genera, after which diversity stabilized in the Ashgill (Williams, 1965, Figure 151).

The relatively stable turnover throughout most of the Ashgill was radically disturbed by the late Ordovician extinction when, according to Raup and Sepkoski (1982), 12% of families became extinct and an average of 20 families per million years disappeared throughout the Ashgill. As will be shown later the extinction of families was concentrated in a short period at the end of the Ashgill.

In the following account an attempt will be made to estimate the severity of the extinction among different groups; constrain the time of extinctions as closely as possible; correlate the extinctions with environmental changes; and discuss the causes of extinction in terms of niche loss.

STRATIGRAPHY

The extinction events which are described in this account occurred in the uppermost stages of the Ashgill (Figure 6.1) and were mostly concentrated around the Hirnantian Stage. Unfortunately, there is currently some confusion about the upper boundaries of the Hirnantian, because a redefinition of the Ordovician–Silurian boundary has entailed raising the upper boundary of the Hirnantian to the position shown in Figure 6.1.

Until 1985, the Ordovician–Silurian boundary had been recognized at the base of the *G. persculptus* Biozone, which approximately corresponds to a time of rapid sea-level rise and hence a change from shallow marine sediments with shelly faunas to deeper-water sediments with graptolites. The British Ordovician stages are mainly based on shelly faunas and the Hirnantian, as originally envisaged, included shallow marine rocks with a distinctive branchiopod fauna. Its upper boundary was marked by a change to graptolitic shales with *G. persculptus*. The redefinition of the Ordovician–Silurian boundary (Cocks, 1985) has entailed repositioning the base of the lowest Silurian stage, the Rhuddanian, at the base of *P. acuminatus*

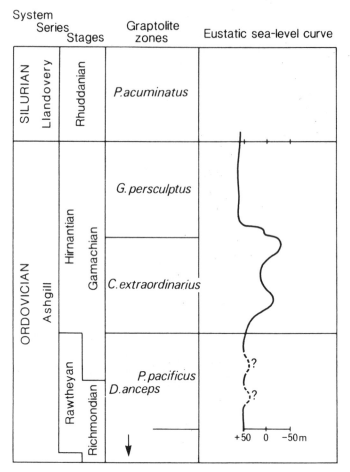

Figure 6.1. Stratigraphy and bathymetric changes close to the Ordovician–Silurian boundary.

Biozone. Hence, by definition, the Hirnantian now includes both the traditional 'Hirnantian' containing shelly fossils and the overlying graptolitic shales of the *G. persculptus* Biozone. The effect of this change is that major environmental and evolutionary events which have previously been referred to as 'end Hirnantian' are now mid-Hirnantian.

The correlation between the standard stages and the graptolite zones is reasonably close, but not completely established. The Rawtheyan–Hirnantian boundary is believed to correlate closely with the base of the *extraordinarius* Biozone and the top of the 'traditional Hirnantian' with the base of the *persculptus* Biozone or a level low within the zone. There are a number of records of *G. persculptus* occurring with a Hirnantian fauna, but most of these have been disputed. The top of the Hirnantian, as now redefined, corresponds to the base of the *P. acuminatus* Biozone.

There is a serious problem in correlating European sequences with some of those in North America, where the stages are best recognized and correlated by their conodont faunas. The uppermost Ordovician stage is the Gamachian, characterized by a Zone 13 conodont fauna. However, the boundaries of the Gamachian do not correspond to those of the Hirnantian. The base of the Gamachian probably corresponds to a level in the middle of the Rawtheyan, and the top probably corresponds to the top of the 'traditional Hirnantian', that is, the base of the *G. persculptus* Biozone.

Correlation of the Hirnantian over a wide area of the globe is possible because the stage is characterized by the distinctive *Hirnantia* fauna which contains several cosmopolitan genera. The fauna has been recognized in Europe, North America, China, central Asia, north Africa and the USSR (Rong, 1984), and more recently in South America (Benedetto, 1986). Where the *Hirnantia* fauna occurs, it establishes the presence of the Hirnantian Stage. However, in many instances it is more difficult to locate the stage boundaries on biostratigraphic grounds, either because the underlying faunas are provincial or through lack of stratigraphically useful fossils throughout the sequence.

Fortunately, in many instances, the lower boundary of the Hirnantian can be identified on the evidence of event stratigraphy. There was a late Ordovician regression, followed by a transgression (Berry and Boucot, 1973), whose effects can be seen on several different plates and which fulfil the criteria for eustatic sea-level changes. The first sedimentary evidence of the regression coresponds with the base of Hirnantian and the transgression corresponds approximately with the base of the *G. persculptus* Biozone (Brenchley and Newall, 1984; Brenchley, 1988). As the sea-level changes were eustatic, they will have had synchronous effects everywhere, except where they are obscured by severe tectonic movements or abundant input of sediment. In clastic sequences, in particular, they provide an effective means of correlation.

The substantial sea-level changes in the Hirnantian (estimated at 50–100 m by Brenchley and Newall, 1980) raise questions about the completeness of late Ordovician sequences. Where the regression affected deep shelf and basinal regions the sequences are likely to be complete, though they can be affected by channelling. On shallow clastic shelves, and particularly on shallow carbonate platforms, as in North America and Greenland, there is a likelihood that shelves became exposed and the Hirnantian will be incomplete or absent.

LATE ORDOVICIAN EXTINCTIONS

Most major fossil groups show a decline in diversity between mid-Ordovician and early Silurian times (Figure 6.2). Maximum diversity in many groups was reached in the Caradoc and declined steadily in the late Caradoc or early Ashgill, as faunal provinces merged as a consequence of plate movements (Williams, 1976). Familial and generic diversity stabilized

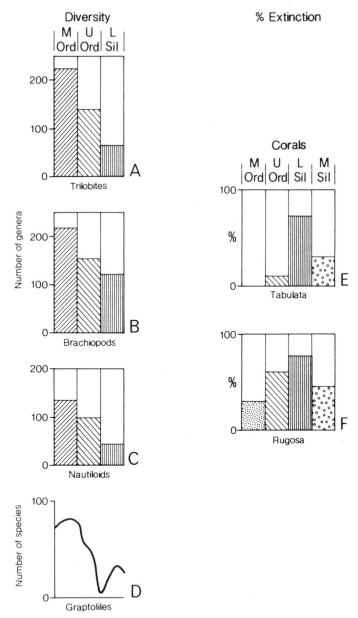

Figure 6.2. (a)–(d) Changes in generic diversity in the Middle and Upper Ordovician and Lower Silurian. (e)–(f) Changes in the extinction of genera, based on the percentage of genera surviving from the preceding unit. Data from (a) Moore (1959); (b) Williams (1965, Figure 149); (c) House (1988, Figure 7.2); (d) Koren and Rickards (1979, Figures 3); (e) Scrutton (1988, Figure 3, 4e).

in the Ashgill, until there was a sharp decrease in diversity near the end of the epoch, now recognized as the late Ordovician extinction.

There were relatively few disappearances of major groups at the level of order, or above, but families, genera and species became extinct in large numbers, though different groups were affected to different degrees.

Extinction among trilobite families was particularly severe, with 38 families in the Ashgill being reduced to about 14 in the early Silurian (Jaanusson, 1979). Much of this extinction was concentrated in the late Ashgill, when 29 late Rawtheyan families are reduced to 15 by the early Silurian (Briggs *et al.*, 1988). A severe extinction among echinoderms is also recorded from the late Ordovician, where 19 Ashgill families of cystoids, edrioasteroids and cyclocystoids are reduced to 11 in the Llandovery. Overall, it has been estimated that 12% of all families became extinct in the late Ordovician (Raup and Sepkoski, 1982).

The late Ordovician extinction radically reduced the number of genera and species among the trilobites, nautiloids, corals, brachiopods, echinoderms (principally crinoids and cystoids), graptolites, conodonts and acritarchs. It affected plants and animals, plankton and benthos. Both the sessile filter feeders and the vagile benthos were affected.

There were regional and local reductions in diversity of bryozoans, ostracods, bivalves and gastropods, but it has not been shown that there was a world-wide extinction among these groups. The patterns of extinction in individual groups are discussed below.

Trilobite extinctions

The number of trilobite genera, as measured from mainly European localities, remained relatively constant throughout most of the Ashgill, but was radically reduced near the Rawtheyan–Hirnantian boundary, when about 75% of genera failed to survive (Figure 6.3(f)). The same sharp drop in diversity has been recognized in North America (Figure 6.3(b); see also Sloan, 1988). Using a world-wide census (Owen, 1987; Briggs *et al.*, 1988), it appears that the extinction at the Rawtheyan–Hirnantian boundary was not quite so extreme as suggested by the regional estimates, but still involved a drop from 113 genera in the Rawtheyan to 71 genera in the Hirnantian, followed by a further decline to 45 genera in the Silurian (Figure 6.3(a)). It is possible that these data underestimate the Rawtheyan–Hirnantian extinction, because certain rich Gamachian faunas (Chatterton and Ludwigsen, 1983; Chatterton *et al.*, 1983) have been included in the Hirnantian, whereas they may well be Rawtheyan.

Many of the Rawtheyan genera are represented by several species, while the Hirnantian genera are often known from a single locality containing one species. In nearly every section throughout the world where trilobites are present there is a major reduction in species diversity at the Rawtheyan–Hirnantian boundary. Range charts for selected Ashgill species in the

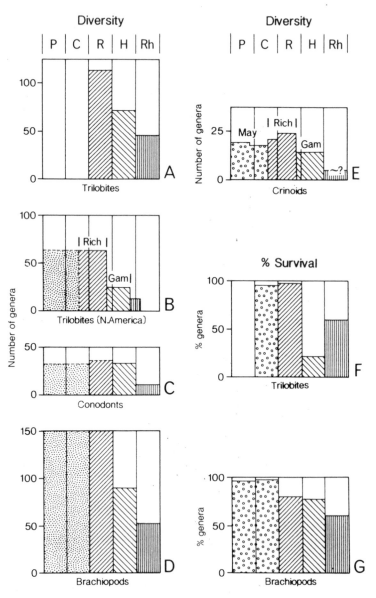

Figure 6.3. (a)–(e) Diversity changes in Ashgill and lower Llandovery stages (P–Pusgillian, C–Cautleyan, R–Rawtheyan, H–Hirnantian, Rh–Rhuddanian, May–Maysvillian, Rich–Richmondian, Gam–Gamachian. (f)–(g) Percentage survival of genera from one stage to the next, based on data from northwest Europe. Data from (a) Briggs *et al.* (1988, Figure 9.5); (b) Sloan (1988); (c) Aldridge (1988, Figure 11a); (d) Williams (1965, Figure 151) and Cocks (1988a).

British Isles (Thomas *et al.*, 1984) show 54 species of Rawtheyan trilobites being reduced to five in the Hirnantian.

The pattern of family survival is interesting in that 29 late Rawtheyan families are reduced by three or four in the Hirnantian, but by a further 10 or 11 before the Silurian. Many of the families which survive into the Hirnantian, but become extinct before the end of the stage, are represented in the Hirnantian by a single genus and often a single species. Some of the families which disappeared had been important since the Cambrian, others such as the Trinucleidae had been abundant throughout the Ordovician (Briggs *et al.*, 1988). Some were typically benthic forms (Asaphidae), others, such as the cyclopygids, were pelagic. Several of the trilobite genera which survived the end-Ordovician extinction were those which lived in carbonate mud-mounds (Mikulic, 1980).

Brachiopod extinctions

The peak of brachiopod generic diversity occurred in the Caradoc and declined into the Ashgill, when it achieved a nearly constant level (Williams, 1965). However, rich and varied brachiopod faunas were present throughout the Ashgill. The diversity was substantially reduced when Rawtheyan endemic faunas (for example, the late Richmondian faunas of North America and European faunas from the Boda Limestone, Sweden) gave way to the more cosmopolitan faunas of the Hirnantian (Cocks, 1988a). It has been estimated that about 25% of genera disappeared at the Rawtheyan–Hirnantian boundary (using data from localities in northwest Europe) and by a further 40% during the mid-Hirnantian extinction event (Figure 6.3(d),(g)). Data for the Rawtheyan–Hirnantian change, based on world-wide localities, are unavailable, but it has been estimated that 90 genera were present in the Hirnantian and this was reduced to 54 in the early Silurian, with 32 genera in common betwen the two lists (Cocks 1988a). These data are not strictly an estimate of extinction, as they take no account of Lazarus taxa, but they are likely to be an approximate estimate of the real level of extinction.

Extinction among echinoderms

The diversity of cystoid, edrioasteroid and cyclocystoid families declined sharply from 19 to 11 near the end of the Ordovician (Paul, 1988). At about the same time 12 families of crinoids became extinct (Eckert, 1988). The stratigraphic distribution of cystoid genera in the Diploporita and Dichoporita shows that a rich and varied Rawtheyan fauna with 26 genera was reduced to a small Hirnantian fauna with about eight genera (Paul, 1973). Cystoids as a whole are represented in the Lower Silurian by less than five genera, reflecting further extinctions during the Hirnantian. The stratigraphic record of the cystoids is patchy and some changes in diversity are likely to be due to preservational effects (Paul, 1988); nevertheless the late Ordovician extinction appears to be severe.

Among the crinoids there was a major late Ordovician extinction in North America. Thirteen families representing 45% of family diversity disappeared and generic diversity was reduced by 70% with the loss of 19 taxa (Eckert, 1988). The extinction is believed to have occurred in the late Richmondian and thus would have preceded the extinctions at the end of the Rawtheyan (Figure 6.3(e)). However, data from the British Isles suggests that a high diversity of crinoids was to be found in the Ashgill and persisted into the Rawtheyan, to be then reduced to about two species in the Hirnantian (Donovan, 1988), suggesting the global extinction might be end Rawtheyan.

Coral extinction

Tabulate and rugose corals show an overall increase in generic diversity from their origins in the Lower and Middle Ordovician, respectively, to a peak in the Middle Devonian. There was, however, a significant extinction in the late Ordovician, when about 75% of all late Ordovician genera became extinct before the early Silurian (Figure 6.2(e),(f); see also Scrutton, 1988). The tabulate corals reached a subsidiary peak of high diversity in Upper Ordovician times, but were particularly affected by late Ordovician extinction, when it has been estimated that 50 from 70 tabulate and heliolitid genera disappeared (Kaljo and Klaamann, 1973). Thereafter, the Rugosa always oustripped the Tabulata in generic diversity (Scrutton, 1988). The exact timing of the coral extinction is difficult to assess on present data. However, varied coral faunas are known from lower Hirnantian bioherms in Norway and bioherms are known from the top of the Gamachian in Anticosti Island, while Lower Silurian bioherms are very rare or absent. On balance, it appears that coral extinction belongs to the mid-Hirnantian extinction event.

Bryozoan extinctions

A review of global family and order level data for bryozoans has failed to reveal any significant late Ordovician extinction (Taylor and Larwood, 1988). On a regional and local scale there were, however, major changes in late Ordovician bryozoan faunas. In North America the endemic genera of the Laurentian Realm, particularly those of the Cincinatti and Reedsville-Lorraine provinces, were severely affected by late Ordovician extinctions (Anstey, 1986). A striking reduction in bryozoan diversity between the late Ordovician and early Silurian has also been recorded by Spjeldnaes (1982) and, in Baltoscandia, by Brood (1981). Unforunately, there is insufficient knowledge of late Ordovician–Silurian bryozoan faunas in general to distinguish regional from global extinctions.

Ostracod extinctions

There is no pronounced extinction of ostracod families at the end of the Ordovician (Moore, 1961). Nevertheless, three studies emphasize the change in the character of ostracod assemblages at the Ordovician–Silurian boundary (Henningsmoen, 1954; Mannil, 1962; Copeland, 1973). These accounts, based on faunas in Norway, Estonia and Anticosti Island, show that ostracods are common and diverse in the uppermost Ordovician and that the major change in the fauna came at the end of the Hirnantian or Gamachian.

Bivalve extinctions

There is no evidence of a major extinction of late Ordovician bivalves (Hallam and Miller, 1988). Two families, Cycloconchidae and Colpomyidae, went extinct in the Ashgill, while the generic diversity of the pteriomorphs was reduced from 25 to 9 and palaeotaxodonts from 15 to 7 (Kriz, 1984).

Cephalopod extinctions

Nautiloids diversified during the lower Ordovician to reach a peak of familial and generic diversity in the middle Ordovician. Diversity decreased into the Ashgill and then again into the Lower Silurian, when there was a particularly sharp decline in generic diversity (Figure 6.2(c)).

Graptolite extinctions

The diversity of graptolite faunas reached a high point in the Caradoc and declined during the Ashgill (Fig. 6.2(d)). At the end of the Ordovician it declined particularly rapidly to a low point in the *extraordinarius* Biozone (that is, in the lower Hirnantian). Koren (1988) has recognized a stepped late Ordovician extinction among the graptolites, with important extinctions in the *pacificus* and *extraordinarius* Biozones, followed by a modest radiation in the *persculptus* Biozone. According to Melchin and Mitchell (1988), the late Ordovician graptolites experienced nearly total extinction and the early Silurian radiation depended on the few species which survived. Both established clades and new forms representing four or five families and about 20 genera became extinct in the upper *supernus* or mid-*bohemicus* Biozones, that is, about the start of the Hirnantian, leaving a residual fauna of one or two genera and a small number of species (Melchin and Mitchell, 1988).

Conodont extinction

There was a very major change in conodont faunas close to the Ordovician–Silurian boundary (Figure 6.3(c); see also Barnes and Bergstrom, 1988). The diversity of conodonts was gradually reduced in diversity from a peak of 75–100 species in the lower–middle Ashgill to about 20 species in the Lower Llandovery (Sweet and Bergstrom, 1984). However, there was a particularly sharp decline in diversity near the end of the Ordovician. Standing diversity of genera dropped from about 33 to 10 across the Ordovician–Silurian boundary (Aldridge, 1988), while, of about 25 late Ordovician genera, only eight cross the boundary into the Silurian (Barnes and Bergstrom, 1988). The result is that late Ordovician and early Silurian conodont faunas are strikingly different.

There are, unfortunately, problems in defining the exact time interval when the extinctions occurred. Most of the conodont faunas come from shallow marine carbonate sequences where there is usually a disconformity near the Ordovician–Silurian boundary caused by the late Ordovician regression. The faunal change can be shown to occur between the Gamachian and overlying rocks on Anticosti Island, but the exact equivalence of this level to the Hirnantian or graptolitic stratigraphy is still uncertain (Barnes and Bergstrom, 1988).

In northwest Europe, where the Hirnantian can be easily recognized, conodont faunas are sparse in the Hirnantian (Orchard, 1980; Barnes and Bergstrom, 1988). It appears likely that in the more temperate parts of the globe the major reduction in conodont diversity was at the base of the Hirnantian, but it could have been later in tropical, carbonate regions.

Chitinozoa

Chitinozoa are generally low in abundance and diversity close to the Ordovician–Silurian boundary. There was a drop in diversity during the Rawtheyan and at the base of the Hirnantian, and an appearance of new forms in the Silurian (Achab and Duffield, 1982; Grahn, 1988). However, it is uncertain whether the low diversity in the lower Hirnantian reflects a true extinction or whether it is caused by the presence of widespread shallow marine facies in which chitinozoa are rare (Grahn, 1988).

Acritarchs

There appear to be important differences between late Ordovician and Silurian acritarchs which have now been recognized in several areas on different plates (Martin, 1988). In Europe, acritarchs are apparently rare in the upper Ordovician, but it is uncertain whether this low diversity is real or a result of lack of investigation. On Anticosti Island the decline in acritarch diversity appears to occur in the Gamachian (Duffield and Legault, 1981; Martin, 1988). There is a striking change between upper Ordovician and Silurian acritarchs in the southern Appalachians (Colbath, 1986), but it is possible that the change occurs here across a disconformity, with the Gamachian missing.

THE PATTERNS OF EXTINCTION

In general the generic diversity of both benthic and pelagic Ordovician faunas was highest in the Caradoc, or early Ashgill, and subsequently declined because of a reduction in provincialism. However, it has been shown in the previous sections that most groups suffered a major extinction of families and genera close to the Ordovician–Silurian boundary.

During the Ashgill most groups still maintained a relatively high standing diversity in spite of previous extinctions. Some groups, such as the trilobites (Figure 6.3(f)) and cystoids, maintained a fairly constant level of generic diversity, while others, such as the conodonts, graptolites and possibly the brachiopods, generally declined in numbers, though not dramatically.

The first major phase of extinction was at the Rawtheyan–Hirnantian boundary, when there was a major reduction in trilobites, cystoids and graptolites, brachiopods were somewhat reduced, and conodonts and acritarchs became rare in temperate regions. Strictly, the data only prove a sharp reduction of diversity during the Rawtheyan, but evidence from individual sections suggests that the greatest change came at the Rawtheyan–Hirnantian boundary itself. For instance, at Cautley in northern England, an uppermost Rawtheyan trilobite fauna of 15 species is reduced to one or two in the Hirnantian. In Norway, an uppermost Rawtheyan trilobite fauna with about 37 species is reduced to about seven in the Hirnantian. Similarly, cystoids change immediately across the boundary in several areas and the varied brachiopod faunas of the upper Rawtheyan are replaced abruptly by the *Hirnantia* fauna. The evidence for an abrupt fauna change at a precise level is strong, but it will require careful collecting through late Rawtheyan sections to determine whether there was also an increased rate of extinction prior to the faunal turnover at the Rawtheyan–Hirnantian boundary.

This first phase of extinction left a residual fauna, the *Hirnantia* fauna, which apparently occupied most temperate regions and has now been recognized in Europe, Kazakhstan, Siberia, China, southeast Asia, north

Africa, North America (Rong, 1984) and Argentina (Benedetto, 1986). The fauna is distinguished by a number of extremely cosmopolitan genera and species, such as *Hirnantia sagittifera, Eostropheodonta hirnantensis, Plectothyrella* and *Dalmanella testudinaria*. It is commonly of low diversity, but deeper-water faunas can be varied. The fauna is unusual because of the extreme cosmopolitanism of some of its species, which are also ecologically eurytopic and may occur in both shallower and deeper marine communities. Another feature of the *Hirnantia* fauna is the rarity of trilobites and echinoderms in any of its component communities.

Hirnantian faunas from tropical regions are less well known because of the widespread stratigraphic gap at this horizon on carbonate platforms. However, carbonate sequences appear to have a different set of brachiopod communities, including the *Brevilamnulella* and *Thebesia* communities (Amsden, 1974; Amsden and Barrick, 1986; Brenchley and Cocks, 1982). Corals appear to have remained common, bryozoans are common in some places and coral/stromatopoid bioherms apparently persisted through the Hirnantian.

In Norway, which probably occupied a marginal-tropical region, the tropical communities and *Hirnantia* fauna occur at different levels in the same sequence, possibly reflecting changing climatic conditions.

It seems likely that, following the first abrupt extinction, there was a progressive phase of extinction among the benthic faunas on the extensive carbonate platforms. As the sea level fell in the early Hirnantian many brachiopods and bryozoans were eliminated from large parts of North America (Sheehan, 1973; 1975; 1982; Anstey, 1986).

The second abrupt phase of extinction occurred in the mid-Hirnantian, at or near the base of the *persculptus* Biozone. This phase coincided with a rapid rise in sea level. At this time several families of trilobites which had just survived into the Hirnantian completely disappeared and similarly residual parts of the cystoid fauna became extinct. Several elements among the brachiopods constituting the *Hirnantia* fauna disappeared at this time. Corals, too, suffered a severe extinction and for some time in the early Silurian there were few or no reefs. This was possibly also the time of maximum extinction of conodonts and acritarchs in tropical regions.

ENVIRONMENTAL CHANGES IN THE LATE ORDOVICIAN

Sea-level changes

Near the end of the Ordovician there was a glacio-eustatic regression which started at the beginning of the Hirnantian, followed by a transgression in the mid-Hirnantian. It has been estimated that the sea level fell by about 70 m and possibly rose by somewhat more than that during the transgression (Brenchley and Newall, 1984). In the unbroken upper Ordovician sequences of the Prague Basin there is evidence of a two-phase regression, with an early Hirnantian phase, followed by a sea-level rise, then a second greater fall in sea level, and finally the major mid-Hirnantian sea-level rise (Brenchley and Storch, 1989). This latter rise apparently occurred in two rapid phases separated by a stillstand with slow sedimentation (Figure 6.5; see also Brenchley and Newall, 1984).

Immediatley prior to the Hirnantian regression there was a short period in the latest Rawtheyan when carbonate deposition was particularly wide-spread (Spjeldnaes, 1981). The carbonate horizons are usually thin (a few centimetres to a few metres) and may consist of carbonate nodules, micritic limestone, bioclastic limestone, or calcareous siltstone. The limestone horizons usually contain a varied Rawtheyan fauna. The horizon could reflect a very early stage in the late Ordovician regression or, more probably, a temporary climatic shift to warmer conditions.

The early Hirnantian regression started slowly and is commonly reflected by only a slight lithological change in shale or siltstone sequences. Sand-stones usually appear rather higher in the sequence and become common at the first regressive maximum. There appears to have been widespread channelling during the later regressive phase and this removed lower Hirnantian beds in some sequences.

The effect of the Hirnantian regression was to drain and expose all those regions which were shallow marine seas in the late Rawtheyan. In particu-lar, the very extensive carbonate platforms of North America and Green-land must have been exposed in the Hirnantian, and karst surfaces have been recognized locally (Kobluk, 1984). Nearly all shelf sequences show evidence of shoaling upwards (Brenchley, 1988). On clastic shelves, muddy outer shelf areas with benthic assemblages 5 to 6 shoal up to above storm wave base (Brenchley and Storch, 1989), while shelves with benthic assemblages 4 to 5 shoal to about the shoreface. Thus, large areas of clastic shelves must have been exposed and much of the rest had shallow marine sediments. Following the downward shift of the shoreline, some basin areas received an influx of turbidites in the Hirnantian (Brenchley and Newall, 1984; Brenchley and Cullen, 1984), while in oceanic regions there was a change to more oxygenated bottom conditions, reflected by a change from black to grey shales (Brenchley, 1988).

Climatic change

In late Ordovician times there was a major Gondwanan ice sheet centred over Saharan Africa, but there is no evidence of a similar development in northern polar regions. The age of the continental ice can only be shown to be between the Caradoc and the early Silurian. However, there is a wide fringe of glacio-marine deposits and these have been shown to be of Hirnantian age in Morocco, the Prague Basin and Portugal (Destombes, 1968; 1981; Brenchley *et al.*, 1988). In the Prague Basin glacio-marine diamictites occur very low in the Hirnantian (Brenchley and Storch, 1989).

The evidence strongly suggests that, even if there was a Gondwanan ice-cap from middle Ordovician times onwards, it expanded rapidly in the early Hirnantian and caused the fall in sea level. The melting of the main part of the ice-cap occurred in the late *extraordinarius* or early *persculptus* Biozone (Legrand, 1986) and coincided with the rise in sea level. The transgression shows a stepped rise in sea level such as might occur when large slabs of grounded ice detach themselves from the margins of an ice-cap and float free (Anderson and Thomas, 1988).

The evidence suggests that cold climates prevailed up to about 60° latitude by the early Hirnantian and there was floating ice up to 45° latitude a little later (Figure 6.4). If the *Hirnantia* fauna is a cool-water fauna, then cool, shallow marine water had spread up to the margins of the tropics early in the Hirnantian and may have periodically made incursions into the tropics during this stage. With a large Gondwanan ice-cap, it would be expected that cold dense bottom waters would have flowed towards the equator from polar regions.

There are a few sequences with Hirnantian carbonates which suggests that the tropics maintained a tropical climate. However, the preserved record is poor and the Hirnantian is widely absent, so any record of Hirnantian cooling of tropical regions might not be preserved.

In addition to cooling, ocean seawater may have undergone changes in its composition in the early Hirnantian. A significant enrichment in $\delta^{13}C$ relative to earlier Ashgill carbonates has been recorded from carbonate mud mounds (Boda Limestone) in Scandinavia. Geochemical changes in shales across the Rawtheyan–Hirnantian boundary have also been recorded by Brenchley (1984). The data are too small to determine whether the changes are local or reflect changes in Ordovician seawater. It has been suggested that substantial changes in the nature of the surface waters could have been caused by oceanic overturn promoted by cold water intruding below a column of unstratified water. The overturn would bring unconditional bottom waters, rich in nutrients, but potentially toxic, to the surface (Wilde and Berry, 1984). An overturn would be likely to produce geologically instant effects.

Continental distribution

The distribution of late Ordovician plates is shown on Figure 6.4. It has been suggested that plate movement was unusually fast in the upper Ordovician (Piper, 1987), which might account for rapid changes in global climate.

The distribution of the continents produced a Northern Hemisphere ocean covering about half the Earth's surface, a band of land and shallow shelf seas partly girdling the earth around the equator, and a major landmass around the pole in the Southern Hemisphere (Wilde *et al.*, 1988). The absence of meridional currents in the Northern Hemisphere would

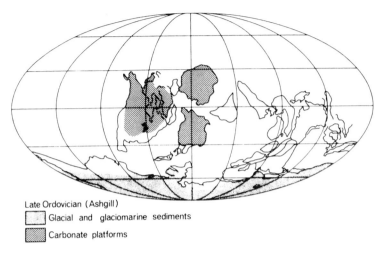

Late Ordovician (Ashgill)

▢ Glacial and glaciomarine sediments

▨ Carbonate platforms

Figure 6.4. Palaeogeography of the late Ordovician. Palaeogeographic base maps, supplied to the Symposium on Palaeozoic Biogeography and Palaeogeography, Oxford, 1988.

produce zonal surface currents with either east or west flow and a low nutrient availability (Wilde *et al.*, 1988). The meridional barriers in the Southern Hemisphere would promote surface flow across lines of latitude and potentially carry cold water towards the equator in some regions.

CAUSES OF EXTINCTION

The causes of mass extinction most commonly cited in the literature are major changes in sea level; climatic changes, particularly temperature change; extraterrestrial impacts; and palaeoceanographic changes. So far it has not been possible to identify any single one of these phenomena as a causal factor in all the mass extinctions identified in the geological record (Jablonski, 1986), though there do seem to have been sea-level changes approximately at the same time as all the extinction events (Newell, 1967). The possible causes of the late Ordovician extinction will now be analysed in turn.

Extraterrestrial impacts

The evidence for a major meteorite impact on the Earth at the end of the Cretaceous is now strong and there is a likelihood that it played a significant role in the extinction at the K–T boundary (Alvarez *et al.*, 1980; Orth, Chapter 3, this volume). Meteorite impacts of 1–10 km are a statistical probability during the Earth's history. Iridium anomalies, which are believed to reflect meteorite impact, are known from five levels other than the K–T boundary, most of which are close to a stratigraphic horizon with high faunal changeover (Raup, 1987). Furthermore, it has been suggested that there is a cyclic extinction pattern in the Mesozoic to Recent, with an approximately 26-million-year periodicity which might be influenced by extraterrestrial events (Raup and Sepkoski, 1986). There is a possibility, therefore, that all major extinctions might be triggered by a common extraterrestrial cause.

So far, the 26-million-year extinction cycle has not been detected in the Palaeozoic, but clearly further investigation is necessary. Two searches have been made for an iridium anomaly about the level of the late Ordovician extinction. One investigation was made through the sequence of graptolitic shales at Dob's Linn, Scotland (Wilde *et al.*, 1986), and the second through the carbonate sequence on Anticosti Island, Canada (Orth *et al.*, 1986). Both sequences are thought to be complete, but both failed to show a significant iridium anomaly. Further sections, however, need investigation before a meteorite impact can be wholly discounted.

Sea-level changes

A major eustatic fall in sea level is thought to have a major effect on faunas by habitat destruction (Newell, 1967) or by reducing habitable area—the species–area effect (Schopf, 1974; Simberloff, 1974). There was a major glacio-eustatic regression followed by a transgression in the late Ordovician, correlating approximately with the two phases of late Ordovician extinction. However, the regression is unlikely to have had a significant influence on plankton, which was severely affected in the first phase of extinction (graptolites, and the conodonts and acritarchs in temperate regions), nor is it thought to have had a major influence on the benthos on the clastic shelves in temperate and cool areas. In several clastic sequences the faunal changeover at the Rawtheyan–Hirnantian boundary occurs at the start of the regression, before there had been any major change in bathymetry. On shelves which sloped into basins or towards the ocean the effect of a regression would be to displace the benthic assemblages outwards and those which might be most affected are the outer shelf communities which would be displaced on to slope (Brenchley, 1984). However, the first phase of extinction was complete in a relatively short period of time, well before the regressive maximum, when the effect of sea-level fall would be most severe.

In contrast to the open clastic shelves, the effect of a major regression on the extensive shallow carbonate platforms is likely to have been much greater (Sheehan, 1973; 1975). Here a fall of about 70 m would have drained most continental interiors, displacing the fauna to a narrow rim around the continental margin. The timing of the extinction relative to the regression in carbonate environments is not well documented, though it might be significant that in the carbonate sequences of Estonia there was a major disappearance of species in the Porkuni (approximately Hirnantian; Kaljo *et al.*, 1988), suggesting the extinction occurred relatively early in the regressive sequence.

The effect of the extinction in tropical carbonate environments was to weed out selectively the more stenotopic elements of the fauna (Sheehan, 1975). Among the brachiopods, of 125 genera which are confined to a single North American province, only ten survive into the Silurian, while of 60 genera known from four or more provinces, 29 survive into the Silurian (Sheehan and Coorough, 1986).

The second peak of extinction coincided very closely with the major mid-Hirnantian transgression. It is estimated that the sea level rose by 70 m or more, and possibly flooded the shelves and platforms to greater depths than were present before the regression. The effect on clastic shelves was widespread deposition of black graptolitic shales, reflecting widespread anoxia. The area of the clastic shelves which could have been inhabited by benthos might have been small immediately after the transgression and a species–area effect might have influenced the extinction of the *Hirnantia* fauna (but see temperature effects, below). On carbonate platforms the corals and stromatoporoids appear to have been severely affected at the time of the transgression. The relationship of the extinctions to sea-level change in the mid-Hirnantian is far from clear.

Temperature changes

Temperature has a major influence on the global distribution of faunas which are broadly correlated with latitudinal climatic belts. Organisms have specific ranges of temperature tolerance and temperature ranges over which they will reproduce. As climate changes, so does the distribution of organisms, with shifts towards the equator during phases of cooling and shifts towards the poles when the climate improves. Substantial extinction might arise when climatic belts contract or are eliminated, though important regional extinctions can occur during climatic change when organisms cannot migrate because of geographic constraints. Climatic change has been postulated as one of the most important causes of extinction (Stanley, 1984), though significant climatic shifts cannot be identified with all major extinctions.

In the Ordovician there was a close relationship between faunal distributions and latitude (Cocks, 1988b), suggesting a strong climatic control on the faunas. At the end of the Ordovician there is a good correlation in time between the first phase of extinction and the spread of cold climates—the first ice-deposited diamictites occur stratigraphically immediately above the Rawtheyan–Hirnantian boundary. The presence of a widespread carbonate horizon immediately below the boundary suggests that temperatures might have been somewhat raised at the end of the Rawtheyan, so that the cooling phase might have been particularly accentuated. At several places in high latitudes, moderate to rich Rawtheyan faunas are succeeded by sediments containing only trace fossils (Brenchley and Storch, 1989), or by a low-diversity *Hirnantia* fauna (Owen *et al.*, 1988). Elsewhere, in mid-latitudes, the Rawtheyan assemblages are succeeded by the cosmopolitan *Hirnantia* fauna which is thought to be descended from cool-water Mediterranean (Gondwanan) faunas (Jaanusson, 1979; Sheehan, 1979). Therefore, the evidence suggests that by early Hirnantian times cool waters were widespread from polar regions to the edge of the tropics. At about the same time as the Rawtheyan benthos suffered extinction, there was a severe extinction among graptolites and probably also among the cool and temperate conodonts and acritarchs. It has been suggested that the cause of the graptolite extinction was the contraction of climatic belts (Skevington, 1974; 1978). A comparison can be made with Cenozoic changes when sea-surface temperatures progressively declined in high and mid-latitudes, but increased in the tropics (Shackleton, 1979). The effect would be for higher-latitude species to suffer range restriction as they moved towards the equator and abutted against the high-temperature water of the tropics. A zone of high extinction would be expected at the margins of the tropics (Valentine, 1984). Whether this model is capable of explaining the extreme extinction among the graptolites is open to some doubt, because there is no reason for high extinction rates in the tropics, too. Only if cool water penetrated tropical regions, possibly only temporarily, during a glacial phase, would tropical extinction be prevalent.

The second peak of extinction also coincided with a climatic change, when the Gondwanan ice-sheets contracted and the sea level rose. Surface seawaters would have generally warmed at this time, though tropical temperatures may have decreased (Valentine, 1984). Extinction would be concentrated towards the poles as the cold-water regions were restricted. The cool-water *Hirnantia* fauna could have suffered extinction at this time. A second area of extinction would be in the outer tropical areas as the carrying capacity of the environment was lowered (Valentine, 1984). It is possible that the conodont extinction in tropical regions might be related to these kinds of climatic change.

Palaeoceanographic changes

The sudden extinction of plankton at the Rawtheyan–Hirnantian boundary, followed possibly by a second abrupt phase in the mid-Hirnantian, suggests the possibility that ocean waters might have deteriorated rapidly beyond the tolerance levels of many species.

Two models have been proposed which outline the kinds of oceanographic change which could have a profound and nearly instantaneous affect on the biota. Wilde and Berry (1984) have described how oceanic overturn might bring unconditional bottom waters to the surface, promoting mass mortality. They envisage that the overturn could occur after a long period of climatic stability, when ocean waters would be weakly stratified. With a shift towards polar climates, cold, dense water would flow outwards from the poles beneath the water column, promoting sufficient instability that some relatively small perturbation, such as internal waves, might cause the body of water to overturn. A second phase of overturn could occur when cold climates gave way rapidly to warmer ones (Wilde and Berry, 1984).

A second oceanographic model was proposed by Thierstein and Berger (1978) and Berger and Thierstein (1979). They envisage the possibility that an influx of dense (cold) water beneath the oceans could lift an oxygen-minimum zone promoting a global upwelling phenomenon. The results would be similar to the oceanic overturn of Wilde and Berry (1984). Alternatively, an influx of freshwater, or very warm water, would isolate the surface waters from waters at depth. The authors developed a model of injection of Arctic fresh or brackish water into the open ocean from an enclosed basin which might account for mass extinctions, including that at the end of the Cretaceous. However, biostratigraphic and isotopic evidence does not support this model for the Cretaceous extinction (Jablonski, 1986).

A palaeoceanographic model which shows how climatic changes could have produced sudden and deleterious changes to the ocean waters and have caused widespread and nearly instantaneous extinction is attractive. The $\delta^{13}C$ isotope excursion described from Hirnantian carbonates by Middleton *et al.* (1988) suggests that there might have been significant changes in seawater composition, but so far there is insufficient evidence to test models involving changes in ocean chemistry.

The conclusion that is reached from the foregoing analysis is that no single cause can be identified for the whole range of end Ordovician extinctions encompassing both plankton and benthos. However, the driving force behind the extinctions was the climatic change which caused the growth and decay on Gondwanan ice-sheets (Berry and Boucot, 1973; Sheehan, 1973, 1975; Brenchley and Newell, 1984), the related changes in sea level and the latitudinal changes in the chemistry of the oceans. All of these perturbations of the environment would have tended to cause extinction, but may have been important at different times and at different locations.

A criticism of the glacial influence on extinction can be raised because extinctions on a similar scale did not occur during the glacial fluctuations of the Pleistocene, which appear to have been on a comparable scale. There are, however, important differences between the conditions at the end of the Ordovician and those in the Pleistocene. First, the groups which were affected in the Ordovician (graptolites, trilobites, conodonts, and so on) are not present in the Pleistocene, so it is not possible to make direct comparisons of the environmental changes in the fauna. Second, the rate of climatic change might have been much greater in the late Ordovician than in the Cenozoic. The evidence for the late Ordovician suggests a rapid climatic change at the Rawtheyan–Hirnantian boundary after a long period of climatic stability, while the Cenozoic record suggests many climatic fluctuations prior to the full glacial period of the Pleistocene (Shackleton and Kennett, 1975). Third, and possibly most important of all, the palaeogeography of the late Ordovician globe included large epeiric seas covering continental interiors. These seas contained long-established, diverse, endemic faunas, many of whose species would have had narrow niches (Sheehan, 1973; 1975). This situation contrasts strongly with the relatively narrow shelves bordering the continents during the Cenozoic. Here, faunas would have been affected by climatic and sea-level changes throughout the Cenozoic and were likely to have been eurytopic and resistant to environmental change.

It was the particular geography of the late Ordovician and its previous history of high sea level and environmental stability which made the stenotopic faunas so vulnerable. The speed of the environmental change might have been the decisive factor which promoted mass extinction.

SYNTHESIS

The late Ordovician extinction involved two peaks of extinction separated by hundreds of thousands of years, probably with other extinctions in the intervening interval when sea level was low.

The first phase of extinction followed a period of several millions of years when the sea level stood high, environments were relatively stable and there was a balance between extinction and new appearances of species. Following a short late Rawtheyan period, when climates might have been slightly warmer than usual for the Ashgill, there was a rapid climatic deterioration when the Gondwanan ice-sheets expanded, temperate climatic belts contracted and sea level started to fall. Coincident in time with these environmental changes there was a short period of major extinction.

The plankton (graptolites, conodonts, acritarchs) was depleted and the varied, relatively endemic benthic communities of late Rawtheyan were replaced in extra-tropical regions by the less varied, cosmopolitan *Hirnantia* fauna (Figure 6.5). The extinction of the plankton and the spread of the cool-water *Hirnantia* fauna is attributed to the spread of cool surface waters to the margins of the tropics and the consequent severe contraction of temperate climatic belts. These changes in ocean temperature may have been accompanied by changes in ocean chemistry, possibly promoted by extreme upwelling. The extinction among the benthos appears to have affected the deeper-water shelf communities in particular. The trilobites and cystoids suffered a greater extinction than the brachiopods. Among

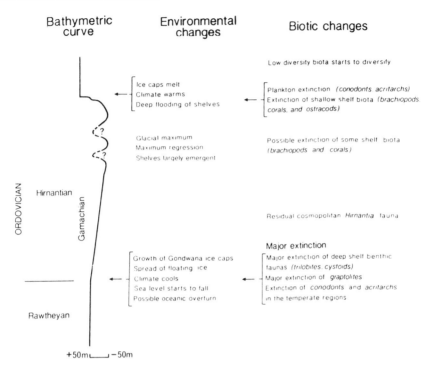

Figure 6.5. Summary of extinctions and environmental changes in the late Ordovician.

the trilobites, it was the deeper-water assemblages which became depleted; among the brachiopods, the deep-shelf *Foliomena* community disappeared. The cause of this selective destruction is uncertain. It could be that deeper shelf species were more stenothermal, or the outer shelf and slope might have been more affected by upwelling of cold bottom waters than areas of the middle and inner shelf.

The fall in sea level which appears to have had relatively little effect on the open clastic shelves of temperate and cool regions probably had a profound effect on the extensive carbonate platforms of the tropics. In these environments, even the small sea-level fall at the start of the Hirnantian would have started to have an effect. The full Hirnantian fall of about 70 m would have had a profound effect, as most of the continental interior became exposed. The diverse and endemic late Ordovician faunas would have suffered a massive reduction as the number of available niches and niche size decreased. It is possible, too, that as the faunas were displaced to the continental margins they may have experienced influxes of cool water during times of glacial advance. The interbedding of the *Hirnantia* fauna with more normal tropical faunas suggests this possibility.

As Sheehan (1973; 1975; 1979) has shown, the effect on the brachiopod faunas was profound and a large proportion of the fauna was lost.

In the period between the two main peaks of extinction, the ice-caps were large, climates were cool (though they probably fluctuated) and the sea level was low. Plankton in extratropical areas had been reduced to a low diversity and the *Hirnantia* fauna typified the benthos. In tropical areas the benthos had been displaced to the continental margins and greatly depleted, though certain elements, such as the corals and stromatoporoids, still flourished in restricted areas.

The second phase of extinction coincided with the rapid rise in sea level in the mid-Hirnantian. The *Hirnantia* fauna survived the first phase of sea-level rise and apparently spread polewards into areas which previously had been too cold for brachiopods. The second, larger, rise in sea level correlates with the disappearance of most of the *Hirnantia* fauna, rapid transgression and the development of widespread anoxia on clastic shelves. Two factors would have contributed to the extinction of the cool water benthos at this time. The spread of warm water would have caused a contraction of the cool-water belt and the spread of anoxia would have restricted the habitable area on the shelves.

In tropical regions the sea returned to the continental interiors and carbonate deposition resumed. However, the post-transgression sequences show a very major change had occurred in the conodonts, acritarchs and chitinozoa, and the coral-stromatoporoid faunas were radically depleted. It is difficult to be sure whether these biotic changes were caused by the preceding lowered sea level, which is thought to have affected the brachiopods, or by the sea-level rise. More precise evidence relating the disappearance of the biota to environmental change is necessary before this question can be answered.

In summary, it appears that the end Ordovician extinction was caused by the growth and decay of the Gondwanan ice-sheet, which promoted a range of environmental perturbations, particularly temperature and sea-level changes, which in turn led to niche destruction and extinction. The particular features of the late Ordovician which led to a mass extinction were the preceding environmental stability and high sea level, promoting the evolution of stenotopic species, and the rapidity of the environmental change which damaged the preceding ecosystem.

The end Ordovician extinction left many empty niches. Some, among the trilobite communities, were probably never replaced. On the carbonate platforms of the tropics the brachiopod niches were filled by descendants of the more eurytopic Old World faunas (Sheehan, 1975). The end Ordovician extinction provided the opportunity for the early Silurian radiation and development of new Silurian communities (Sheehan, 1982).

The broad pattern of the late Ordovician extinction and its likely causes is becoming quite well constrained. However, we need to know more about the exact timing of species extinction, especially in the tropical carbonate environments. The analysis of selective extinctions among different communities and the niches which disappeared is still in an early stage. Many of the associated environmental changes are now well known and well constrained stratigraphically, but knowledge of chemical changes is still sparse and could be important.

The late Ordovician extinction provides one of the best examples of ecological collapse followed by re-establishment of the ecological structure. It should prove instructive to all those interested in the flux of ecosystems.

REFERENCES

Achab, A. and Duffield, S.L., 1982, Palynological changes at the Ordovician–Silurian boundary on Anticosti Island, Quebec. In D.L. Bruton and S.H. Williams (eds), *Abstracts, 4th International Symposium on the Ordovician System, Paleontological Contributions from the University of Oslo*, **280**: 3.

Aldrige, R.J., 1988, Extinction and survival in the Conodonta. In G.P. Larwood (ed.), *Extinction and survival in the fossil record, Systematics Association Special Volume* **34**, Clarendon Press, Oxford: 231–56.

Alvarez, L.W., Alvarez, W., Asaro, F. and Michel, H.V., 1980, Extraterrestrial causes for the Cretaceous–Tertiary extinction, *Science*, **208** (4448): 1095–1108.

Amsden, T.W., 1974, Late Ordovician and early Silurian articulate brachiopods from Oklahoma, southwestern Illinois and eastern Missouri, *Bulletin of the Oklahoma Geological Survey*, **119**: 1–154.

Amsden, T.W. and Barrick, J.E., 1986, Late Ordovician–early Silurian strata in the central United States and the Hirnantian Stage, *Bulletin of the Oklahoma Geological Survey*, **139**: 1–95.

Anderson, J.B. and Thomas, M.A., 1988, Collapse of marine ice sheets as a mechanism for rapid episode transgressions, *Abstracts, SEPM Conference on Shelf Sedimentation Events and Rhythms, University of California at Santa Cruz*: 1.

Anstey, R.L., 1986, Bryozoan provinces and patterns of generic evolution and extinction in the late Ordovician of North America, *Lethaia*, **19** (1): 33–51.

Barnes, C.R. and Bergstrom, S.M., 1988, Conodont biostratigraphy of the uppermost Ordovician and lowermost Silurian. In L.R.M. Cocks and R.B. Rickards (eds), *A global analysis of the Ordovician–Silurian boundary, Bulletin of the British Museum (Natural History)*, *Geology*, **43**: 325–43.

Benedetto, J.L., 1986, The first typical *Hirnantia* fauna from South America (San Juan Province, Argentine Precordillera). In P.R. Racheboeuf and C.C. Emig (eds), *Les brachiopodes fossiles et actuels, Actes du 1er Congrès International sur les Brachiopodes, Brest, 1985, Biostratigraphie du Paléozoïque*, **4**: 439–47.

Berger, W.H. and Thierstein, H.R., 1979, On Phanerozoic mass extinctions, *Naturwissenschaften*, **66** (1): 46–7.

Berry, W.B.N. and Boucot, A.J., 1973, Glacio-eustatic control of late Ordovician–early Silurian platform sedimentation and faunal changes, *Bulletin of the Geological Society of America*, **84** (1): 275–84.

Brenchley, P.J., 1984, Late Ordovician extinctions and their relationship to the Gondwana glaciation. In P.J. Brenchley (ed.), *Fossils and climate*, Wiley and Sons, Chichester: 291–315.

Brenchley, P.J., 1988, Environmental changes close to the Ordovician–Silurian boundary. In L.R.M. Cocks and R.B. Rickards (eds), *A global analysis of the*

Ordovician–Silurian boundary, Bulletin of the British Museum (Natural History), Geology, **43**: 377–85.

Brenchley, P.J. and Cocks, L.R.M., 1982, Ecological associations in a regressive sequence: the latest Ordovician of the Oslo-Asker district, Norway, *Palaeontology*, **25** (4): 783–815.

Brenchley, P.J. and Cullen, B., 1984, The environmental distribution of associations belonging to the *Hirnantia* fauna—evidence from Wales and Norway. In D.L. Bruton (ed.), *Aspects of the Ordovician System, Paleontological Contributions from the University of Oslo*, **295**: 113–25.

Brenchley, P.J. and Newall, G., 1980, A facies analysis of upper Ordovician regressive sequences in the Oslo region, Norway: a record of glacio-eustatic changes, *Palaeogeography, Palaeoclimatology, Palaeoecology*, **31** (1): 1–38.

Brenchley, P.J. and Newall, G., 1984, Late Ordovician environmental changes and their effect on faunas. In D.L. Bruton (ed.), *Aspects of the Ordovician System, Palaeontological Contributions from the University of Oslo*, **295**: 65–79.

Brenchley, P.J., Romano, M., Young, T.P. and Storch, P., 1988, Hirnantian diamictites—evidence for the spread of glaciation and its affect on late Ordovician faunas, *Abstracts, 5th International Symposium on the Ordovician System, St John's, Newfoundland*: 13.

Brenchley, P.J. and Storch, P., 1989, Environmental changes in the Hirnantian (upper Ordovician) of the Prague Basin, Czechoslovakia, *Geological Journal*, **24**.

Briggs, D.E.G., Fortey, R.A. and Clarkson, E.N.K., 1988, Extinction and the fossil record of the arthropods. In G.P. Larwood (ed.), *Extinction and survival in the fossil record, Systematics Association Special Volume* **34**, Clarendon Press, Oxford: 171–209.

Brood, K., 1981, Hirnantian (upper Ordovician) Bryozoa from Baltoskandia. In G.P. Larwood, and C. Nielson (eds), *Recent and fossil Bryozoa*, Olsen and Olsen, Friedensburg, PA: 19–27.

Chatterton, B.D.E., Lespérance, P.J. and Ludvigsen, R., 1983, Trilobites from the Ordovician–Silurian boundary of Anticosti Island, eastern Canada, *Papers from the Symposium on the Cambrian–Ordovician and Ordovician–Silurian boundaries, Nanjing, China, October 1983*, Nanjing Institute of Geology and Palaeontology, Academica Sinica: 144–5.

Chatterton, B.D.E and Ludvigsen, R., 1983, Trilobites of the Ordovician–Silurian boundary of the Mackenzie Mountains, northwestern Canada, *Papers from the Symposium on the Cambrian–Ordovician and Ordovician–Silurian boundaries, Nanjing, China, October 1983*, Nanjing Institute of Geology and Palaeontology, Academica Sinica: 146–7.

Cocks, L.R.M., 1985, The Ordovician–Silurian boundary, *Episodes*, **8** (2): 98–100.

Cocks, L.R.M., 1988a, Brachiopods across the Ordovician/Silurian boundary. In L.R.M. Cocks and R.B. Rickards (eds), *A global analysis of the Ordovician–Silurian boundary, Bulletin of the British Museum (Natural History)*, Geology, **43**: 311–16.

Cocks, L.R.M., 1988b, Biogeography of Ordovician and Silurian faunas, *Abstracts, Symposium on Palaeozoic Biogeography and Palaeogeography, Oxford*: 18.

Cocks, L.R.M. and Fortey, R.A., 1982, Faunal evidence for oceanic separation in the Palaeozoic of Britain, *Journal of the Geological Society of London*, **139** (4): 465–78.

Colbath, G.K., 1986, Abrupt terminal Ordovician extinction in phytoplankton associations, southern Appalachians, *Geology*, **14** (11): 943–6.

Copeland, M.J., 1973, Ostracoda from the Ellis Bay Formation (Ordovician), Anticosti Island, Quebec, *Geological Survey of Canada Paper*, **72-43**: 1–49.

Destombes, J., 1968, Sur la présence d'une discordance générale de ravinement d'âge Ashgill supérieur dans Ordovician terminal de l'Anti Atlas (Maroc), *Comptes Rendu des Séances de l'Academie des Sciences de Paris*, **267** (5 August): 565–7.

Destombes, J., 1981, Hirnantian (upper Ordovician) tillites on the north flank of the Tindouf Basin, Anti-Atlas, Morocco. In J. Hambrey and W.B. Harland (eds), *Earth's pre-Pleistocene glacial record*, Cambridge University Press, Cambridge: 84–8.

Donovan, S.K., 1988, The British Ordovician crinoid fauna, *Lethaia*, **21** (4): 424.

Duffield, S.L. and Legault, J.A., 1981, Acritarch biostratigraphy of upper Ordovician-lower Silurian rocks, Anticosti Island, Quebec: preliminary results. In P.J. Lespérance (ed.),. *Field meeting, Anticosti-Gaspé, Quebec, 1981, 2 (stratigraphy and palaeontology), IUGS Subcommission on Silurian stratigraphy, Ordovician–Silurian boundary working group*, IUGS, Montreal: 91–9.

Eckert, J.D., 1988, Late Ordovician extinction of North American and British crinoids, *Lethaia*, **21** (2): 147–67.

Grahn, Y., 1988, Chitinozoan stratigraphy in the Ashgill and Llandovery. In L.R.M. Cocks and R.B. Rickards (eds), *A global analysis of the Ordovician–Silurian boundary, Bulletin of the British Museum (Natural History)*, Geology, **43**: 317–24.

Hallam, A. and Miller, A.I., 1988, Extinction and survival in the Bivalvia. In G.P. Larwood (ed.), *Extinction and survival in the fossil record, Systematics Association Special Volume* **34**, Clarendon Press, Oxford: 121–38.

Henningsmoen, G., 1954, Upper Ordovician ostracods from the Oslo region, Norway, *Norsk Geologisk Tidsskrift*, **33** (1): 69–108.

House, M.R., 1988, Extinction and survival in the Cephalopoda. In G.P. Larwood (ed.), *Extinction and survival in the fossil record, Systematics Association Special Volume* **34**, Clarendon Press, Oxford: 139–45.

Jaanusson, V., 1979, Ordovician. In R.A. Robison and C. Teichert (eds), *Treatise on Invertebrate Paleontology, Part A, Introduction*, Geological Society of America and University of Kansas Press, New York and Lawrence: A136–66.

Jablonski, D., 1986, Causes and consequences of mass extinctions: a comparative approach. In D.K. Elliott (ed.), *Dynamics of extinction*, Wiley and Sons, New York: 183–230.

Kaljo, D. and Klaaman, E., 1973, Ordovician and Silurian corals. In A. Hallam (ed.), *Atlas of palaeobiogeography*, Elsevier, Amsterdam: 37–45.

Kaljo, D., Nestor, M. and Polma, L., 1988, East Baltic region. In L.R.M. Cocks and R.B. Rickards (eds), *A global analysis of the Ordovician–Silurian boundary, Bulletin of the British Museum (Natural History)*, Geology, **43**: 85–91.

Kobluk, D.R., 1984, Coastal palaeokarst near the Ordovician–Silurian boundary, Manitoulin Island, Ontario, *Canadian Bulletin of Petroleum Geology*, **32**: 398–407.

Koren, T.N., 1988, Evolutionary crisis of Ashgill graptolites, *Abstracts, 5th International Symposium on the Ordovician System, St John's, Newfoundland*: 48.

Koren, T. and Rickards, R.B., 1979, Extinction of the graptolites. In A.L. Harris,

C.H. Holland and B.E. Leake (eds), *The Caledonides of the British Isles*, *reviewed*, Scottish Academic Press, Edinburgh: 457–66.

Kriz, J., 1984, Autecology and ecogeny of Silurian Bivalvia, *Special Papers in Palaeontology*, **32**: 183–95.

Legrand, P., 1986, The Lower Silurian graptolites of Oned in Djerane: a study of populations at the Ordovician–Silurian boundary. In C.P. Hughes and R.B. Rickards (eds), *Palaeoecology and biostratigraphy of graptolites*, *Geological Society Special Publication* **20**: 145–53.

McKerrow, W.S. and Cocks, L.R.M., 1976, Progressive faunal migration across the Iapetus Ocean, *Nature*, **263** (5575): 304–6.

Mannil, R., 1962, A faunistic characterisation of the Porkuni Stage, *Akadeemia Geoloogia-Instituudi Uurimused*, **10**: 115–29 (in Russian with English summary).

Martin, F., 1988, Late Ordovician and early Silurian acritarchs. In L.R.M. Cocks and R.B. Rickards (eds), *A global analysis of the Ordovician–Silurian boundary*, *Bulletin of the British Museum (Natural History), Geology*, **43**: 299–310.

Melchin, M.J. and Mitchell, C.E., 1988, Late Ordovician extinction among the Graptoloidea, *Abstracts, 5th International Symposium on the Ordovician System, St John's, Newfoundland*: 58.

Middleton, P., Marshall, J.D. and Brenchley, P.J., 1988, Isotopic evidence for oceanographic changes associated with the late Ordovician glaciation, *Abstracts, 5th International Symposium on the Ordovician System, St John's, Newfoundland:* 59.

Mikulic, D.G., 1980, Trilobites of Palaeozoic carbonate buildups, *Lethaia*, **14** (1): 45–56.

Moore, R.C. (ed.), 1959, *Treatise on Invertebrate Paleontology, Part O, Arthropoda* 1, Geological Society of America and University of Kansas Press, New York and Lawrence.

Moore, R.C., 1961, Summary of classification and stratigraphic distribution. In R.C. Moore (ed.), *Treatise on Invertebrate Paleontology, Part Q, Arthropoda 3*, Geological Society of America and University of Kansas Press, New York and Lawrence; Q92–9.

Newell, N.D. 1967, Revolutions in the history of life, *Special Paper of the Geological Society of America*, **89**: 63–91.

Orchard, M.J., 1980, Upper Ordovician conodonts from England and Wales, *Palaeontographica*, **175**: 1–88.

Orth, C.J., Gilmore, J.S., Quintana, L.R. and Sheehan, P.M., 1986, Terminal Ordovician extinction: geochemical analysis of the Ordovician–Silurian boundary, Anticosti Island, Quebec, *Geology*, **14** (5): 433–6.

Owen, A.W., 1987, The uppermost Ordovician (Hirnantian) trilobites of Girvan, S.W. Scotland, with a review of coeval trilobite faunas, *Transactions of the Royal Society of Edinburgh, Earth Sciences*, **77** (3): 231–9.

Owen, A.W. Harper, D.A.T. and Rong, J.-Y., 1988, Hirnantian brachiopods in space and time, *Abstracts, 5th International Symposium on the Ordovician System, St John's, Newfoundland*: 69.

Paul, C.R.C., 1973, British Ordovician cystoids, part 1, *Palaeontographical Society Monograph* London, **127** (536): 1–64.

Paul, C.R.C., 1988, Extinction and survival in the echinoderms. In G.P. Larwood (ed), *Extinction and survival in the fossil record*, *Systematics Association Special Volume* **34**, Clarendon Press, Oxford: 156–70.

Piper, J.D.A., 1987, *Palaeomagnetism and the continental crust*, Open University Press, Milton Keynes.

Raup, D.M., 1987, Mass extinction: a commentary, *Palaeontology*, **30** (1): 1–13.

Raup, D.M. and Sepkoski, J.J., Jr, 1982, Mass extinctions in the marine fossil record, *Science*, **215** (4539): 1501–3.

Raup, D.M. and Sepkoski, J.J., Jr, 1986, Periodic extinction of families and genera, *Science*, **231** (4740): 833–6.

Rong, J.-Y., 1984, Distribution of the *Hirnantia* fauna and its meaning. In D.L. Bruton (ed.), *Aspects of the Ordovician System, Palaeontological Contributions of the University of Oslo*, **295**: 101–12.

Schopf, T.J.M., 1974, Permo-Triassic extinctions: relations to sea-floor spreading, *Journal of Geology*, **82** (2): 129–43.

Scrutton, C.T., 1988, Patterns of extinctions and survival in Palaeozoic corals. In G.P. Larwood (ed.), *Extinction and survival in the fossil record*, Systematics Association Special Volume **34**, Clarendon Press, Oxford: 65–88.

Sepkoski, J.J., Jr and Sheehan, P.M., 1983, Diversification, faunal change and community replacement during the Ordovician radiations. In M.J.S. Tevesz and P.L. McCall (eds), *Biotic interactions in Recent and fossil benthic communities*, Plenum, New York: 673–717.

Shackleton, N.J., 1979, Evolution of the Earth's climate during the Tertiary Era. In D. Gautier (ed.), *Evolution of planetary atmospheres and climatology of the Earth*, Centre Nationale d'Etudes Spatiales, Toulouse: 49–58.

Shackleton, N.J. and Kennett, J.P., 1975, Palaeotemperature history of the Cenozoic and the initiation of Antarctic glaciation: oxygen and carbon isotope analyses in DSDP Sites 277, 279 and 281, *Initial Reports of the Deep Sea Drilling Project*, **29**: 745–55.

Sheehan, P.M., 1973, The relation of late Ordovician glaciation to the Ordovician–Silurian changeover in North American brachiopod faunas, *Lethaia*, **6** (2): 147–54.

Sheehan, P.M., 1975, Brachiopod synecology in a time of crisis (late Ordovician–early Silurian), *Paleobiology*, **1** (2): 205–12.

Sheehan, P.M., 1979, Swedish late Ordovician marine benthic assemblages and their bearing on brachiopod zoogeography. In J. Gray and A.J. Boucot (eds), *Historical biogeography, plate tectonics and the changing environment*, Oregon State University Press, Corvallis: 61–73.

Sheehan, P.M., 1982, Brachiopod macroevolution at the Ordovician–Silurian boundary, *Proceedings of the 3rd North American Paleontological Convention*, **2**: 477–81.

Sheehan, P.M. and Coorough, P.J., 1986, Brachiopod zoogeography at the terminal Ordovician extinction, *Abstracts, 4th North American Paleontological Convention, Boulder, Colorado*: A42.

Simberloff, D.S., 1974, Permo-Triassic extinctions: effects of area on biotic equilibrium, *Journal of Geology*, **82** (2): 267–74.

Skevington, D., 1974, Controls influencing the composition and distribution of Ordovician graptolite faunal provinces. In R.B. Rickards, D.E. Jackson and C.P. Hughes (eds), *Graptolite studies in honour of O.M. Bulman, Special Papers in Palaeontology*, **13**: 59–73.

Skevington, D., 1978, Latitudinal surface water temperature gradients and Ordovician faunal provinces, *Alcheringa*, **2** (1): 21–6.

Sloan, R.E., 1988, a chronology of North American Ordovician trilobite genera, *Abstracts, 5th International Symposium on the Ordovician System, St John's, Newfoundland*: 94.

Soper, N.J. and Hutton, D.H.W., 1984, Late Caledonian sinistral displacement in Britain: implications for a three-plate collision model, *Tectonics*, **3** (7): 781–94.

Spjeldnaes, N., 1981, Lower Palaeozoic climatology. In C.H. Holland (ed.), *Lower Palaeozoic of the Middle East, eastern and southern Africa, and Antarctica*, Wiley and Sons, Chichester: 199–256.

Spjeldnaes, N., 1982, Ordovician bryozoan fauna, *Abstract, 4th International Symposium on the Ordovician System, Paleontological Contributions of the University of Oslo*, **280**: 49.

Stanley, S.M., 1984, Temperature and biotic crises in the marine realm, *Geology*, **12** (4): 205–8.

Sweet, W.C. and Bergstrom, S.M., 1984, Conodont provinces and biofacies in the late Ordovician, *Special Paper of the Geological Society of America*, **196**: 69–87.

Taylor, P.D. and Larwood, G.P., 1988, Mass extinctions and the pattern of bryozoan evolution. In G.P. Larwood (ed.), *Extinction and survival in the fossil record, Systematics Association Special Volume 34*, Clarendon Press, Oxford: 99–119.

Thierstein, H.R. and Berger, W.H., 1978, Injection events in ocean history, *Nature*, **276** (5687): 461–6.

Thomas, A.T., Owens, R.M. and Rushton, A.W.A., 1984, Trilobites in British stratigraphy, *Special Report of the Geological Society of London*, **16**: 1–78.

Valentine, J.W., 1984, Climate and evolution in the shallow seas. In P.J. Brenchley (ed), *Fossils and climate*, Wiley and Sons, London: 265–77.

Wilde, P. and Berry, W.B.N., 1984, Destabilisation of the oceanic density structure and its significance to marine 'extinction' events, *Palaeogeography, Palaeoclimatology, Palaeoecology*, **48** (2–4): 143–62.

Wilde, P., Berry, W.B.N. and Quinby-Hunt, M.S., 1988, Oceanography in the Ordovician, *Abstracts, 5th International Symposium on the Ordovician, St John's, Newfoundland*: 112.

Wilde, P., Berry, W.B.N., Quinby-Hunt, M.S., Orth, C.J., Quintana, L.R. and Gilmore, J.S., 1986, Iridium abundances across the Ordovician–Silurian stratotype, *Science*, **233** (4761): 339–41.

Williams, A., 1965, Stratigraphic distribution. In R.C. Moore (ed.), *Treatise on Invertebrate Paleontology, Part H, Brachiopoda*, Geological Society of America and University of Kansas Press, New York and Lawrence: H237–50.

Williams, A., 1976, Plate tectonics and biofacies evolution as factors in Ordovician correlation. In M.G. Bassett (ed.), *The Ordovician System*, University of Wales Press and National Museum of Wales, Cardiff: 29–66.

Chapter 7

THE FRASNIAN–FAMENNIAN EXTINCTION EVENT

George R. McGhee, Jr

INTRODUCTION

The ultimate measure of biological crisis is the loss of species diversity from one geological time interval to another. Catastrophic decreases in the diversity of life on the Earth are referred to as 'mass extinctions', to distinguish these events from the normal extinctions which occur constantly in time, but which do not markedly affect the diversity of life, as the species which are lost are replaced by new species originations.

The simplest measure of the magnitude and duration of mass extinction events is a plot of the diversity of life as a function of geological time. Plotting the diversity of families of marine animals versus time reveals five major diversity crises during the past 600 million years (Raup and Sepkoski, 1982). These crises occurred at the end of the Ordovician, Permian, Triassic, and Cretaceous Periods—and within the Late Devonian Epoch.

The Late Devonian event is particularly controversial at present. It is referred to as the 'Frasnian–Famennian' event because the major diversity crisis appears not to have taken place in the terminal stage of the Devonian (the Famennian), but rather in the preceding Frasnian. Even this point has been disputed, however. Widespread disagreement exists as to whether it was in fact an 'event' or a series of events, and an even wider diversity of opinion exists as to the cause of the Frasnian–Famennian crisis.

A major loss in global biological diversity clearly occurred during the Late Devonian. As many as 21% of all families of marine organisms (Sepkoski, 1982), and approximately 50% of all genera (Sepkoski, 1986), were eliminated from the biosphere in this event. Analyses of regional ecosystems indicate the loss of at least 70% of local species numbers (McGhee, 1982).

TEMPORAL DURATION OF THE EXTINCTION EVENT

The temporal duration of the Frasnian–Famennian (Late Devonian) mass extinction has been the subject of considerable debate. The duration debate has often also been linked with the discussion concerning the causation of the extinction event. Those favouring gradual mechanisms of extinction have argued for a protracted event, while those supporting catastrophic mechanisms have reasoned that the event must have been abrupt. The two questions are, however, at least semi-independent. For example, if a gradual process of global climatic deterioration were rapidly accelerated by an asteroidal impact, the result would be an event of significant temporal duration. To use this duration result to rule out a catastrophic component of causation would, of course, be mistaken, as the causation (as hypothesized) was multifactorial and not simple.

Estimates of the duration of the mass extinction have varied widely. A sample of some published duration estimates include 15 million years (Raup and Sepkoski, 1982), 10 million (House, 1967), 6–8 million (Farsan, 1986), 7 million (McGhee, 1982), 3–5 million (Copper, 1984), 1 million (two conodont zones; Ziegler, 1984), 0.5 million (one conodont zone; Sorauf and Pedder, 1984), to less than 0.5 million (McLaren, 1982; 1984). Among those workers who have maintained that the event was not of short duration, disagreement exists as to whether the event records a series of episodic extinctions ('stepdown extinctions', Copper, 1986; 'episodic gradualism', Kalvoda, 1986), a series of multiple extinctions with distinctive and recurrent phases (House, 1986; Becker, 1986), the simple accumulation of successive extinctions (Farsan, 1986), or the gradual disappearance of species (Dutro, 1984). The great variety of opinion seen in the published literature has prompted at least one reviewer to exclaim: 'The Late Devonian extinction is one where not even the major facts are agreed on yet' (Van Valen, 1984, p. 130).

Data resolution and duration estimates

Duration estimates are strongly influenced by the completeness of the stratigraphic record and the minimum interval of stratigraphic correlation. Luckily, abundant stratigraphic sequences exist around the globe for the Late Devonian interval. Rather than stratigraphic completeness, the main problem in determining the duration of the Frasnian–Famennian event has been in the resolution of the correlation data.

Series and stage level patterns of extinction

An early work which pointed out that a major extinction event had taken place during the Devonian is that of Newell (1967). Several metrics exist to measure the severity of 'extinctions' (as opposed to diversity loss). Newell chose to use the metric of proportional number of extinctions, which is the number of extinctions (E) scaled against the total number of taxa present (N), per geological interval. His analyses indicated that approximately 30% of all families of marine organisms became extinct in the Late Devonian (Figure 7.1(a)), a peak which rises considerably above other extinction levels in the Devonian and Carboniferous.

The data available to Newell (1967) could only be obtained at the series level of temporal resolution, and at the familial level of taxonomic resolution. Plotting these data as points at series termini (Figure 7.1(a)), the maximum extinction appears to occur at the very end of the Devonian; thus, the extinction appears to mark the close of the Devonian and the boundary to the Carboniferous.

In the intervening years data of finer temporal resolution, though still at the familial taxonomic level, have been gathered by Sepkoski (1982) and analysed by Raup and Sepkoski (1982). These data are largely resolvable to the stage level, though some data remain which are only known to the series level (Sepkoski, 1982). Raup and Sepkoski chose a slightly different measure of extinction severity in using the absolute rate of extinction, which is the number of extinctions (E) for a given time interval (Δt), usually a million years. In these analyses the Frasnian Stage emerges with the maximum extinction rate for the entire Devonian, though the points for the temporally flanking Givetian and Famennian are close to the Frasnian peak (Figure 7.1(b)). In their analyses, Raup and Sepkoski (1982) concluded that all three stages—the Givetian, Frasnian and Famennian— were above the computed 'background' extinction rate, with the Frasnian at the maximum.

One could seemingly conclude from Figure 7.1(b) (and from Raup and Sepkoski, 1982) that extinction rates are elevated for the entire span of time from the Givetian though the Famennian, or a period of some 25 million years. Thus, though the temporal duration of the extinction event in the Newell (1967) data remained unresolvable—the event could have been of very short duration somewhere within the Late Devonian, or of extended duration spread throughout the Late Devonian—the stage-level data of Sepkoski (1982) would appear to indicate that there was no short-duration extinction event, but rather an extended period of high extinction rates in the latter half of the Devonian Period, with the Frasnian at the peak.

Substage-level patterns of extinction

Finer-scale temporal data are scarce at present, particularly on a global geographic scale. Such data which do exist are generally confined to taxonomic groups which are important for international geological correlation, such as ammonoids and conodonts. Multispecies data, resolvable to substage levels, usually exist only for restricted geographic areas. On the other hand, finer-scale temporal data are generally obtainable at the species-taxonomic level.

Substage data indicate that the single broad extinction peak seen in

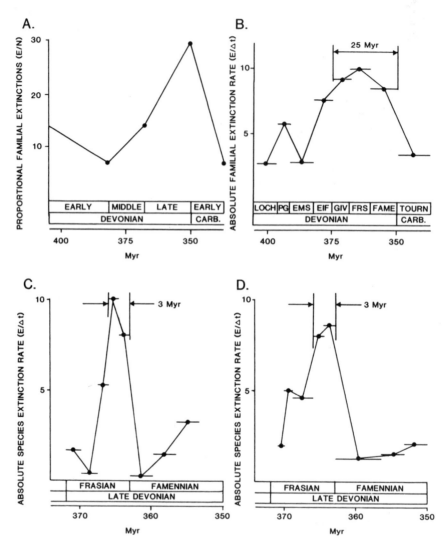

Figure 7.1. Effect of the level of data resolution on temporal duration estimates of the Frasnian–Famennian extinction event: (a) gives the temporal pattern of familial extinctions during the Devonian with resolution at the series level, while (b) gives the pattern of familial extinctions with resolution at the stage level. These are contrasted with the temporal pattern of extinction rates of brachiopod species, with resolution at the substage level, in New York (c) and the southern Urals, USSR (d). (a), (b) adapted from McGhee (1988a); (c), (d) from McGhee (1988b).

Figure 7.1(b) is an artefact of the level of resolution. In fact, three distinct peaks in absolute extinction rates occur *within* the Givetian, Frasnian, and Famennian Stages (Bayer and McGhee, 1986, Figure 10); however, these are 'blurred' into an overall peak spanning the entire Middle and Late Devonian in Figure 7.1(b). At the substage level of resolution, the late Frasnian stands out sharply as a period of elevated extinction rates (Figure 7.1(c),(d)).

Of importance is the fact that *multiple* periods of elevated extinction rates can be discerned within the Frasnian—there is no single isolated peak occurring at the very end of the Frasnian (Figure 7.1(c),(d)). Data at the substage level of resolution still indicate that a geologically significant span of time (approximately 3 million years) during the late Frasnian was characterized by elevated extinction rates. Thus the 'extinction event', as measured by absolute extinction rates, appears not to have been geologically instantaneous.

Biozone level patterns of extinction
Quantitative data currently do not exist at the finest level of international temporal resolution, that of the biozone. However, the data which do exist indicate that Late Devonian extinctions were not a single event (Table 7.1), but were spread over the span of six biozones in the latest Frasnian and earliest Famennian. The Late Devonian conodont biozones are estimated to have had an average duration of 0.5 million years (Sandberg *et al.*, 1988), thus the temporal spread of the extinctions outlined in Table 7.1. would represent the passage of approximately 3 million years. This figure agrees with the duration estimates previously discussed.

Careful analyses of the conodont fauna (Sandberg *et al.*, 1988) reveal that the most severe diversity crisis experienced by these elements of the global biota occurred during the *linguiformis* Zone (Table 7.1). Thus, a particularly severe, and perhaps short-term, diversity loss occurred at the very end of the Frasnian. This pattern is also seen in other elements of the ecosystem, as will be demonstrated below.

Evolutionary dynamics of the extinction event

There exists a persistent observation among field geologists, going back over one hundred years (see Chadwick, 1935, for a summary of New York State field studies), that a marked drop in species diversity occurred at the end of the Frasnian. The fact that extinction rates were elevated long before the end of the Frasnian (Figure 7.1(c),(d)) suggests that the drop in species diversity at the end of this stage was not a simple function of extinction-rate magnitudes (see McGhee, 1982, p. 494). In fact, Frasnian marine ecosystems in New York State appear to have been flourishing, in terms of standing species diversity, during the same interval of time

Table 7.1. *Biozonal position of some major biological events during the Late Devonian.*

Stage	Zone	Biological event
	crepida	Final extinction of corals and stromatoporoids in Moravia (Hladil *et al.*, 1986)
	Upper *triangularis*	
	Middle *triangularis*	Decimation of reefs in the Urals (Kalvoda, 1986) Final and total extinction of all atrypoid brachiopods (Copper, 1986)
FAMENNIAN	Lower *triangularis*	Global decimation of calcareous foraminifera (Kalvoda, 1986) Global decimation of rugose corals (Sorauf and Pedder, 1986)
FRASNIAN	*linguiformis*	Conodont diversity crisis horizon (Sandberg *et al.*, 1988)
	Upper *gigas*	Beginning of coral and stromatoporoid extinctions in Moravia (Hladil *et al.*, 1986) Decimation of reefs in western Europe (Tsien, 1980) Beginning of atrypoid brachiopod extinctions (Copper, 1986)
	Lower *gigas*	

corresponding to the maximum extinction interval in Figure 7.1(c),(d) (McGhee and Sutton, 1983; Sutton and McGhee, 1985).

At this point it should be re-emphasized that the Frasnian–Famennian biological crisis is actually recognized and defined by the sharp drop in global species diversity which occurred during this interval of geological time. Much attention has been focused upon extinction as the mechanism driving the decline in species diversity during periods of biotic crisis, and the equally important mechanism of speciation cessation has in general been neglected. Clearly one way in which diversity can be abruptly lost is by greatly increasing the number of species which go extinct (E) in a time interval Δt. Alternatively, extinction can remain more or less constant and abrupt diversity loss can still occur if the number of new species originations (O) is greatly decreased in the same time interval Δt. Maximum diversity loss can be obtained by combining these two effects.

An alternative to plotting diversity (N) as a function of time (in order to locate the intervals of diversity decrease) is to plot the rate of change of diversity per time interval ($\Delta N/\Delta t$). This metric, known as the 'turnover rate', pinpoints intervals of diversity loss as periods of negative turnover magnitudes. It is further useful in that it allows us to examine the underlying forcing mechanism of diversity decline in terms of the two effects discussed above:

$$\Delta N/\Delta t = O/\Delta t - E/\Delta t$$

Sharp drops in brachiopod species diversity occurred in both New York and the southern Urals (USSR), in the very latest Frasnian (Figure 7.2(a),(b)). This abrupt net loss of species diversity was not, however, a direct function of elevated extinction rates (cf. Figures 7.1(c) and 7.2(a), and Figures 7.1(d) and 7.2(b)). Extinction rates were high for approximately 3 million years during the late Frasnian; the sharp negative pulses in species turnover rates were of much shorter duration, with an estimated maximum of 1.5 million years. The actual interval of diversity loss may have been less than this figure.

Frasnian marine ecosystems flourished during the same time intervals characterized by elevated extinction rates, because origination rates of new

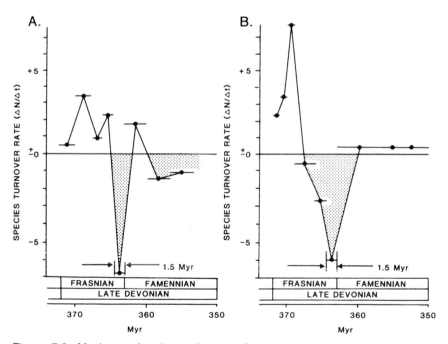

Figure 7.2. Maximum duration estimates of the period of brachiopod species diversity loss in (a) New York and (b) the southern Urals. In both regions an abrupt loss in diversity occurs at the very end of the Frasnian.

species were higher than corresponding extinction rates. It was the abrupt reversal in this pattern of relative origination/extinction rates which precipitated the rapid loss of species diversity seen at the very end of the Frasnian. In viewing the Frasnian–Famennian diversity crisis from an ecological perspective, the most important question is not what triggered the elevated extinction rates, but rather what was the inhibiting factor that caused the cessation of new species originations.

ECOLOGICAL SIGNATURE OF THE EXTINCTION EVENT

The massive deterioration in ecosystems which occurred throughout the world during this event can correctly be described as catastrophic in effect (McLaren, 1982). Frasnian ecosystems were ecologically very diverse and equitable in structure. Early Famennian ecosystems, in contrast, were impoverished in ecological diversity and in overall species richness. The effect of the biotic crisis can easily be seen in the 'bottleneck' constriction of ecological complexity which occurred in the Appalachian region of eastern North America (McGhee, 1982), where the diverse Frasnian ecosystem was replaced by an ecologically depauperate Famennian ecosystem proportionately overdominated by reduced species numbers of brachiopods, bivalves and glass sponges.

The analysis of the fossil remains of organisms which perished around the globe, as well as those which survived, during the Frasnian–Famennian interval reveals several differential, and non-random, ecological effects of the extinction event.

Latitudinal effect

Tropical reefal and peri-reefal marine ecosystems were particularly hard hit during the Frasnian–Famennian biotic crisis. The low-latitude, geographically widespread and massive stromatoporoid-tabulate reefal ecosystems vanished, and peri-reefal rugose coral-tabular stromatoporoid bioherms were decimated. Though the stromatoporoids suffered a severe reduction in biomass during the extinction event, they did not become totally extinct, nor did they totally lose their reef-building potential (Stearn, 1987). Post-Frasnian stromatoporoid structures were of small dimensions, and are generally found in the warm-water equatorial region of the Palaeotethys. Famennian stromatoporoids found outside of this geographic area are generally labechiids, which are believed to have been better adapted to cool water than the majority of Frasnian species, which were tropical and low-latitude in distribution.

Differential survival of high-latitude species adapted to cool water is also exhibited by the Brachiopoda, which are the dominant form of shellfish in Frasnian benthic ecosystems (Copper, 1977). Of the total brachiopod fauna, approximately 86% of Frasnian genera did not survive into the Famennian. However, 91% of brachiopod families whose species were generally confined to low-latitude, tropical regions perished in the extinction event, in contrast to a loss of 27% of the brachiopod families with species members which ranged into high-latitude, cool-water regions.

Other elements of the marine benthos which exhibit latitudinal patterns of survival include the foraminifera (Kalvoda, 1986). The foraminifera suffered major losses in species diversity with the substantial reduction in the global belt of carbonate sedimentation which occurred at the Frasnian–Famennian boundary. Species of the high-latitude regions differentially survived the event, and species of the cool-water Siberian Realm expanded their geographic ranges into low-latitude regions concurrent with the latitudinal contraction in range of the Palaeotethys species.

Bathymetric effect

In general, shallow-water marine ecosystems were much more severely affected during the Frasnian–Famennian interval than deeper-water systems. The bathymetric selectivity in extinction is seen most dramatically within the rugose cnidarians, a group which suffered a massive loss in biomass at the Frasnian–Famennian boundary. Only 4% of the shallow-water species survived the biotic crisis. Deeper-water species suffered a 60% extinction in their numbers and, while this reduction was severe, it pales in comparison with the 96% loss of species in the shallow waters. The decimation of the shallow-water cnidarians was actually more severe than that of the stromatoporoids (Stearn, 1987).

A particularly intriguing bathymetric pattern of selective extinction and diversification occurs across the Frasnian–Famennian boundary in the Appalachian marine ecosystems of eastern North America (McGhee, 1982). Simultaneous with the extinction of many shallow-water benthic species, the hyalosponges (glass sponges) migrated from deeper-water regions into the shallows and underwent a burst of diversification in species numbers. Modern glass sponges are generally found in water depths in excess of 200 m, and are considered to be better adapted to colder waters than most other invertebrate species. Blooms in other siliceous organisms, most notably the radiolarians, are also reported during the Frasnian–Famennian interval.

Marine versus terrestrial habitat effects

A marked habitat effect in selective survival can be observed in the Devonian fish groups which had both marine and freshwater species members. During the Frasnian–Famennian crisis only 35% of marine placoderm species survived the event, in contrast to a 77% survival of the placoderm species which lived in terrestrial freshwater regions. A similar pattern can be seen in the differential survival of acanthodian fishes: only 12% of the marine species survived, in contrast to 70% survival of freshwater species (McGhee, 1982).

A key environmental parameter which differentiated the two habitat regions (other than salinity) is temperature. In general, terrestrial freshwater species are adapted to seasonal and diurnal fluctuations in temperature, in contrast to the temperature-buffered shallow-water marine regions. The differential survival of terrestrial freshwater fish species may possibly reflect their greater tolerance to temperature changes. Other elements of the terrestrial ecosystem appear to have been unaffected by the Frasnian–Famennian event. Terrestrial floras exhibit no major disruptions, and plant biomass productivity appears to have been either unchanged, or even perhaps enhanced, during the Frasnian–Famennian interval.

Within the shallow-water marine benthos, epifaunal filter-feeding organisms appear to have been most affected by the extinction event; infauna and detritus feeders were relatively unaffected.

In common with other extinction events in the Earth's history, the upper oceanic water habitat of the marine plankton was massively disrupted (Tappan, 1982). Approximately 90% of the preservable phytoplankton was affected during the Frasnian–Famennian crisis, and massive biomass reductions also occurred among the zooplankton.

Summary

The ecological signature preserved in the fossil record of the Frasnian–Famennian extinction event appears to indicate a significant drop in global temperatures during the crisis interval. The decimation of low-latitude tropical reef ecosystems and of warm-water shallow marine faunas, combined with the relatively higher survival of high-latitude faunas, deep-water faunas, and the terrestrial faunas and floras seems best compatible with lethal temperature decline as a primary signal at the global level. At the local and regional level, the extinction event doubtless records the additional complicating factors of specific local environmental conditions during this interval of time.

THE CAUSE OF THE EXTINCTION EVENT

Hypothesized terrestrial mechanisms

Thermal
A proximate cause for the Frasnian–Famennian event which is most compatible with the observed ecological patterns of selectivity is lethal temperature decline. Significant global cooling would decimate reefal and

tropical ecosystems, as species which inhabit these regions have no refuge against cold. Higher-latitude ecosystems would be less affected, as many species could simply migrate to lower latitudes and thus maintain their tolerable temperature ranges. Global cooling would most adversely affect the warm-water, temperature-buffered, epicontinental sea benthos. Deeper-water faunas are generally adapted to cooler water conditions and would therefore be expected to survive differentially. Shallow oceanic planktic ecosystems would also be decimated, with greater survival of the plankton in high-latitude regions. Lastly, terrestrial ecosystems are more resistant to temperature fluctuations than marine. Each of these expected patterns of ecological selectivity in survival is seen in the Frasnian–Famennian event; thus, lethal temperature decline is a prime candidate for the cause of the event.

A problem arises in determining the ultimate cause of global temperature decline, in that the same temperature effect may be produced by a variety of forcing mechanisms, both terrestrial and extraterrestrial. The most obvious correlate with global cooling is continental glaciation, yet even here it can be debated whether the onset of glaciation has an extraterrestrial trigger (cycles in orbital eccentricities of the Earth) or a terrestrial trigger (positioning of a continental landmass over a pole), or both.

Caputo and Crowell (1985) and Caputo (1985) have recently argued that the Frasnian–Famennian extinctions were triggered by the onset of glaciation in Gondwana. Their argument is based on two lines of reasoning: reinterpretations of polar wandering paths for Gondwana, which have the South Pole positioned on continental South America rather than over the ocean in the Late Devonian; and new biostratigraphic miospore dates for glacial sediments in South America, which move these previously Carboniferous dated sediments back in time to the early Famennian.

Several problems are posed by the postulated Frasnian–Famennian glaciations. First, significant glaciation is usually correlated with marine regression due to ice-volume build-up, yet the late Frasnian and early Famennian were periods of maximum global sea level (Johnson *et al.*, 1985). Only late in the Famennian did significant marine regression occur. Second, the early Famennian dates for South American glacial sediments have been questioned by Streel (1986), who argues that the reported early Famennian miospores have been reworked upwards into younger sediments of latest Famennian age. Late Famennian to early Carboniferous glaciation would be consistent with the observed pattern of marine regression during this interval of time (McGhee and Bayer, 1985; Johnson *et al.*, 1985). Lastly, new palaeomagnetic data suggest that the South Pole remained positioned over continental Gondwana for the entire interval of time from the middle Ordovician to the Late Devonian (Van der Voo, 1988). Such a positioning brings into question the glacial model of Caputo and Crowell (1985) and Caputo (1985), in which glaciation is initiated only when a continent moves over a pole.

An alternative model of global cooling has been proposed by Copper (1986). In this model, Laurussia and Gondwana connected in the Frasnian–Famennian interval, disrupting the low-latitude circumequatorial flow of warm water which had previously existed in the ocean between these two supercontinents. Cessation of easterly equatorial currents would have brought high-latitude cold water into equatorial regions along the western margins of this continental configuration, and restricted circulation with euxinic conditions in the surviving warm-water basins along the eastern margins. The resultant cooler Famennian oceans and global climates are postulated to have eventually initiated glaciation on Gondwana in the Famennian–Carboniferous transition.

This palaeogeographic model of climatic cooling has, however, been questioned both in terms of Late Devonian palaeomagnetic data (Hurley and Van der Voo, 1987; Van der Voo, 1988) and biogeographic data (Raymond *et al.*, 1987), neither of which support a Frasnian–Famennian collision of Laurussia and Gondwana. Rather, both lines of evidence suggest a much later collision of these two supercontinents in the Carboniferous, and indicate the continued existence of an open ocean between Laurussia and Gondwana in the Frasnian–Famennian interval.

Marine regression

It is clear from the geological record that sea-level fall is often associated with periods of biotic crisis (Hallam, 1981; Jablonski, 1986). However, the causal link (if there is one) between lowered sea levels and elevated extinction levels still remains problematical (Jablonski, 1986; Stanley, 1987).

The Frasnian–Famennian event is one of the few known in the geological record where a major biotic crisis occurs during an interval of global sea-level high stand. Thus, major sea-level fall cannot be invoked (regardless of the proposed killing mechanism of such a regression) to account for the Frasnian–Famennian diversity crisis. Johnson (1974) suggested that a geologically rapid regressive-transgressive sea-level pulse may have occurred during the latest Frasnian, which may have extinguished the 'perched' faunas produced by the overall high stands of the Late Devonian oceans. A minor regression does appear to have occurred near the Frasnian–Famennian boundary (McGhee and Bayer, 1985; Johnson *et al.*, 1985; Sandberg *et al.*, 1988), but many sea-level fluctuations of similar magnitude occurred within the span of the Frasnian, none of which precipitated the massive disappearance of species seen during the latest phase of the stage (McGhee, 1982; Sutton and McGhee, 1985).

Marine transgression

Periods of sea-level high stand are generally viewed as producing equitable global climates favourable to both marine and terrestrial organisms. However, there are several proposals that the reverse might be true, at least for marine benthic ecosystems. Schlager (1981) has proposed that the extended

Late Devonian transgression may have actually led to the 'drowning' of major reef ecosystems, and House (1985) suggested that poisonous euxinic conditions were created with the spread of shallow stagnant seas, conditions which would decimate bottom-dwelling marine organisms.

It is clear that euxinic conditions were associated with many broad shallow seas produced by transgressive phases in the past. It is not clear, however, how the entire global marine ecosystem can become euxinic (Stanley, 1987). Anoxic black shale facies are indeed widespread in the Late Devonian deposits of much of the world, and at least one model suggests that climatic cooling may have triggered a major oceanic overturn during this period of time (Wilde and Berry, 1984; 1988), an event of such magnitude as to bring deep anoxic waters into the surface mixed layers on a global scale. Alternatively, it has been suggested that an overturn event of similar magnitude could have been triggered by the impact of a large asteroid with the Earth during the latest Frasnian (Geldsetzer *et al.*, 1987; Sandberg *et al.*, 1988; but see also Wilde and Berry, 1988).

Hypothesized extraterrestrial mechanism

Asteroidal or cometary impact
Perhaps the most interesting and controversial proposal made in palaeobiology in the past decade is the proposal that the Late Cretaceous mass extinction was triggered by the impact of a large asteroid on the Earth (Alvarez *et al.*, 1980). There is an entire class of asteroid-sized bodies, the Apollo objects, which have orbits that cross that of the Earth and it is only a matter of time before a collision takes place. Thus, it is entirely reasonable to propose that such events have taken place in the past, and that the catastrophic results of such a collision could trigger global ecosystem collapse. In fact, such a scenario was proposed specifically for the Frasnian–Famennian event by McLaren (1970) almost two decades ago, and though the suggestion was not taken seriously at the time, it is now (McGhee, 1981).

A problem with the impact hypothesis is the geologically instantaneous, but short-term, effect of such a collision. The Frasnian–Famennian event was not instantaneous, in that multiple periods of high extinction rates existed in the late Frasnian, spanning an interval of 2–4 million years (McGhee, 1988a). This does not rule out asteroidal or cometary impact as the cause of some of the extinctions, but it does indicate that a single impact is not sufficient in and of itself to explain the multiple periods of elevated extinction rates. Alternatively, it could be proposed that the multiple periods of high extinction in the Frasnian were triggered by multiple impacts distributed over a geologically significant period of time, in a sort of 'asteroidal shower'.

The climatic effects of an asteroid or comet impact are difficult to predict. Large amounts of debris from the vapourized bolide and impact site would be ejected into the atmosphere. Meteorological models suggest such a global dust cloud would result in the total blockage of sunlight from the Earth's surface for weeks, and light intensities too low to support photosynthesis for several months. Surface temperatures would quickly drop to below freezing, where they would remain for a period of several months.

Global temperature decline is consistent with the ecological signature of the Frasnian–Famennian event, thus the impact hypothesis provides a possible forcing mechanism for the biological crisis. The chief difficulty with the hypothesis is its geologically short-term climatic effect, which is not consistent with the 2–4 million year period of high extinction rates seen in the late Frasnian. However, it should be noted that the climatic predictions of the impact hypothesis are strongly model-dependent. The climatic effects of an asteroid or comet impact could be more long-term than currently predicted, and may produce global changes on a time-scale of tens of thousands, or perhaps even millions, of years. Such long-term climatic effects would be more in concert with the pattern of species extinctions seen in the fossil record. The only other alternative for the impact scenario is to invoke a series of impacts distributed over 2–4 million years in the late Frasnian.

The search for physical evidence of impact
One obvious physical sign of impact is the existence of a crater produced by the collision of a bolide with the earth. Interestingly, the Late Devonian does appear to have been a time when several large objects impacted the Earth, in that there exist several craters which are generally dated as late Frasnian to early Famennian in age (McGhee, 1982, p. 498). None can be precisely dated to exactly the Frasnian–Famennian boundary at present, though the existence of several craters might be used to support the scenario of an extended shower of impacts, as outlined above.

The other, more precise, types of evidence currently used to support impact hypotheses are: the presence of anomalous levels of the element iridium, which is differentially enriched in some extraterrestrial bodies (Alvarez *et al.*, 1980); the presence of impact-produced structures, such as shocked minerals (Bohor *et al.*, 1984), or impact melt structures, such as tektites (Glass, 1982); or anomalous carbon concentrations, hypothesized to have been produced by global wildfires triggered by the energy of the bolide impact (Wolbach *et al.*, 1988).

Extensive geochemical analyses to date have failed to reveal an iridium anomaly at the Frasnian–Famennian boundary. Strata which span this horizon have been examined for iridium in Belgium (McGhee *et al.*, 1984), Germany (McGhee *et al.*, 1986a; 1986b), the eastern United States (McGhee *et al.*, 1984), and western Canada (Geldsetzer *et al.*, 1985; 1987), and all have produced negative geochemical evidence for a bolide impact.

It is entirely possible that the Earth could have been impacted by a bolide which would have not produced an iridium anomaly, such as a highly fractionated body or an ice-rich comet. The search for other impact structures at the Frasnian–Famennian boundary, such as shocked minerals or tektites, has also given negative results (McGhee, 1986a).

Several geochemical anomalies do exist at the Frasnian–Famennian boundary (Figure 7.3), but are not those normally considered to be impact-produced. An iridium anomaly has been found in early Famennian strata in Australia (Playford *et al.*, 1984), but this anomaly is thought to have been produced by terrestrial bacteriological concentration (Playford *et al.*, 1984; Donovan, 1987; Kyte, 1988; Orth, Chapter 3, this volume). It also occurs two biozones above the conodont crisis horizon (Figure 7.3), thus postdating the event by some 1.0–1.5 million years. In any event, a sharp positive shift occurs in both $\delta^{13}C$ and $\delta^{18}O$ at the conodont diversity crisis horizon in Europe. The $\delta^{13}C$ data suggest a sudden environmental shift to anoxic conditions in bottom waters during the latest Frasnian. While an equivalent positive shift in $\delta^{13}C$ is not seen in Australia (Playford *et al.*, 1984) or in western Canada (Geldsetzer *et al.*, 1987), sulphur anomalies at the approximate Frasnian–Famennian boundary do occur in western Canada and may signal the same onset of ocean anoxia.

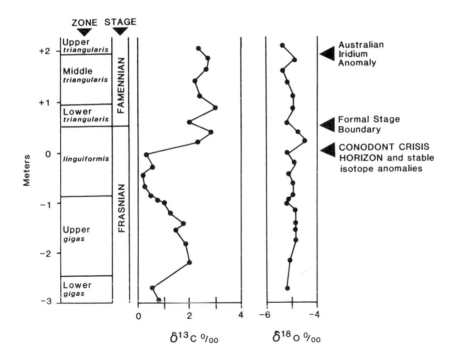

Figure 7.3. Stratigraphic position of the conodont crisis horizon and of stable isotope anomalies in the Steinbruch Schmidt section, Federal Republic of Germany. For contrast, the stratigraphic position of the iridium anomaly reported from Australia (Playford *et al.*, 1984) is also given. Modified from McGhee *et al.* (1986b).

The $\delta^{18}O$ data are more difficult to interpret, both in Europe and western Canada. A sharp positive shift of about 1 per mille does occur at the crisis horizon (Figure 7.3), which may signal a brief period of temperature decline. An extended period of global cooling, however, is not indicated from these geochemical data. Given the sensitivity of oxygen isotopes to diagenetic alteration, the oxygen data may in fact have been 'homogenized' to the extent that any meaningful signal has been subsequently lost in the stratigraphic sequence.

CONCLUSION

The ultimate cause of the Frasnian–Famennian diversity crisis remains enigmatic. The finest data at present available indicate that the event was of at least 3 million years' duration, through the latest Frasnian and early Famennian, with abrupt and severe losses in organic diversity occurring at the very end of the Frasnian. The ecological signature of the event indicates lethal temperature decline as the proximate cause. The stratigraphic record indicates widespread anoxia in many of the world's epeiric seas throughout much of the Late Devonian. It has been postulated that global cooling itself may have triggered a massive overturn event in the world's oceans during the Late Devonian, creating widespread anoxia in the surface waters of the oceans around the world (Wilde and Berry, 1984; 1988).

The ultimate cause of the postulated global cooling remains unknown. Tectono-glacial, palaeogeographic, and bolide impact hypotheses have been proposed, each of which predicts global temperature decline as a resultant. Serious discrepancies also exist for each model. The extraterrestrial impact hypothesis has generated the most debate, yet to date all attempts to provide independent physical evidence for such a forcing mechanism for the Frasnian–Famennian biological crisis have proven negative.

REFERENCES

Alvarez, L.W., Alvarez, W., Asaro, F. and Michel, H.V., 1980, Extraterrestrial cause for the Cretaceous–Tertiary extinction, *Science*, **208** (4448): 1095–1108.

Bayer, U. and McGhee, G.R., Jr, 1986, Cyclic patterns in the Paleozoic and Mesozoic: implications for time scale calibrations, *Paleoceanography*, **1** (4): 383–402.

Becker, R.T., 1986, Ammonoid evolution before, during and after the 'Kellwasser Event'—review and preliminary new results. In O.H. Walliser (ed.), *Global bioevents*, Springer-Verlag, Berlin: 181–8.

Bohor, B.F., Foord, E.E., Modreski, P.J. and Triplehorn, D.M., 1984, Mineralogic evidence for an impact event at the Cretaceous–Tertiary boundary, *Science*, 224 (4651): 867–9.

Caputo, M.V., 1985, Late Devonian glaciation in South America, *Palaeogeography, Palaeoclimatology, Palaeoecology*, 51 (1–4): 291–317.

Caputo, M.V. and Crowell, J.C., 1985, Migration of glacial centers across Gondwana during the Paleozoic Era, *Bulletin of the Geological Society of America*, 96 (8): 1020–36.

Chadwick, G.H., 1935, Faunal differentiation in the Upper Devonian, *Bulletin of the Geological Society of America*, 46 (2): 305–42.

Copper, P., 1977, Paleolatitudes in the Devonian of Brazil and the Frasnian–Famennian mass extinction, *Palaeogeography, Palaeoclimatology, Palaeoecology*, 21 (3): 165–207.

Copper, P., 1984, Cold water oceans and the Frasnian–Famennian extinction crisis, *Geological Society of America Abstracts with Programs*, 16 (1): 10.

Copper, P., 1986, Frasnian–Famennian mass extinction and cold-water oceans, *Geology*, 14 (10): 835–9.

Donovan, S.K., 1987, Iridium anomalous no longer?, *Nature*, 326 (6111): 331–2.

Dutro, J.T., 1984, The Frasnian–Famennian extinction event as recorded by Devonian articulate brachiopods in New Mexico, *Geological Society of America Abstracts with Programs*, 16 (1): 14.

Farsan, N.M., 1986, Frasnian mass extinction—a single catastrophic event or cumulative? In O.H. Walliser (ed.), *Global bio-events*, Springer-Verlag, Berlin: 189–97.

Geldsetzer, H.H.J., Goodfellow, W.D., McLaren, D.J. and Orchard, M.J., 1985, The Frasnian–Famennian boundary near Jasper, Alberta, Canada, *Geological Society of America Abstracts with Programs*, 17 (7): 589.

Geldsetzer, H.H.J., Goodfellow, W.D., McLaren, D.J. and Orchard, M.J., 1987, Sulfur-isotope anomaly associated with the Frasnian–Famennian extinction, Medicine Lake, Alberta, Canada, *Geology*, 15 (5): 393–6.

Glass, B.P., 1982, Possible correlations between tektite events and climatic change? In L.T. Silver and P.H. Schultz (eds), *Geological implications of impacts of large asteroids and comets on the Earth, Special Paper of the Geological Society of America*, 190: 251–6.

Hallam, A., 1981, *Facies interpretation and the stratigraphic record*, Freeman, Oxford.

Hladil, J., Kesslerova, Z. and Friakova, O., 1986, The Kellwasser Event in Moravia. In O.H. Walliser (ed.), *Global bio-events*, Springer-Verlag, Berlin: 213–17.

House, M.R., 1967, Fluctuations in the evolution of Palaeozoic invertebrates. In W.B. Harland *et al.* (eds), *The fossil record, a symposium with documentation*, Geological Society, London: 41–54.

House, M.R., 1985, Correlation of mid-Palaeozoic ammonoid evolutionary events with global sedimentary perturbations, *Nature*, 313 (5997): 17–22.

Hurley, N.F. and Van der Voo, R., 1987, Paleomagnetism of Upper Devonian reefal limestones, Canning Basin, Western Australia, *Bulletin of the Geological Society of America*, 98 (2): 138–46.

Jablonski, D., 1986, Causes and consequences of mass extinctions: a comparative approach. In D.K. Elliott (ed.), *Dynamics of extinction*, Wiley and Sons, New York: 183–229.

Johnson, J.G., 1974, Extinction of perched faunas, *Geology*, 2 (10): 479–82.

Johnson, J.G., Klapper, G. and Sandberg, C.A., 1985, Devonian eustatic fluctuations in Euramerica, *Bulletin of the Geological Society of America*, **96** (5): 567–87.

Kalvoda, J., 1986, Upper Frasnian and lower Tournaisian events and evolution of calcareous foraminifera—close links to climatic changes. In O.H. Walliser (ed.), *Global bio-events*, Springer-Verlag, Berlin: 225–36.

Kyte, F.T., 1988, The extraterrestrial component in marine sediments: description and interpretation, *Paleoceanography*, **3** (2): 235–47.

McGhee, G.R., Jr, 1981, The Frasnian–Famennian extinctions: a search for extraterrestrial causes, *Bulletin of the Field Museum of Natural History*, **52** (7): 3–5.

McGhee, G.R., Jr, 1982, The Frasnian–Famennian extinction event: a preliminary analysis of Appalachian marine ecosystems. In L.T. Silver and P.H. Schultz (eds), *Geological implications of impacts of large asteroids and comets on the Earth, Special Paper of the Geological Society of America*, **190**: 491–500.

McGhee, G.R., Jr, 1988a, The Late Devonian extinction event: evidence for abrupt ecosystem collapse, *Paleobiology*, **14** (3): 250–7.

McGhee, G.R., Jr, 1988b, Evolutionary dynamics of the Frasnian–Famennian extinction event. In N.J. McMillan, A.F. Embry and D.J. Glass (eds), *Devonian of the world, Memoir of the Canadian Society of Petroleum Geologists*, **14**.

McGhee, G.R. Jr, and Bayer, U., 1985. The local signature of sea-level changes. In U. Bayer and A. Seilacher (eds), *Sedimentary and evolutionary cycles*, Springer-Verlag, Berlin: 98–112.

McGhee, G.R., Jr, Gilmore, J.S., Orth, C.J. and Olsen, E., 1984, No geochemical evidence for an asteroid impact at late Devonian mass extinction horizon, *Nature*, **308** (5960): 629–31.

McGhee, G.R., Jr, Orth, C.J., Quintana, L.R., Gilmore, J.S. and Olsen, E.J., 1986a, The Late Devonian 'Kellwasser Event' mass-extinction horizon in Germany: no geochemical evidence for a large-body impact, *Geology*, **14** (9): 776–9.

McGhee, G.R., Jr, Orth, C.J., Quintana, L.R., Gilmore, J.S. and Olsen, E.J., 1986b, Geochemical analyses of the Late Devonian 'Kellwasser Event' stratigraphic horizon at Steinbruch Schmidt (F.R.G.). In O.H. Walliser (ed.), *Global bio-events*, Springer-Verlag, Berlin: 219–24.

McGhee, G.R., Jr and Sutton, R.G., 1983, Evolution of late Frasnian (Late Devonian) marine environments in New York and the central Appalachians, *Alcheringa*, **7** (1): 9–21.

McLaren, D.J., 1970, Time, life, and boundaries, *Journal of Paleontology*, **44** (5): 801–15.

McLaren, D.J., 1982, Frasnian–Famennian extinctions. In L.T. Silver and P.H. Schultz (eds), *Geological implications of impacts of large asteroids and comets on the Earth, Special Paper of the Geological Society of America*, **190**: 477–84.

McLaren, D.J., 1984, An Upper Devonian event: Frasnian–Famennian extinctions, *Geological Society of America Abstracts with Programs*, **16** (1): 49.

Newell, N.D., 1967, Revolutions in the history of life, *Special Paper of the Geological Society of America*, **89**: 63–91.

Playford, P.E., McLaren, D.J., Orth, C.J., Gilmore, J.S. and Goodfellow, W.D., 1984, Iridium anomaly in the Upper Devonian of the Canning Basin, Western Australia, *Science*, **226** (4673): 437–9.

Raup, D.M. and Sepkoski, J.J., Jr, 1982, Mass extinctions in the marine fossil record, *Science*, **215** (4539): 1501–3.

Raymond, A., Kelley, P.H. and Lutken, C.B., 1987, Comment on 'Frasnian/ Famennian mass extinction and cold-water oceans', *Geology*, **15** (8): 777.

Sandberg, C.A., Ziegler, W., Dreesen, R. and Butler, J.L., 1988, Part 3: Late Frasnian mass extinction: conodont event stratigraphy, global changes, and possible causes, *Courier Forschungsinstitut Senckenberg*, **102**: 263–307.

Schlager, W., 1981, The paradox of drowned reefs and carbonate platforms, *Bulletin of the Geological Society of America*, **92** (4): 197–211.

Sepkoski, J.J., Jr, 1982, A compendium of fossil marine families, *Milwaukee Public Museum Contributions to Biology and Geology*, **51**: 1–125.

Sepkoski, J.J., Jr, 1986, Phanerozoic overview of mass extinction. In D.M. Raup and D. Jablonski (eds), *Patterns and processes in the history of life*, Springer-Verlag, Berlin: 277–95.

Sorauf, J.E. and Pedder, A.E.H., 1984, Rugose corals and the Frasnian–Famennian boundary, *Geological Society of America Abstracts with Programs*, **16** (1): 64.

Sorauf, J.E. and Pedder, A.E.H., 1986, Late Devonian rugose corals and the Frasnian–Famennian crisis, *Canadian Journal of Earth Sciences*, **23** (9): 1265–87.

Stanley, S.M., 1987, *Extinction*, Scientific American Books, New York.

Stearn, C.W., 1987, Effect of the Frasnian–Famennian extinction event on the Stromatoporoids, *Geology*, **15** (7): 677–9.

Streel, M., 1986, Miospore contribution to the upper Famennian–Strunian event stratigraphy, *Annales de la Société Géologique de Belgique*, **109** (1): 75–92.

Sutton, R.G. and McGhee, G.R., Jr, 1985, The evolution of Frasnian marine 'community-types' of south-central New York, *Special Paper of the Geological Society of America*, **201**: 211–24.

Tappan, H., 1982, Extinction or survival: selectivity and causes of Phanerozoic crises. In L.T. Silver and P.H. Schultz (eds), *Geological implications of impacts of large asteroids and comets on the Earth*, Special Paper of the Geological Society of America, **190**: 265–76.

Tsien, H.H., 1980, Les régimes récifaux Devoniens du Ardenne, *Bulletin de la Société Géologique de Belgique*, **89** (1): 71–102.

Van der Voo, R., 1988, Paleozoic paleogeography of North America, Gondwana, and intervening displaced terrenes: comparisons of paleomagnetism with paleo-climatology and biogeographical patterns, *Bulletin of the Geological Society of America*, **100** (3): 311–24.

Van Valen, L.M., 1984, Catastrophes, expectations, and the evidence, *Paleobiology*, **10** (2): 121–37.

Wilde, P. and Berry, W.B.N., 1984, Destabilization of the oceanic density structure and its significance to marine 'extinction' events, *Palaeogeography, Palaeoclimatology, Palaeoecology*, **48** (2–4): 143–62.

Wilde, P. and Berry, W.B.N., 1988, Comment on 'Sulfur-isotope anomaly associated with the Frasnian–Famennian extinction, Medicine Lake, Alberta, Canada', *Geology*, **16** (1): 86.

Wolbach, W.S., Gilmour, I., Anders, E., Orth, C.J. and Brooks, R.R., 1988, Global fire at the Cretaceous–Tertiary boundary, *Nature*, **334** (6184): 665–9.

Ziegler, W., 1984, Conodonts and the Frasnian/Famennian crisis, *Geological Society of America Abstracts with Programs*, **16** (1): 73.

Chapter 8

THE END PERMIAN MASS EXTINCTION

W. Desmond Maxwell

INTRODUCTION

The greatest crisis to affect the Earth's biota occurred approximately 250 million years ago, at the end of the Permian period. Marine life was devastated, with a 57% reduction in the number of families (Sepkoski, 1986) and an estimated 96% extinction at the species level (Raup, 1979). Terrestrial forms were similarly affected, with a 77% reduction in the number of tetrapod families (Maxwell and Benton, 1987). All major groups of marine organisms were affected with the crinozoans (98%), anthozoans (96%), brachiopods (80%) and bryozoans (79%) suffering the greatest extinction (McKinney, 1987). Other severely affected groups included the cephalopods, corals, ostracodes and foraminiferans, all predominantly tropical groups or members of the reef-building community.

There has been a constant expression of doubt regarding the scale of the end Permian extinction, especially in the terrestrial realm. Two major factors must be taken into account: the presence of Lazarus taxa (see Donovan, Chapter 2, this volume); and the incomplete stratigraphic record across the Permian–Triassic (P–Tr) boundary.

Many taxa which are known from the late Permian and Middle Triassic have no fossil record in the early Triassic. It seems reasonable to expect that, in a number of cases, taxa suspected of extinction during the end Permian event actually spanned the P–Tr boundary and survived into the Triassic. An absence of suitable facies in the early Triassic, however, means that any taxa suffering extinction prior to Middle Triassic times remain unknown from the post-Permian.

A few examples of complete marine stratigraphic records across the

P–Tr boundary are observable in China. Complete sequences are reported for Kashmir, the Salt Range, central and northern Iran, and the Kap Stosch area of East Greenland, but some of these sequences may be disconformable (Nakazawa *et al.*, 1980). No complete terrestrial sections are known, causing problems with the interpretation of the extinction of non-marine forms.

The most famous area for gathering data regarding the extinction of terrestrial tetrapod families is the Permo-Trias Karroo sequence of South Africa. The sequence is notoriously incomplete and the facies of the late Permian and early Triassic are in no way comparable (Hotton, 1967). Changes in sedimentation in the Karroo Basin may exaggerate the abruptness and magnitude of the terrestrial extinctions (Pitrat, 1973) and this tends to detract from the validity of any conclusions regarding the elimination of terrestrial South African forms. When combined with an undoubted taxonomic inflation of therapsids across the boundary (Padian and Clemens, 1984), serious doubts arise regarding the conclusions drawn from extinction data. Recent work has helped to resolve some of the taxonomic inflation with the advent and application of cladistic analysis. These thorough re-evaluations of taxonomy have served to eliminate some of the many synonymous and poorly defined false genera and species erected by various authors over the past 150 years.

This still leaves the problem of incomplete sections in the classic South African localities. As with marine invertebrate genera, it seems reasonable to assume that a number of terrestrial genera, whose last record is in the late Permian, may actually have extended into the Triassic, but, because of the change in lithology and facies in the rock record across the P–Tr boundary, 'survivors' are not preserved. Because of this, quantitative estimates of tetrapod extinction are based on an assumption of no 'missing' Triassic survivors and, as a result, may be slightly anomalous. The end Permian extinction of terrestrial forms was real, none the less, and must be borne in mind when considering a causal mechanism for the larger extinction in the marine realm.

Problems with dating the P–Tr boundary should be recognized when examining the scale of the extinction. To state the obvious, the position of the boundary denotes the transition from Permian to Triassic, thus affecting the number of species that are seen to terminate within the Permian and the number that continue into the Triassic. A date of 248 million years BP seems to be most widely accepted for the P–Tr boundary at the moment, with closest tie-points of 238 and 268 million years BP (Harland *et al*, 1982), but any alteration of the boundary date, as a result of stratigraphic or palaeontological revision, may change the observed pattern of faunal depletion. The interpretation of time-related diversity may change in the future but it will not alter the end Permian's status as the greatest extinction in history.

The two main aspects of any mass extinction that should be considered before proposing a mechanism of causation are the taxa affected and the timing of their extinction. The late Permian and early Triassic faunas have been extensively documented, revealing that it was mainly the warm-adapted pelagic and benthic stenohaline taxa from low latitudes which suffered the greatest extinction. The timing of their elimination is not so well documented and there is dispute regarding the timing of the event as a whole.

TIMING OF THE EXTINCTION

The question of a catastrophic or gradual end Permian extinction has been considered by a number of workers during the 1980s. The recent interest was generated by the proposed extraterrestrial causes of periodic mass extinctions which were revealed by a statistical analysis of the marine fossil record (Raup and Sepkoski, 1982; 1984). The extraterrestrial cause, be it a single bolide, cometary phenomenon, asteroid shower, or whatever, implies a short-term, catastrophic extinction as outlined originally by Alvarez *et al.* (1980) for the Cretaceous–Tertiary (K–T) boundary, and subsequently by numerous other authors.

Geochemical evidence

The 'fingerprint' of any impact, or series of impacts, a concentration of iridium in sediments, has been reported from China (Sun *et al.*, 1984; Xu *et al.*, 1985). Sun *et al.* reported an iridium concentation of 8 parts per billion (ppb) for the boundary clay at Meishan, Changxing, China, while Xu *et al.* detected a 2 ppb concentration from the same clay. Xu *et al.*'s preliminary report gives a detailed account of quantitative variations in trace-element concentrations in a continuous marine sequence 4 m thick, at Meishan. Ten samples taken from the sequence and spanning the P–Tr boundary were analysed and, in all but one, iridium was found to be absent. The sample which revealed the presence of iridium in detectable quantities was that taken from a clay regarded as marking the transition from Permian to Triassic times (Figure 8.1). This anomalous concentration was measured at 2.48 ppb. Samples of the K–T boundary clay from western Europe were also examined by Xu *et al.* (1985) and the results

---

Figure 8.1. P–Tr boundary stratigraphical section at Shangsi, Guangyuan, Meishan. The iridium concentration was recorded from division 3. Divisions: 1, silty shale; 2, grey shale; 3, grey-black calcareous shale; 4, light-coloured shale; 5, grey-black calcareous shale; 6, grey siliceous limestone; 7, grey shale; 8, dark grey calareous shale; 9, grey shale; 10, dark-grey siliceous shale; 11, grey shale; 12, dark-grey siliceous limestone; 13, grey shale; 14, dark-grey siliceous limestone. After Xu *et al.* (1985).

Series	Stages	Formation	Division	Lithology	Thickness (cm)	Major Fossils
Lower Triassic	Induan	Fiexianguan Formation	1			*Hypophiceras* sp. *Glyptophiceras* sp. *Claraia griesbachi* *Bakevillia* sp. *Anchignathodus decresense* *Neogondolella changxiangensis*
			2		4	
			3		4	*Pseudotirolites asiaticus*
			4		6	*Pleuronodoceras* sp.
			5		4	
			6		22	*Neogondolella subcarinata*
			7		2	
			8		2	*Pleuronodoceras* sp.
			9		2	
			10		2	*Rotodiscoceras*
			11		6	
Upper Permian	Changxingian	Dalong Formation	12		52	*Pleuronodoceras* sp. *Xenodiscus* *Neogondolella subcarinata* *Porodiscus* sp. *Cenellipsis* sp.
			13		6	
			14		70	Many Fossils

concurred. Variations noted in the concentration of other members of the platinum metal group throughout the section were also compared and their pattern of variation was found to be similar to that seen in the K–T clays analysed.

Evidence supporting an extraterrestrial origin for the iridium is provided by a consideration of the microspherules present in the section. A plot of the nickel/iridium (Ni/Ir) ratio, as outlined by Ganapathy (1983), suggests that the microspherules are of extraterrestrial, rather than terrestrial, origin. All in all, the evidence gathered and presented from the Meishan P–Tr section by these two groups of workers appears to support the occurrence of an extraterrestrial event at the end of the Permian.

Clark *et al.* (1986) carried out analyses of a number of P–Tr sections in China, including that at Meishan originally examined by Xu *et al.* (1985). Consideration of trace-element concentrations and conodont faunas produced results at variance with those of Sun *et al.* (1984) and Xu *et al.* (1985). In all the sections examined, no trace of any iridium concentration was recorded and indeed, a depletion of iridium (0.002 ppb, compared with average earth crustal abundances of 0.005–0.008 ppb) was observed for the boundary clay at Changxing, with slightly higher levels above and below (0.004 ppb and 0.034 ppb respectively). Hence, Clark *et al.* (1986) found no evidence that would indicate an extraterrestrial event at the end of the Permian.

The elemental abundances in boundary clays across China suggest that there is a remote possibility that the predominantly illite boundary clay resulted from the alteration of ejecta dust from a comet impact, but the most likely source was ash from a massive volcanic eruption (Clark *et al.*, 1986). The trace elements suggest that the dust was highly acidic and the ratios of TiO_2 and Al_2O_3 are low enough to support the volcanic dust scenario. Evidence from Clark *et al.*'s (1986) examination of conodont diversity and abundance across the boundary also supports rejection of an extraterrestrial event.

A geochemical study of well-preserved and nearly continuous P–Tr sections in the southern Alps utilized carbon-isotope data (Magaritz *et al.*, 1988). Carbon-isotope ratios are known to shift at some stratigraphic boundaries associated with extinction events, although the connection between mass extinctions and carbon-isotope shifts is ambiguous. Drops in the level of $\delta^{13}C$ may result from a cessation of biological production in the 'Strangelove ocean' following a meteorite impact (Hsü *et al.*, 1985), or from a large drop in sea level that reduces shelf area, exposing the shelf and its accumulated organic carbon to erosion.

The sections examined in the southern Alps of Italy and Austria are part of the late Permian Bellerophon Formation and the early Triassic Tesero Horizon of the Werfen Formation. Magaritz *et al.* (1988) reported a gradual change in the $\delta^{13}C$ content of marine carbonates across the P–Tr boundary within the sections, the carbon isotopes showing no dramatic shift that can be associated with the mass extinction. The gradual change of

$\delta^{13}C$ across the boundary corresponds to 10–20 m of carbonate sedimentation, equivalent to a time interval of 0.1–2 million years at expected rates of sedimentation (Wilson, 1975). The carbon-isotope data suggest the possibility of a stepwise extinction in the late Permian and a final gradual decline crossing the boundary.

Considering the findings of Clark *et al.* (1985) and Magaritz *et al.* (1988), it appears that geochemical evidence points to a gradual, rather than a catastrophic, extinction event.

Faunal evidence

Evidence gleaned from P–Tr faunas appears to be accumulating to support a gradual and protracted end Permian event, rather than a cataclysmic end to the Palaeozoic. Late Permian generic extinctions began to reach inordinately high levels in the Leonardian, peaked in the Guadalupian, and declined in the Tatarian (Sepkoski, 1986) (Figure 8.2). The continuation of the peak into the early Triassic is thought to reflect high rates of evolutionary turnover among groups that radiated after the extinction event (Sepkoski, 1986) and should, perhaps, not be interpreted as a continuation of the above-background level of extinction dominant in the latest Permian.

Sepkoski (1986), perhaps seeking to maintain faith in a catastrophic extraterrestrial cause for periodic extinctions, proposed that the breadth of the late Permian extinction peak is a result of inadequate sampling, because of the virtual absence of complete late Permian sections and complete sections across the P–Tr boundary. Nevertheless, Sepkoski may be essentially correct in attributing the supposedly protracted nature of the extinction to an insufficiency of late Permian sections. Extensive regressions of the late Permian seas are known from lithostratigraphical studies and may present a false picture of a gradual extinction (Signor and Lipps, 1982).

Lazarus taxa, those that in this case crop up in the Triassic record having apparently suffered extinction in the early late Permian, must have extended right up to, and over, the P–Tr boundary. These 'false' extinctions of the early late Permian serve to extend the apparent time-span of the event. There is no doubt that this Lazarus phenomenon is real for the late Permian–early Triassic interval; Batten (1973) recorded a greater number of 'Palaeozoic' gastropod genera in the Middle Triassic than the late Permian. Similar patterns for other invertebrate groups such as bivalves (Nakazawa and Runnegar, 1973) and articulate brachiopods (Waterhouse and Bonham-Carter, 1976) suggest strong preservational biases in the late Permian fossil record. The possibility therefore exists that the prolonged generic extinction peak is in part an artefact (Jablonski, 1986) and that the end Permian event was of shorter duration than a 'blind' examination of the fossil record would suggest, although the numbers of Leonardian and

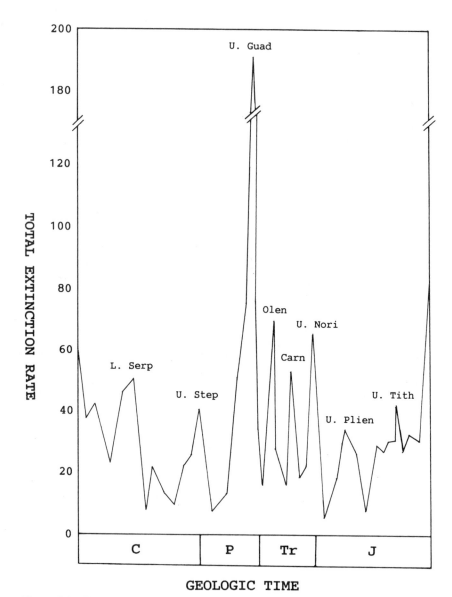

Figure 8.2. Total extinction rate (genera per million years) for Carboniferous to Jurassic times. The total extinction rate for any stage is given by the number of extinctions divided by the estimated duration of the stage. Abbreviations: l = lower, u = upper, Serp = Serpukhovian, Step = Stephanian, Guad = Guadalupian, Olen = Olenekian, Carn = Carnian, Nori = Norian, Plein = Pleinsbachian. After Sepkoski (1986).

Guadalupian generic extinctions appear to be an overwhelming testimony to a protracted extinction event.

The faunal evidence is open to interpretation, considering the widespread late Permian regressions and the bias they may have introduced to the fossil record. Most authors today favour the idea of a gradual event. Detailed biostratigraphical studies of the complete P–Tr marine sequences in China, and perhaps elsewhere (see above), should help to resolve the debate.

CAUSES OF THE END PERMIAN EXTINCTION

The end Permian extinction has long been recognized as the most comprehensive devastation of Phanerozoic life. Thankfully, however, the proposed reasons for such an event are neither as numerous, nor as varied, as those that have been put forward for the end Cretaceous extinction. The elimination of the dinosaurs assured the K–T event of a place in the public consciousness and there was a knock-on effect that produced many research papers, not necessarily by palaeontologists or geologists. The end Permian event, 'famed' for its reduction of the diversity and numbers of marine invertebrate taxa, holds no attraction for a public conditioned to marvelling at the 'fantastic' dinosaurs. A similar, but opposite, knock-on effect restricted research and speculation on possible causes of the P–Tr extinction to serious palaeontologists and geologists/geochemists.

Recent postulations of causal mechanisms include reduction in the area of shallow epicontinental seas (Newell, 1963; Valentine, 1973) and hence reduction of species-habitable areas, one cause of which may have been sea-floor spreading (Schopf, 1974); trace-element poisoning (Cloud, 1959); increased cosmic radiation (Hatfield and Camp, 1970); high temperatures (Waterhouse, 1973); low temperatures (Stanley, 1984; 1988); increased salinity (Bowen, 1968), and reduced salinity (Beurlen, 1956; Fischer, 1963; Stevens, 1977). The notion of sea-level fluctuations appears to have held sway over the past few decades, while many workers have considered the possible effects of varying temperatures and varying salinity. A summary of the essential facts of each possible causal mechanism, as outlined by various authors, is presented below, followed by a discussion utilizing the latest data and publications.

Sea-level fluctuations

The effects of sea-level fluctuations are outlined by Newell (1963), without elaborating as to why the fluctuations would occur. During Palaeozoic and Mesozoic times land surfaces were generally much lower than they are today. Approximately 30 major oscillations of sea level have occurred in the Phanerozoic (along with hundreds of minor ones) and, with each, a small increase or decrease in sea level was sufficient to flood or expose large, flat-lying, near-shore areas, producing major environmental changes. Repeated expansion and contraction of habitats by way of response to draining or . flooding has been envisaged as having created ecological disturbances among offshore and lowland communities, the repercussions of which probably extended inland and out to sea. Draining of continents reduced or eliminated shallow inland seas and many organisms adapted to the special estuarine conditions of these seas failed to survive in more exposed ocean margins. So it was that draining of continents induced extinction and natural selection. Flooding of continents increased the number of habitats, possibly leading to increased diversification, although this has been disputed because of the biogeographical complexities involved (Boucot and Gray, 1978).

Newell (1963) outlined the concept of key species, those, not necessarily represented in the fossil record, whose disappearance would lead directly to the extinction of many ecologically dependent species higher in the pyramid of community organization. A simple example would be a primary food plant. Through its removal a wave of extinction, beginning in a shrinking coastal habitat, could extend inland and out to deeper seas.

A decrease in area and the number of biotopes, consequences of a reduction in area of shallow epicontinental seas, satisfies the unique features of the late Permian–early Triassic interval which are discussed below. There is a direct correlation between the number of biotopes and the size of the available area, as biotopes exist in relation to many physical variables. Another consequence of significant regression is an increase in rates of evolution brought about by reduction of populations to smaller reproducing units. Rapid evolution is evident in many groups throughout the Triassic.

Fluctuations in ocean salinity

A hypothesis of reduced ocean salinity as the cause of the end Permian marine extinctions (Beurlen, 1956) was based on the observation that largely stenohaline groups, such as the bryozoans, ostracodes and corals, were severely reduced at the end of the Permian, but those which were least affected were groups such as gastropods and freshwater fishes, which, in Recent faunas, contain many brackish water representatives. He postulated that salinity was progressively reduced during the second half of the Permian, reaching critically low values at the P–Tr boundary, before persisting into the early Triassic. As noted above, early Triassic marine faunas are sparse with cosmopolitan distribution, and many groups that were both diverse and abundant before and after that time are absent. Beurlen suggested that this reduced marine fauna was derived from brackish-water lagoonal and estuarine faunas of the Permian which could exist in an ocean of then greatly reduced salinity. Stenohaline species were envisaged as having suffered extinction, except for those existing in a few places of the world where normal salinities were maintained. A return to normal salinities world-wide in later Triassic times allowed these few stenohaline forms to repopulate the seas and, as a result, crop up once more in the fossil record after their temporary absence.

However, what would cause such a large reduction in ocean salinity world-wide? Fischer (1963) proposed a model of large-scale evaporite sedimentation accompanied by the formation of large quantities of dense brine which becomes stored on the deep-sea floor. Circulating waters were thus reduced in salinity, possibly to a value of 30 parts per thousand. The geological consequences of such a process would be the deposition of enormous volumes of evaporites including anhydrite, gypsum, and halite. Beurlen (1956) estimated the quantity of existing Permian evaporites at 5×10^{14} tonnes, a figure which he doubled to take into account salts lost by erosion in the past 225 million years. The resulting figure of 1×10^{15} tonnes represents approximately 15% of the minimum salt loss required for the hypothesis to work. Fischer (1963), who criticized Beurlen's theory, used estimates and assumptions to produce a figure of 200 000 km^3, less than 10% of the quantity of salts required. It appeared, therefore, that the Permian evaporite deposits, even when taking erosion into account, were wholly inadequate to have lowered oceanic salinities by the required amount.

Over a decade later new data revealed that Permian halite deposits were of a much greater volume than previously suspected and amounted to at least 10% of the volume of salt presently in solution in the oceans (Stevens, 1977). This is well over one-half of the volume that would result in brackish conditions if removed from modern oceans. There may be considerably more Permian halite than has yet been discovered or dated as Permian, and, taking into account the amounts of halite lost to erosion, Stevens (1977) considered it probable that sufficient salt was removed for the Permian oceans to produce brackish conditions, and that this was an important factor in the Permian extinctions.

The idea of extinction resulting from an increase in salinity was put forward by Bowen (1968). Based on an assumption that the Louann salt mass extends below the Gulf of Mexico, Bowen calculated its volume to be approximately 4×10^6 km^3. If this amount of salt was added to the present ocean water along with other post-Palaeozoic salts of Africa and the Middle East, the ocean would be roughly 20% saltier than at present. Extrapolating back, Bowen envisaged the end Permian oceans to be just as salty and the extinction of many forms of marine life was a reflection of their inability to survive in such salty waters. The proposed increase in concentration of salts in the oceans would not be considered typical of Palaeozoic oceans, but rather a consequence of Permian climatic and weathering conditions. A return to normal salinity throughout the course of the Triassic resulted from the deposition of the Louann Salt mass, caused by a prolonged period of highly evaporative conditions.

Fluctuations in temperature

A dominant role for temperature change in extinction events has been forcefully argued by Stanley (1984; 1988). The late Permian was cited as a protracted event during which there was a major decline in tropical marine biotas including stenothermal calcareous algae. The reef community suffered substantial extinction and warm-adapted taxa were displaced towards the equator (Stanley, 1988). Global cooling would have its greatest impact in the tropics. Severe cooling, which would eliminate the tropical zone, would lead to major extinctions of tropical taxa. However, climatic zones in higher latitudes would migrate towards the equator with their constituent nontropical taxa, thus allowing them to survive. There appears to be a progressive contraction of the tropical zone of major taxa such as the bryozoans, fusulinacean foraminiferans and rugose corals, with all of these groups restricted to the Tethyan by Dzhulfian time (Stanley, 1988).

Stanley (1988) utilized the sparse and cosmopolitan faunas of the early Triassic to support his theory. A severe global refrigeration would be expected to leave its mark in the rock record as a depletion of limestones (a rock of predominantly tropical origin), greatly reduced reef growth, and reduced provinciality of faunas. It has already been observed that the faunas of the early Triassic were cosmopolitan and Kummel (1973) noted an uncommon dearth of carbonate sediments in this interval. Reefs are unknown for this time, yet important reef builders of the late Permian, such as sponges and calcareous algae, reappeared and expanded to build reefs in the Middle Triassic (Flugel and Stanley, 1984). The reappearance of these taxa is envisaged as being indicative of two things. Reef growth in the late Permian was not halted by extinction alone, but the onset of ecological conditions that suppressed reef taxa, and these conditions persisted for several million years (Stanley, 1988). Stanley found it difficult to imagine that any factor other than temperature could limit and suppress organisms as simple as calcareous algae for so long. Cool conditions at low latitudes could easily account for the pattern.

The suggestion that an increase in temperature may have caused the extinction (Waterhouse, 1973) has received little support. With the onset of cooling, it is impossible for tropical taxa to move to a warmer environment, but there seems little reason why these taxa could not have migrated towards cooler polar regions if temperature increased. One possible explanation, which utilizes both warming and cooling, is that in late Permian times latitudinal temperature gradients became more gentle without net global cooling, and tropical regions may have briefly suffered cooling while there was a warming at high latitudes (Stanley, 1984).

Increased cosmic radiation

There is a crude correlation between the galactic position of the solar system and the major faunal extinctions (Hatfield and Camp, 1970). Our galaxy undergoes one revolution in the vicinity of the Sun around the galactic centre once every 200 million years and, while this occurs, the Sun completes between two and three vibrations perpendicular to the galactic plane, that is, once every 80–90 million years. The most intense lines of force in the magnetic field are probably orientated in the galactic plane and this maximum magnetic energy density may be coincident with the most intense concentration of cosmic radiation within the galaxy. On this basis, Hatfield and Camp (1970) speculated that major magnetic events for the Earth might be expected as a result of its movement perpendicular to the galactic plane, and that these events could have produced faunal reductions.

Greatly increased cosmic radiation is envisaged as having the potential to produce mutations within organisms and cause extinctions. Because the end Permian extinction was such a protracted event, a considerable period of increased radiation has been speculated. The probability of radiation-induced mutation increases not only with increased radiation dosage, but also with the increased time over which a given dose is applied, so there was an 'immensity of time' available for the accumulation of lethal genetic changes.

Trace-element poisoning

The undeniably high level of marine extinctions and the apparently low level of terrestrial extinctions suggested the possibility of a chemical causal mechanism in large or interconnected water systems (Cloud, 1959). The introduction of poisonous elements into the world's seas may cause extinctions as a result of direct assimilation of the elements in lethal quantities, death of essential dietary components, or the intake of overdoses via a concentrator dietary component. Permian terrestrial floras and faunas were not significantly affected because there was but a slow diffusion of lethal substances in the nonmarine environment, especially under late Permian desert conditions (Cloud, 1959).

The result of the high level of marine extinctions was a reintroduction of potassium, phosphate, vanadium, and other biogenic elements to seawater and sediment in excess of organic withdrawal, and local concentrations in favourable sites. This is seen as a contributing factor to the vanadium-rich Permian phosphates, and the world's greatest potash reserves which are held in late Permian salt deposits.

Decline in origination rates

A decline in the origination rate for any given period of time will result in an increased number of extinctions, provided that the background extinction rate remains close to its previous level, for the duration of that period. Hüssner (1983) advanced evidence for dominant low origination rates in the end Permian marine environment and concluded that the low diversity at the P–Tr boundary did not result from a catastrophic event, but from reduced speciation. The formation of Pangaea in Upper Palaeozoic times is envisaged as the cause with continental convergence resulting in less differentiated environments and a reduction in the possibility of allopatric speciation. Lowering of species origination rates is therefore seen as the cause of the enormous reduction in marine invertebrate family numbers.

A similar theory, of reduction in the number of isolated environments leading to reduced speciation, has been forwarded for the terrestrial realm (Bakker, 1977). There is an observed synchroneity, throughout the geologic record, of large-scale tetrapod extinctions and widespread regressions which are thought to coincide with periods of reduced orogenic activity. This is the Haug Effect (Johnson, 1971) which links maximum transgression with maximum orogeny and hence a high topographic–geographic diversity of species with high speciation rate and standing diversity. Bakker (1977) saw regression and diminution of orogenic activity reducing the number of continental habitats. This would result in a reduction in speciation rate below extinction rate, and, hence, a significant decrease in species numbers and diversity.

DISCUSSION OF POSSIBLE CAUSAL MECHANISMS

Before considering the causal mechanisms outlined above, a cautionary note regarding reduced origination rates. A decrease in the number of environments and habitats, and an increased equability of climate, as proposed by Hüssner (1983) and Bakker (1977), would cause a decrease in speciation, and, hence, a reduction in family numbers if extinction rates remain at background levels. However, extinction rates, far from remaining steady in the end Permian, soared, as shown by numerous analyses over the past few decades. The elimination of hundreds of families in such a relatively short period of time cannot be attributed to very low origination and background extinction, but more likely resulted from the pressures exerted by one or more external physical factors.

Of the five proposed causal mechanisms outlined above, the first three (drop in sea level, salinity fluctuation and temperature decrease) account for virtually all the research time spent on the pursuit of a satisfactory model for the end Permian marine and terrestrial mass extinctions. The notion of regressions causing a reduction in the area of shallow epicontinental seas has attracted most attention (Newell, 1963; Valentine, 1973; Schopf, 1974; Simberloff, 1974; Boucot and Gray, 1978), while Stanley (1988) has recently argued strongly in favour of global refrigeration as a causal factor. However, before discussing further the relative merits of regressions, global cooling, and salinity changes, we should consider the radiation and trace-element poisoning theories.

Radiation and trace-element poisoning

These theories fall into the category of untestable scenarios. Newell (1963) effectively dismissed the radiation theory, stating that the late Permian floras were largely unaffected, but Hatfield and Camp (1970) cited experimental work which revealed that the degree of resistance to radiation damage varies as much as 100 000-fold from taxon to taxon (Platt, 1963), although they added that plants are generally more susceptible to radiation damage than animals. The criticism that any increase in cosmic radiation intensity would be too slight to cause extinctions is countered with the claim that the amount of time involved would be sufficient for genetic mutations to accumulate and have a lethal affect.

The classic argument against any theory invoking the effects of radiation, from whatever source, upon the Earth's surface, is the occurrence of widespread marine extinctions despite the shielding effect of the oceans. If the end Permian marine benthos was protected by the water in which it lived, why was it devastated? Hatfield and Camp (1970) argued for the elimination of marine plankton, which were not so well protected. Their removal from the food chain would have resulted in extinctions in the marine benthos. Observed extinction patterns show that the foraminiferids were severely (but preferentially) reduced (Brasier, 1988), but claims that waves of late Permian extinction were related to the extinction of phytoplankton (Tappan, 1968) were disputed by Pitrat (1970). The radiation theory, as its proposers admit, is based largely on scant data and has received little support in recent times.

Poisoning in ancient environments is not unknown, for example, the fossil fish of the Kupferschiefer are thought to have been poisoned by an influx of copper and silver (Moret, 1948). However, the notion that such a poisoning could occur on a world-wide scale and be selective in the taxa it affected (Cloud, 1959) is one that few people have given serious consideration. It seems inconceivable that the quantities of metallic trace elements, such as copper and vanadium, required to effect a world-wide extinction could be circulated throughout the world's oceans after initial introduction, before being removed as a result of accumulation by organisms, chemical reactions, or depositional processes.

Fluctuations of sea level

The effective elimination of radiation and poisoning as serious contenders for the cause of the greatest single reduction of Phanerozoic life leaves the three more plausible mechanisms, those which have received most attention. Before considering these, it would perhaps be best to list a few facts regarding the late Permian–early Triassic interval that are relevant to a discussion of causal factors (taken from Boucot and Gray, 1978).

1. The greatest level of regression of shallow seas from continents of any Phanerozoic interval occurred at this time.
2. Reef environments are unknown during the latest Permian and early Triassic.
3. There are few taxa up to class level of early Triassic benthic and pelagic organisms, contrasting with large numbers before and after this time.
4. Early Triassic taxa were organized into a small number of shelly invertebrate communities with very low species diversity.
5. Biogeographic diversification was at a Phanerozoic low in the early Triassic.
6. There are abundant late Permian evaporite deposits.

The first of these points would seem to support the theory of regression acting as an agent of extinction by effecting a reduction in the areal extent of shallow epicontinental seas, with the subsequent loss of species-habitable area. Shallow continental shelves would be reduced in width and limited to continental margins. The near-shore area, rich in nutrients and over which primary production is highest, would also be reduced, causing a reduction in food supply and increasing the possibliity of the extinction of forms higher up the food chain (Schopf, 1974). Other hypotheses have invoked alterations in climate or oceanic circulation accompanying regression, but the mainstay of the regression theory is the dynamic equilibrium theory of island biogeography, in which the reduction in diversity (with reduction in area) results from decreased population size and thus increased vulnerability to stochastic extinction processes (MacArthur and Wilson, 1967; Simberloff, 1972).

The hypothesis of reduced diversity resulting from reduced habitat area has been questioned recently (Stanley, 1984; Jablonski, 1985). It would perhaps be more accurate to say that the scale of the reduction in the number of species has been questioned; the fact that a contraction of habitable area will cause extinctions of some degree is not disputed. Stanley (1984) adopted the premise that if the species–area hypothesis is valid, it can be regarded as 'a fundamental ecological rule for a particular kind of ecosystem' and response to the rule should be universal, in that every large sustained global regression should result in the mass extinction of shallow-water benthos. Mass extinction has not accompanied every regression (Jablonski, 1986). Further, Stanley (1984) used data from the modern molluscan faunas of the Hawaiian islands and the Panamic-Pacific Province to show that the species–area effect is too weak on the sea-floor to have resulted in anything more than a minimal amount of excessive extinctions, as very small areas of shallow sea-floor currently support enormous faunas.

A similar study of modern faunas leads to the conclusion that the late Permian extinction 'is far too large to be accounted for by simple area effects due to contraction of shelf seas' (Jablonski, 1985). A survey of oceanic island faunas world-wide revealed that of the 267 families considered suitable for analysis, only 13% would suffer extinction if the continental shelf biota were completely eliminated, the remaining 87% persisting on the oceanic islands. The modern faunas indicate that most shallow-water families are spread over many habitat patches and possess large populations that are maintained over many localities well distributed both longitudinally and latitudinally (Jablonski, 1985). This suggests that the majority of shallow-marine families would be resistant to even prolonged regressions.

The fact that the Pleistocene sea-level fluctuations, which were among the largest ever, were not accompanied by mass extinctions is commonly utilized as an argument against the regression-extinction theory. Too much credit should not be given to this, however, as Pleistocene continents stood high above sea level and the expansion and contraction of shelf seas was minimal and in no way comparable to that which occurred in the late Permian–early Triassic interval (Newell, 1963; Jablonski, 1985).

Salinity decrease

The basis for the decreased salinity hypothesis (Beurlen, 1956; Fischer, 1963) is found in the selective extinction of marine families. The stenohaline forms such as corals and ostracodes were severely affected, while organisms with some tolerance of salinity variations survived and proliferated in the early Triassic. The claim that stenohaline forms lived in local pockets of normal salinity throughout the early Triassic, before emerging to repopulate the seas after a return to normal salinities world-wide (Beurlen, 1956), has been questioned. It is possible to picture small regions of normal salinity surrounded by brackish environments, but to suggest that these enclaves between the brackish world-ocean reservoir and the hypersaline areas undergoing intensive evaporation could support a sufficiently great diversity of marine life to account for the survivors which supposedly radiated in the Middle Triassic seems slightly optimistic (Boucot and Gray, 1978). A number of Permo-Triassic epicontinental sea habitats were extreme environments such as those of high salinity, but they supported relatively few taxa (Jablonski, 1985).

If, for whatever reasons, these areas of normal salinity could have supported a sufficiently diverse fauna, what mechanism could have removed such an enormous quantity of salts from the world's oceans? Salinity is not easily changed dramatically on a world-wide scale (Stanley, 1984) and the only likely model proposed to date is Fischer's (1963) brine-reflux hypothesis. Evaporation of seawater and the subsequent deposition of salts in evaporite basins produced dense brines which sank to the deep-sea floor, leaving circulating oceanic waters depleted of salts. Consideration of ostracode faunas shows that 75% of known Permian forms became extinct in the late Permian extinction and the survivors can be divided into two groups, those whose recent representatives live in freshwater and marginal environments, and those whose marine distribution is either very broad or typical of the deep-sea (Benson, 1984). Constituent genera of the deep-sea group such as *Cytherella* and *Paracypris* are found in today's oceans at depths exceeding 2000 m, and Benson considered it unlikely that this fauna could have survived had a salt brine existed at these depths.

Temperature decrease

There is a rough correlation between climatic changes and most, but not all, extinction events. There is a distinct tendency for researchers to attempt to tie in periods of postulated global cooling with large-scale glaciations, but the relationship is not so simple. If it were, we could immediately reject a global refrigeration scenario for the late Permian, as most of this protracted extinction event corresponds to deglaciation and approximately 5–8 million years of unglaciated time before the end of the Palaeozoic (Jablonski, 1986). In contrast to this widely held view of a direct link between cooling and glaciation, it is probable that cooling can cause mass extinction at low latitudes without also causing glaciation (Stanley, 1988) and, to reverse the implication, polar glaciation does not necessarily result in refrigeration and mass extinction of tropical taxa.

The most recent description of the end Permian climatic conditions and their implications (Stanley, 1988) reveals that, although Pangaea was moving off the South Pole in late Permian time and glaciation was waning, the Ross Sea area of Antarctica was glaciated and may have provided the source for sea-ice that reached regions off the Australian coast (Crowell and Frakes, 1971) (Figure 8.3). Also, as Pangaea moved off the South Pole and northward, it encroached on the North Pole. Glacial and glacial marine deposits are abundant in Australia, Siberia, and the Kolyma block, a terrain which was separated from Siberia in the Permian, but became sutured to it in the Mesozoic. A spread of cool temperatures, from high latitudes towards the equator, would have forced the geographical ranges of many major taxa towards zero, leading, as discussed above, to the extinction of tropical taxa. The observed pattern of the extinction, both in terms of its protracted duration and geographical affect, 'makes the most parsimonious explanation for the mass extinction a simple extrapolation of the cooling trend to low latitudes' (Stanley, 1988).

Overview of plausible causes

There is no doubt that regressions of the sea, fluctuations in salinity, and periods of global refrigeration cause extinctions. The problem with applying these concepts to the end Permian extinction is their inability, when considered separately, to explain the scale and selectivity of the extinction. The reduced salinity hypothesis certainly fits the faunal data and there

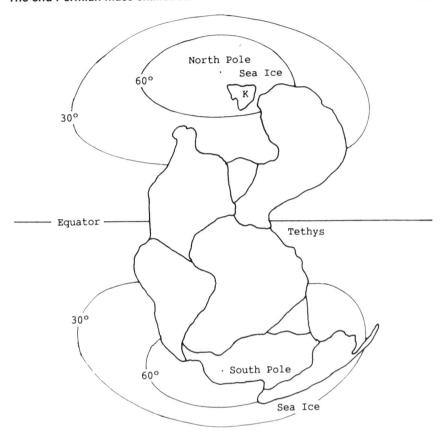

Figure 8.3. Disposition of Pangaea in the late Permian. The frigid condition of the poles, and the resulting bipolar cooling of continental margins, would have accentuated the climatic affects of sea-level lowering. After Stanley (1984). K = Kolyama block.

appears to be support in the rock record with the existence of enormous quantities of Permian halite (Stevens, 1977), but to expect stenohaline forms to survive in considerable, but undetected, numbers in local pockets of normal salinity seems to be asking too much. The concepts of regression, and hence reduction of species area, and temperature reduction appear most attractive. The recent data suggest that the species–area effect would be of relatively minor significance because of survivorship on oceanic islands (Jablonski, 1985). The proposal that sea-floor spreading caused the end Permian regression (Schopf, 1974; Bambach *et al.*, 1980) allows the postulate that a reduction in a sea-floor spreading rate could, via failure of heat flow and contraction of the lithosphere, cause destruction of the photic zone of oceanic islands and thus heighten the species–area effect. This concept may be rejected after consideration of late Permian subduction rates and the possibility that the regression may have resulted, not from sea-floor spreading, but from increased depth and enlarged capacity of the world ocean basin (Jablonski, 1985).

It appears, then, that regression alone could not account for the scale of the marine extinctions, although climatic changes resulting from the elimination of warm shallow seas may be evoked to explain the terrestrial extinctions. A scenario which may satisfy most of the data, be it faunal, lithological, or palaeogeographical, is one in which a period of global refrigeration, accompanied by a major regression, resulted in the observed marine extinctions, mainly through temperature reduction, and the terrestrial extinctions through climatic changes. The position of Pangaea in the late Permian (Figure 8.3) lends itself to the suggestion that the seas of its western margin were refrigerated by bipolar cooling, while the late Permian regressions not only accentuated the situation by eliminating warm shallow seas in this area, but also reduced the area of warm seas in the Tethyan embayment on the eastern margin (Stanley, 1988). Global cooling explains the selective extinction of tropical warm-adapted taxa, and the large-scale regressions evident from lithostratigraphy assisted in altering the atmospheric climate which may have played a significant role in terrestrial extinctions. Reduced speciation rates, associated with a reduction in the number of habitats which accompanies regression, may have contributed to the reduction in tetrapod diversity.

SUMMARY

The greatest reduction in diversity of Phanerozoic life occurred as the Permian Period and the Palaeozoic Era closed. Terrestrial faunas were badly affected, while marine faunas were devastated. Such an enormous reduction in the biota invites speculation of a catastrophic event, but geochemical and faunal evidence favours a gradual process of reduction in species numbers. The most plausible causal mechanism would appear to be a sustained period of global refrigeration, accompanied by widespread regressions and reduction of warm shallow seas on the western and eastern margins of Pangaea.

ACKNOWLEDGEMENTS

Thanks are due to Dr M.J. Benton and Dr S.K. Donovan for helpful comments and criticisms. This chapter was completed under tenure of a Science and Technology Award granted by the Department of Education for Northern Ireland.

REFERENCES

Alvarez, L.W., Alvarez, W., Asaro, F. and Michel, H.V., 1980, Extraterrestrial cause for the Cretaceous–Tertiary extinction, *Science*, **208** (4448): 1095–1108.

Bakker, R.T., 1977, Tetrapod mass extinctions—a model of the regulation of speciation rates and immigration by cycles of topographic diversity. In A. Hallam (ed.), *Patterns of evolution, as illustrated by the fossil record*, Elsevier, Amsterdam: 439–68.

Bambach, R.K., Scotese, C.R. and Ziegler, A.M., 1980, Before Pangaea: the geographies of the Paleozoic world, *American Scientist*, **68** (1): 26–38.

Batten, R.L., 1973, The vicissitudes of the gastropods during the interval of Guadalupian–Ladinian time. In A. Logan and L.V. Hills (eds), *The Permian and Triasssic Systems and their mutual boundary, Memoir of the Canadian Society of Petroleum Geologists*, **2**: 596–607.

Benson, R.H., 1984, The Phanerozoic 'crisis' as viewed from the Miocene. In W.A. Berggren, and J.A. Van Couvering (eds), *Catastrophes and Earth history*, Princeton University Press, Princeton, NJ: 437–46.

Beurlen, K., 1956, Der Faunenschnitt an der Perm-Trias Grenze, *Zeitschrift der Deutschen Geologischen Gesellschaft*, **108** (1): 88–99.

Boucot, A.J. and Gray, J., 1978, Comment on 'Catastrophe theory: application to the Permian mass extinction', *Geology*, **6** (11): 646–7.

Bowen, R.L., 1968, Paleoclimatic and paleobiologic implications of Louaan salt deposition, *Bulletin of the American Association of Petroleum Geologists*, **52** (9): 1833.

Brasier, M.D., 1988, Foraminiferid extinction and ecological collapse during global biological events. In G.P. Larwood (ed.), *Extinction and survival in the fossil record, Systematics Association Special Volume* **34**: 37–64.

Clark, D.L., Wang, C.-Y., Orth, C.J. and Gilmore, J.S., 1986, Conodont survival and the low iridium abundances across the Permian–Triassic boundary in south China, *Science*, **223** (4767): 984–6.

Cloud, P.E., Jr, 1959, Paleoecology—retrospect and prospect, *Journal of Paleontology*, **33** (5): 926–62.

Crowell, J.C. and Frakes, L.A., 1971, Late Palaeozoic glaciation of Australia, *Journal of the Geological Society of Australia*, **17** (2): 115–55.

Fischer, A.G., 1963, Brackish oceans as the cause of the Permo-Triassic marine faunal crisis. In A.E.M. Nairn (ed.), *Problems in paleoclimatology*, Wiley and Sons, New York: 566–74.

Flugel, E. and Stanley, G.D., 1984, Reorganization, development, and evolution of post-Permian reefs and reef organisms, *Palaeontographica Americana*, **54**: 177–86.

Ganapathy, R., 1983, The Tunguska explosion of 1908: discovery of meteoritic debris near the explosion site and at the Southern Pole, *Science*, **220** (4602): 1158–61.

Harland, W.B., Cox, A.V., Llewellyn, P.G., Pickton, C.A.G., Smith, A.G. and Walters, R., 1982, *A geologic time scale*, Cambridge University Press, Cambridge.

Hatfield, C.B. and Camp, M.J., 1970, Mass extinctions correlated with periodic galactic events, *Bulletin of the Geological Society of America*, **81** (3): 911–14.

Hotton, N., III, 1967, Stratigraphy and sedimentation in the Beaufort Series (Permian–Triassic), South Africa. In C. Teichert and E.L. Yochelson (eds), *Essays in paleontology and stratigraphy, Special Publication of the Department of Geology, University of Kansas*, **2**: 390–428.

Hüssner, V.H., 1983, Die Faunenwende Perm/Trias, *Geologische Rundschau*, **72** (1): 1–22.

Hsü, K.J., Oberhänsli, H., Gao, J.Y., Shu, S., Haihong, C. and Krähenbuhl, U.,

1985, 'Strangelove ocean' before the Cambrian explosion, *Nature*, **316** (6031): 809–11.

Jablonski, D., 1985, Marine regressions and mass extinctions: a test using the modern biota. In J.W. Valentine (ed.), *Phanerozoic diversity patterns: profiles in macroevolution*, Princeton University Press, Princeton, NJ: 333–54.

Jablonski, D., 1986, Causes and consequences of mass extinctions: a comparative approach. In D.K. Elliott (ed.), *Dynamics of extinction*, Wiley and Sons, New York: 183–229.

Johnson, J.G., 1971, Timing and coordination of orogenic, epeirorogenic and eustatic events, *Bulletin of the Geological Society of America*, **82** (11): 3263–98.

Kummel, B., 1973, Lower Triassic (Scythian) molluscs. In A. Hallam (ed.), *Atlas of palaeobiogeography*, Elsevier, Amsterdam: 225–33.

MacArthur, R.H. and Wilson, E.O., 1967, *The theory of island biogeography*, Princeton University Press, Princeton, NJ.

Magaritz, M., Bar, R., Baud, A. and Holser, W.T., 1988, The carbon-isotope shift at the Permian/Triassic boundary in the southern Alps is gradual, *Nature*, **331** (6154): 337–9.

McKinney, M.L., 1987, Taxonomic selectivity and continuous variation in mass and background extinctions of marine taxa, *Nature*, **325** (6100): 143–5.

Maxwell, W.D. and Benton, M.J., 1987, Mass extinctions and data bases: changes in the interpretation of tetrapod mass extinction over the past 20 years. In P.J. Currie and E.H. Koster (eds), *4th Symposium on Mesozoic Terrestrial Ecosystems, Occasional Paper of the Tyrrell Museum of Palaeontology, Alberta*, **3**: 156–60.

Moret, L., 1948, *Manuel de paléontologie animale*, Masson, Paris.

Nakazawa, K., Bando, Y. and Matsuda, T., 1980, *Geology and Palaeontology of SE Asia*, **21** (1): 75.

Nakazawa, K. and Runnegar, B., 1973, The Permian–Triassic boundary: a crisis for bivalves? In A. Logan and A.V. Hills (eds), *The Permian and Triassic Systems and their mutual boundary, Memoir of the Canadian Society of Petroleum Geologists*, **2**: 608–21.

Newell, N.D., 1963, Crises in the history of life, *Scientific American*, **208** (2): 76–92.

Padian, K. and Clemens, W.A., 1985, Terrestrial vertebrate diversity: episodes and insights. In J.W. Valentine (ed.), *Phanerozoic diversity patterns: profiles in macroevolution*, Princeton University Press, Princeton, NJ: 41–96.

Pitrat, C.W., 1970, Phytoplankton and the late Palaeozoic wave of extinction, *Palaeogeography, Palaeoclimatology, Palaeoecology*, **8** (1): 49–66.

Pitrat, C.W., 1973, Vertebrates and the Permo-Triassic extinctions, *Palaeogeography, Palaeoclimatology, Palaeoecology*, **14** (4): 249–64.

Platt, R.B., 1963, Ecological effects of ionising radiation on organisms, communities and ecosystems. In V. Schultz and A.W. Klement, Jr (eds), *Radioecology*, Reinhold, New York: 243–55.

Raup, D.M., 1979, Size of the Permo-Triassic bottleneck and its evolutionary implications, *Science*, **206** (4415): 217–18.

Raup, D.M. and Sepkoski, J.J., Jr, 1982, Mass extinctions in the marine fossil record, *Science*, **215** (4539): 1501–3.

Raup, D.M. and Sepkoski, J.J., Jr, 1984, Periodicity of extinctions in the geologic past, *Proceedings of the National Academy of Science U.S.A.*, **81** (3): 801–5.

Schopf, T.J.M., 1974, Permo-Triassic extinction: relation to sea-floor spreading, *Journal of Geology*, **82** (2): 129–43.

Sepkoski, J.J., Jr, 1986, Phanerozoic overview of mass extinctions. In D.M. Raup and D. Jablonski (eds), *Patterns and processes in the history of life*, Springer-Verlag, Berlin: 277–95.

Signor, P.W., III and Lipps, J.H., 1982, Sampling bias, gradual extinction patterns, and catastrophes in the fossil record. In L.T. Silver and P.H. Schultz (eds), *Geological implications of impacts of large asteroids and comets on the Earth*, *Special Paper of the Geological Society of America*, **190**: 291–6.

Simberloff, D.S., 1972, Models in biogeography. In T.J.M. Schopf (ed.), *Models in paleobiology*, Freeman, Cooper, San Francisco: 160–91.

Simberloff, D.S., 1974, Permo-Triassic extinctions: effects of area on biotic equilibrium, *Journal of Geology*, **82** (2): 267–74.

Stanley, S.M., 1984, Marine mass extinctions: a dominant role for temperature. In M.H. Nitecki (ed.), *Extinctions*, University of Chicago Press, Chicago: 69–117.

Stanley, S.M. 1988, Paleozoic mass extinctions: shared patterns suggest global cooling as a common cause, *American Journal of Science*, **288** (4): 334–52.

Stevens, C.H., 1977, Was development of brackish oceans a factor in Permian extinctions? *Bulletin of the Geological Society of America*, **88** (1): 133–8.

Sun, Y.-Y., Chai, Z.-F., Ma, S.-L., Mao, X.-Y., Xu, D.-Y., Zhang, Q.-W., Yang, Z.-Z., Sheng, J.-Z., Chen, C.-Z., Rui, L., Liang, X.-L. and Hi, and J.-W., 1984, The discovery of iridium anomaly in the Permian–Triassic boundary clay in Changxing, Zhijing, China, and its significance. In G. Tu (ed.), *Developments in geoscience*, Science Press, Beijing: 235–45.

Tappan, H., 1968, Primary production, isotopes, extinctions and the atmosphere, *Palaeogeography, Palaeoclimatology, Palaeoecology*, **4** (3): 187–210.

Valentine, J.W., 1973, *Evolutionary paleoecology of the marine biosphere*, Prentice Hall, Englewood Cliffs, NJ.

Waterhouse, J.B., 1973, The Permian–Triassic boundary in New Zealand and New Caledonia and its relationship to world climatic changes and extinction of Permian life. In A. Logan and L.V. Hills (eds.), *The Permian and Triassic Systems and their mutual boundary*, *Memoir of the Canadian Society of Petroleum Geologists*, **2**: 445–64.

Waterhouse, J.B. and Bonham-Carter, G.F., 1976, Global distribution and character of Permian biomes based on brachiopod assemblages, *Canadian Journal of Earth Science*, **12** (7): 1085–1146.

Wilson, J.L. 1975, *Carbonate facies in geologic history*, Springer, New York.

Xu, D.-Y., Ma, S.-L., Chai, Z.-F., Mao, X.-Y., Sun, Y.-Y., Zhang, Q.-W. and Yang, Z.-Z., 1985, Abundance variation in iridium and trace elements at the Permian/Triassic boundary at Shangsi in China, *Nature*, **314** (6007): 154–6.

NOTE ADDED IN PROOF

Two recent papers have made significant contributions to this debate. Studies of a core from an almost complete P–Tr section in Austria reveal two widely spaced iridium anomalies and Au/Ir ratios similar to those found in basalts and kimberlites rather than chondrites (Holser *et al*, 1989). Similarly, carbon isotope data from the core provides no evidence for a sharp event or events at the P–Tr boundary. Further isotope studies, this time from a brachiopod fauna in West Spitsbergen (Gruszczynski *et al.*, 1989), reject a catastrophic scenario and suggest a decline in atmospheric oxygen levels and nutrient deficiency as the main cause of the extinction.

Holser *et al*, 1989, A unique geochemical record at the Permian/Triassic boundary, *Nature*, **337** (6202): 39–44.

Gruszczynski, G., Halas, S., Hoffman, A. and Malkowski, K., 1989, A brachiopod calcite record of the oceanic carbon and oxygen isotope shifts at the Permian/Triassic transition, *Nature*, **337** (6202): 64–68.

Chapter 9

THE TIMING AND CAUSE OF LATE TRIASSIC MARINE INVERTEBRATE EXTINCTIONS: EVIDENCE FROM SCALLOPS AND CRINOIDS

Andrew L.A. Johnson and Michael J. Simms

INTRODUCTION

According to Sepkoski (1984) and Benton (1986), some 23% of both marine and non-marine animal families became extinct in the late Triassic, making this an interval of mass extinction comparable in importance to the late Cretaceous. Sponges, gastropods, bivalves, cephalopods, brachiopods, insects and a variety of vertebrate groups all suffered heavy losses of families. Among those groups disappearing completely were strophomenid brachiopods, conodonts, conulariids, placodonts, nothosaurs and rhynchosaurs. It has commonly been regarded (see, for example, Hallam, 1981; Raup and Sepkoski, 1984; 1986) that extinctions were concentrated at or near the end of the Triassic, in the Norian stage. However, Hoffman (1985) considered that a significant amount of extinction also occurred in the preceding Carnian stage and Benton (1986) has shown that the single late Triassic extinction peak recognized by Raup and Sepkoski (1984; 1986) is divisible into at least two, with one or more Carnian extinction peaks in addition to a peak in the Norian.

The idea of multiphase late Triassic extinctions has been criticized by Newton *et al.* (1987), principally on the basis of information from a study of bivalves. However, their evidence is not entirely conclusive (see below) and their arguments cannot necessarily be applied to other organisms. It is the main object of the present study to assess the evidence for multiphase extinctions, principally through a detailed analysis of patterns of extinction shown by late Triassic scallops and crinoids. Although these groups represent but a small subset of the total fauna, we believe that any patterns common to the two—phylogenetically so distant and autecologically so dissimilar—may be taken as an indication of patterns of extinction in a much broader range of taxa, at least many other marine invertebrates.

The existence of multiple extinction peaks in the late Triassic has a bearing on the possible causes of extinction. The single peak recognized by Raup and Sepkoski (1984; 1986) formed part of their evidence for a regular (26-million-year) extinction cyclicity, and thus of a possible extraterrestrial forcing mechanism. As pointed out by Benton (1986), multiple late Triassic peaks reduce the credibility of this proposition. In this contribution we attempt an independent assessment of the likelihood of involvement of extraterrestrial agencies through an appraisal of fine-scale patterns of extinction, and of correlations between extinction and environmental change. The latter raises an important philosophical point: if the disappearance of a species coincides with an environmental change, how are we to be sure that it has not simply emigrated from the area for which information is available, rather than become extinct? This problem is particularly acute for the Triassic, where the occurrence of fossiliferous rock is more patchy than in some systems (for example, compared with the Jurassic). Obviously, a knowledge of palaeoautecology may suggest whether we should expect a species to emigrate upon a given environmental change, but then the same environmental change might reasonably be regarded as a likely agency to cause the extinction of the species. There is no easy solution to this problem. Probably the most parsimonious general approach is to regard last occurrences that coincide with the onset of unfavourable facies as marking terminations of lineages. We discuss this point further below in connection with our data and, in some cases, are able to provide additional arguments in favour of an 'extinction' interpretation.

ANALYSIS OF PREVIOUS WORK

Although the main thrust of this contribution is directed towards an understanding of extinctions among marine invertebrates, part of the evidence concerning the involvement of extraterrestrial agencies is derived from vertebrates, and must be considered. Benton (1986) has shown that non-marine tetrapods suffered major end Carnian, as well as end Norian, extinctions and has used this information (together with the evidence from ammonoids, and marine and non-marine organisms in general) to call into question the regular extinction cyclicity, and possibility of extraterrestrial forcing, recognized by Raup and Sepkoski (1984; 1986). However the abrupt disappearance (within 1 million years) of non-marine tetrapod families at the Triassic–Jurassic boundary in eastern North America, together with the possible contemporaneity of an impact crater in Quebec (Olsen *et al.*, 1987), restores some credibility to the notion of extraterrestrial forcing, at least for the later of the two tetrapod extinction episodes identified by Benton.

For previously studied marine invertebrates the dating of last occurrences is generally too imprecise to give any clues as to the possible occurrence of catastrophic extinctions at the very end of the Triassic. Hallam (1981) has claimed a 92% extinction of bivalve species in Europe between the Sevatian (latest Triassic substage) and the Hettangian (earliest Jurassic stage). However, the Penarth Group of England and Wales and the Kössen and Zlambach Beds of the Northern Calcareous Alps, which are the source of data on latest Triassic bivalves, probably span the last two Sevatian ammonite zones (more than 1 million years; Hallam, 1981) and for most species available information on occurrence is no better than, for example, 'Kössen Beds'. Much the same can be said of latest Triassic brachiopod species in Europe, which suffered 80% extinction before the Jurassic (Pearson, 1977). For those bivalve and brachiopod species whose ranges have been plotted in relation to a measured section in the Kössen Beds (see Morbey, 1975), highest occurrences are fairly regularly distributed through some 100 m of strata (possibly spanning two ammonite zones) beneath the Triassic–Jurassic boundary. There is no reason to think that this reflects a progressive decline in sampling quality (cf. Signor and Lipps, 1982).

Stratigraphic resolution is rather better for latest Triassic Ammonoidea. Eight ceratitine genera became extinct in the last (Crickmayi) zone of the Sevatian (Tozer, 1979). However, there had been a progressive decline before this. The demise of the conodonts before the Jurassic appears to have been similarly somewhat gradual: of ten genera known in the Sevatian Kössen Beds, only five definitely survived into the Crickmayi Zone (recognized by the occurrence of the ammonite *Choristoceras*) in the Austrian Tyrol (Mostler *et al.*, 1978).

For groups other than the above, stratigraphic resolution of late Triassic occurrences is rarely better than the substage. Clearly, therefore, there is only any real evidence of catastrophic extinctions, and thus of the possible involvement of extraterrestrial agencies, for non-marine tetrapods at the end of the Triassic. However, this should not be taken as an indication of the non-occurrence of such extinctions among other organisms—better stratigraphic resolution (potentially achievable) is required to test this possibility adequately. Nevertheless, Hallam (1981) has argued strongly for terrestrial causation of latest Triassic extinctions, drawing attention to major environmental changes—including a regression and the loss of reefal facies, succeeded by a major transgression and widespread sea-floor anoxicity—over the Triassic–Jurassic boundary. In addition, Hallam and El Shaarawy (1982) have shown that the salinity of the late Triassic sea in northwestern Europe was probably somewhat reduced. The palaeoautecology of the bivalve orders which, according to Hallam (1981), underwent especially heavy losses (Unionoida, Trigonoida) is not such as to suggest that they would have had a particular propensity to emigrate upon these environmental changes (in comparison, for instance, to the Pholadomyoida, which suffered relatively few losses). That they suffered genuine extinctions seems, therefore, fairly certain, although paradoxically this form of reasoning renders it difficult, logically, to account for the extinctions in terms of the observed environmental changes. Laws (1982) has shown that the succession of molluscan assemblages (most including a number of unique species) in the late Norian Gabbs Formation of Nevada may be related to a general regression through the sequence; he thus adopts an 'ultra-emigrationist' stance concerning faunal changes over the Triassic–Jurassic boundary. However, this begs the question as to whether some species might have actually become extinct (as a result of the change in environment), an interpretation which is at least equally plausible (see above).

The most detailed published information relating to Carnian extinctions concerns tetrapods (as noted above) and ammonoids (Benton, 1986). The latter show a mid-Carnian, rather than end Carnian, extinction peak, the precise position of which is probably in the interval of the Upper Cordevolian and Lower Julian substages (Simms and Ruffell, in press). Newton *et al.* (1987) have criticized Benton's concept of major Carnian extinctions, suggesting that apparent disappearances at this time may be an artefact of inadequate information from the early part of the succeeding Norian stage. They show that some species, previously known only from the late Ladinian to early Carnian Cassian Formation of Italy, occur in the early Norian of Oregon. While this may mean that a higher proportion of extinctions took place in the Norian, it does not necessarily imply that there was, in fact, but a single phase of extinction in the late Triassic. Newton *et al.* (1987) did not consider the possibility that the Oregon fauna may differ from the bivalve fauna of the late Norian in Europe. Of the 42 named species listed from the latter by Hallam (1981), not one is quoted among the 50 named species recognized by Fürsich and Wendt (1977) from the Cassian Formation. While this may well reflect parochial taxonomy and the concern of these authors with matters other than the possibility of major Carnian or early Norian extinctions, it does at least suggest a need for further work before the idea of two phases of extinction in the late Triassic can be entirely discounted. Whilst Newton (in Newton *et al.*, 1987) compared late Norian bivalves from elsewhere in the world (for example, the Gabbs Formation) with species in the early Norian fauna, only 32% of the latter (10 of 31 species assigned to a genus) are even remotely suggested to be conspecific with late Norian forms.

It is the evidence that a major phase of extinction occurred between the early Carnian and late Norian that we examine particularly below through a detailed analysis of scallops and crinoids. Evidence concerning extinctions during this interval among bryozoa, conodonts, reef-building organisms and plants, as well as among ammonoids and non-marine tetrapods (see above), has been recently compiled by Simms and Ruffell (in press). They provide evidence of an alteration in climate during the mid- to late Carnian, specifically increased rainfall, which is believed to be causally linked to the extinctions. We discuss this, and other possible causes of extinction, following our analysis of the scallop and crinoid data below.

ANALYSIS OF LATE TRIASSIC SCALLOP AND CRINOID EXTINCTIONS

Scope and stratigraphic background

A compelling argument for the extinction of a species is its absence from sedimentary facies in which it was formerly present. It is the greatest good fortune that in the Alpine region of Europe sediments of early Carnian and late Norian age—thus bracketing the interval in which major extinctions may have occurred—are developed in a similar range of facies and are, moreover, richly fossiliferous. An excellent opportunity thus exists to test

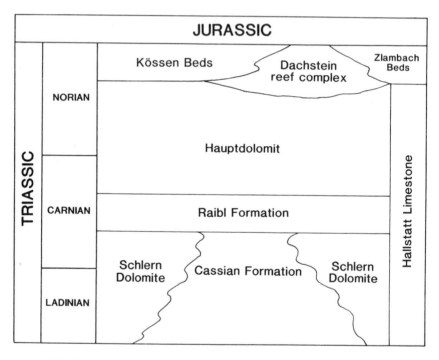

Figure 9.1. Summary stratigraphy of the late Triassic in the eastern and southern Alps, based on Zankl (1971), Fürsich and Wendt (1977), Urlichs (1977) and Zapfe (1974). The Rhaetian stage is included in the Norian (cf. Hallam, 1981). 'Dachstein reef complex' includes semilagoonal carbonate platform deposits (Lofer facies) representing the inferred backreef zone (reef front itself adjacent to Zlambach Beds). Deposition of the Raibl Formation apparently continued to the end of the Carnian in some areas. Similar facies to the units in the chart are present at equivalent horizons, but have different names, in some parts of the region; they may also occur further afield (see text).

the reality of the supposed extinction episode. The units concerned, the Cassian Formation (late Ladinian–early Carnian, southern Alps, Italy) and the Kössen and Zlambach Beds (late Norian, N. Calcareous Alps, Bavaria and Austria), accumulated in a variety of shallow (and rather deeper) basinal settings adjacent to sizeable carbonate platforms (represented respectively by the Schlern Dolomite and the Dachstein reef complex; Figure 9.1). At both times patch reefs were locally developed in shallow subtidal settings (Fürsich and Wendt, 1977; Zankl, 1971). Scallops and crinoids, investigated respectively by ALAJ and MJS, both occur commonly in the upper (early Carnian) part of the Cassian Formation; scallops are more common in the Kössen Beds of the late Norian units, whilst crinoids are more richly represented in the Zlambach Beds.

Figure 9.1 indicates the main stratigraphic units representing the interval between the early Carnian and late Norian in the Alpine region. The Raibl Formation (and correlative units such as the Lunz Sandstone) represents an influx of terrigenous clastic material, known widely in the Alps of Italy and Austria. The Hauptdolomit represents a variety of peritidal carbonate environments found in Bavaria and Austria, but also developed at the same horizon elsewhere in the Alps and further afield (Hallam, 1981). 'Hallstatt Limestone' is a locally developed red limestone facies whose name derives from a sequence of Ladinian–Norian age in the northern Calcareous Alps of Austria, but which is also developed much further afield (Hallam, 1981), and at earlier and later times. It represents clastic-starved sedimentation, probably in fairly deep water (Zankl, 1971). While Alpine tectonics have, to a greater or lesser extent, shifted all of the above units from their original positions of deposition, we can be sure that their faunas were not part of different biogeographic provinces in the Triassic from such evidence as the occurrence of typical 'Kössen' faunas in the late Norian of the southern Alps (Allasinaz, 1962).

We have supplemented our information from the Cassian and Kössen/Zlambach faunas with data on all European late Triassic scallops and crinoids so as to facilitate a comparison with Jurassic forms from the same region (Johnson, 1984; 1985; Simms, 1988a, 1988b; 1989, in press).

Scallop extinctions

Late Triassic scallop species and their ranges
The term 'scallop' implies an overall shell shape, characteristic of the bivalve superfamily Pectinacea (*sensu* Waller, 1978), for which group the term is usually reserved. Forms possibly representing the Aviculopectinacea and Buchiacea, but otherwise exhibiting typical scallop morphology, are included in the present analysis. The following 'scallop' species can thus be recognized in the late Triassic of Europe (see Appendix 9.1 for further

details):

(A)   'Leptochondria' subalternans (d'Orbigny)
(B)   Antijanira auristriata (Münster)
(C)   Chlamys (Granulochlamys) nodulifera (Bittner)
(D)   Chlamys (Granulochlamys) tubulifera (Münster)
(E)   Chlamys (Chlamys) interstriata (Münster)
(F)   Chlamys (Praechlamys) ?stenodictya (Salomon)
(G)   Filopecten filosus (Hauer)
(H)   Entolioides deeckei (Parona)
(I)   Propeamussium schafhäutli (Winkler)
(J)   Chlamys (Chlamys) mayeri (Winkler)
(K)   Chlamys (Chlamys) coronata (Schafhäutl)
(L)   Chlamys (Chlamys) valoniensis (Defrance)
(M)   Chlamys (Chlamys) pollux (d'Orbigny)
(N)   'Chlamys' subcutiformis (Kittl)

Illustrations (identified by the appropriate letter) and plots of stratigraphic range are provided in Figure 9.2. With the exception in each case of 'C'. subcutiformis (N), all the species represented in the Cordevolian are known in the upper part of the Cassian Formation and all those occurring in the Sevatian are known in the Kössen Beds (some species also occur in age-equivalent formations elsewhere). From Figure 9.2 it is evident that there was 100% extinction of Cassian species before the Sevatian. Although the possibility of pseudoextinction has not been fully assessed through a cladistic analysis, the general contrast in the styles of ornamentation (both internal and external; see Appendix 9.1) of Cassian and Kössen species makes it unreasonable to invoke a significant amount of pseudoextinction. Moreover, if the generic and subgeneric assignments (based on Allasinaz, 1972; Johnson, 1984) are accepted as valid, then only one of the Cassian species, C. (C.) interstriata (E), may be considered to have possibly given rise to Kössen forms. The number of species listed above is far fewer than the number recognized by Allasinaz (1972) from the same interval in Europe. In most cases these can be regarded as variants of the above species (Johnson, in preparation), but, since a high proportion are from the upper part of the Cassian Formation, the recognition of some as additional valid species would serve only to enhance the scale of extinctions between the Cordevolian and Sevatian.

   The scale of extinction of Kössen species before the Jurassic was evidently rather smaller than the earlier extinction of Cassian forms, but nevertheless still apparently considerable (60%).

Ecology and extinction
Many of the Cassian species are not known after the Cordevolian (considered here to be the age of the upper part of the Cassian Formation; see legend of Figure 9.2). However, we cannot assume uncritically that their

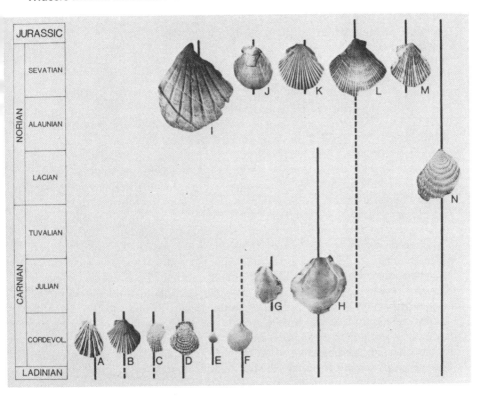

Figure 9.2. Ranges of European Carnian and Norian scallop species recognized in this study (dashes indicate uncertainty). See text for names corresponding to letters and Appendix 9.1 for further details. Species first known in the Raibl Formation, or equivalent (G, ?L), and those last known in the upper part of the Cassian Formation (A–E), are herein assumed for the sake of simplicity to respectively appear and disappear at the Cordevolian–Julian boundary. There is now evidence that the Cassian Formation extends into the early Julian (Bizzarini *et al.*, 1986). Species last known in the Raibl Formation (G, ?F) are assumed not to range beyond the Julian (cf. Figure 9.1). The earliest occurrences of species appearing in the Sevatian (I–K, M) are probably in the Amoenum ammonite zone (middle of three). All illustrations to same scale (× 0.045).

absence from the Julian Raibl Formation was due to their extinction because, as noted above, this unit marks a major facies change. The same sort of considerations apply to the apparent extinction of Kössen species by the Hettangian stage (earliest Jurassic).

All the scallops occurring in the Cassian Formation are very small (height rarely more than 20 mm) and those with thin shells would have required byssal attachment to maintain stability. A deep byssal notch, attesting a byssate mode of life, is known, or can be reasonably inferred, in all such species (see Appendix 9.1). *Entolioides deeckei* (H) had only a small byssal notch and, being relatively thick shelled, was probably a recliner, at least when adult. It is significant that *E. deeckei* is the only Cassian species which definitely survived into the Raibl Formation. The latter marks the loss of reefal environments—in which many byssate species (for example, *Chlamys (Granulochlamys) tubulifera* (D); see Fürsich and Wendt, 1977) probably found attachment sites—and their replacement by terrigenous clastic facies, often of high energy (Bosellini *et al.*, 1978). Such environments would have afforded few suitable sites for attachment. The loss of byssate species and the survival of *E. deeckei* (which, moreover, underwent considerable phyletic size increase, presumably as a further adaptation for reclining) is thus fully intelligible. Nevertheless, we still cannot be certain from this evidence whether the byssate species became extinct in the Julian or simply migrated elsewhere, becoming extinct at some later juncture before the Sevatian. Considerable light is thrown on this problem by the recent description of examples of *E. deeckei* (as *Crenamussium concentricum* Newton) from the early Norian of Oregon (Newton *et al.*, 1987). This is the only Cassian scallop species which can be recognized in the Oregon fauna. Notwithstanding the presence of other Cassian bivalves (only four more, however, among a total of 19 named species) and the possible failure of some species to migrate this far, it is surely significant that the only Cassian scallop represented in the Oregon fauna is the single species known to have survived into the Raibl Formation in Europe. The strong suggestion is that the other Cassian scallops were forced into extinction by the environmental changes associated with the passage into the Raibl Formation. The ecological congruence (and thus presumed similar palaeoenvironment) of the total Cassian and Oregon bivalve faunas (Newton *et al.*, 1987) makes it at least fair to suggest that the byssate Cassian scallops were extinct by the early Norian. The absence of the European Julian species *Filopecten filosus* (G) from the Oregon fauna can be reasonably explained by the strong environmental contrast and does not necessarily imply any impediment to colonization derived from geographic separation.

The distribution of *'Chlamys' subcutiformis* (N) strongly supports the idea that the extinction of all Cassian scallops (including *E. deeckei*) before the Sevatian was due to habitat loss. *'C.' subcutiformis* is the only scallop known to range through the Cordevolian–Sevatian interval and occurs exclusively in 'Hallstatt Limestone' facies, which are developed continuously over this period in the type area (Figure 9.1). Elsewhere in Europe after the Julian very shallow-water carbonate facies, subject to frequent emergence, were developed universally (Hauptdolomit and equivalents). Such environments would have been inimical to all the Cassian scallops.

We have already remarked upon the major environmental changes, including the loss of reefal facies, at the Triassic–Jurassic boundary. Reefal environments, not known north of the Alps in Europe during the Sevatian, were almost certainly the habitat of *Propeamussium schafhäutli* (I), and possibly also of *Chlamys (C.) mayeri* (J) and *C. (C.) coronata* (K), which are unknown north of the Alps. The fact that these three species disappeared before the Hettangian might therefore reflect emigration upon habitat loss, rather than extinction. Certainly they were extinct by the late Pliensbachian, as they are absent from the reef-associated fauna of Jbel Bou-Dahar in Morocco (Dubar, 1948; Johnson, 1984). It is worth noting here, however, that *Pseudopecten dentatus* (J. de C. Sowerby), one of the species present at Jbel Bou-Dahar, bears a strong resemblance to '*Pecten' coronatiformis* Krumbeck, a species described from the late Triassic of Timor (Krumbeck, 1924). On the basis of material recently collected by P. Wignall from the unit (Hallam, 1981) marking the Triassic–Jurassic boundary in the Kendelbachgraben (Austria), it appears that *P. schafhäutli* survived at least until the very end of the Triassic. Further detailed collecting is required at higher horizons before it can be said that this marks the last occurrence of the species.

Comparison with Jurassic scallops

A fair assumption is that six out of seven Cassian scallop species were extinct by the early Norian. However, there remains the possibility that this represents no more than the total which might have been expected to result from 'background' extinctions during the Carnian. In order to gain an estimate of 'background' rate for comparison, stage-by-stage extinction rates for Jurassic scallops have been calculated (Johnson, 1984; 1985), using three different measurements, and these have been plotted in conjunction with data from Triassic species (Figure 9.3). For this analysis it is assumed that the stratigraphic disappearance of a species coincides with its extinction. It is evident that Carnian extinctions were at a rate far higher than developed during most of the Jurassic—which may be described as the 'background' rate—and only matched at the end of the Jurassic, which has been recognized previously as a time of mass extinction for bivalves (Hallam, 1986). This term also seems appropriate to describe events amongst scallops in the Carnian.

Crinoid extinctions

Figure 9.4 shows ranges of crinoid lineages represented in the Carnian to Pliensbachian interval in Europe (data from Simms, 1988a; 1989; in press), together with extinction rates calculated as in Figure 9.3. For reasons of sparse preservation Comatulidina and Millericrinida have been excluded (although see below). A few other forms are also excluded (see Appendix 9.2). The range chart is organized in a slightly different way from Figure 9.3 in order to emphasize patterns of extinction within certain higher taxa. Species constituting the lineages (synonyms or those related by phyletic evolution) are listed in Appendix 9.2.

It is immediately obvious that many lineages disappeared in the Carnian, resulting in values for apparent extinction rate which are much higher than

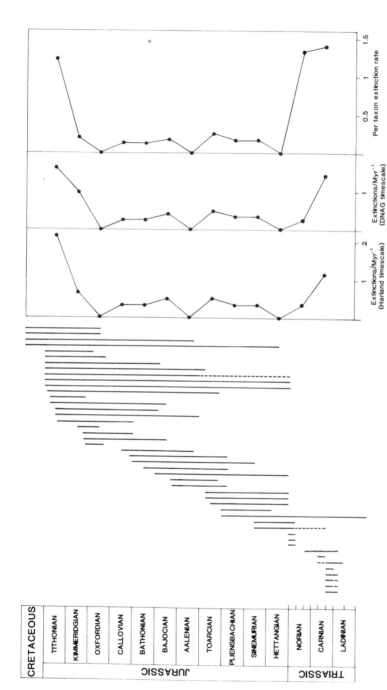

Figure 9.3. Ranges of European late Triassic and Jurassic scallop species (dashes indicate uncertainty). Data from Johnson (1985) and the present study; first 14 species from left in order of letters (A–N) in Figure 9.2. Corresponding extinction rates calculated assuming presence of species in dashed intervals. Timescales from Harland *et al.* (1982) and DNAG (Palmer, 1983). Per-taxon extinction rates calculated according to the method described in Benton (1986). Subdivisions of Triassic stages are substages (names in Figure 9.2).

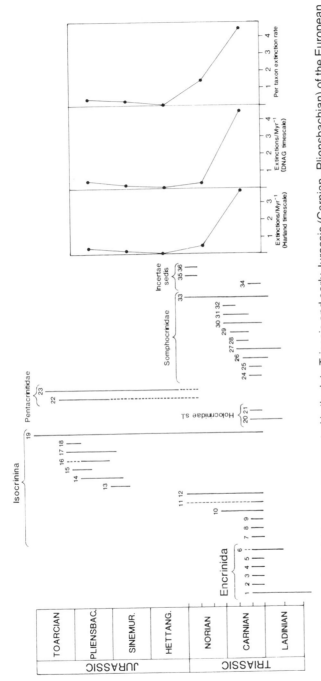

Figure 9.4. Ranges of crinoid lineages represented in the late Triassic and early Jurassic (Carnian–Pliensbachian) of the European area (dashes indicate uncertainty). See Appendix 9.2 for species comprising lineages, and for taxa excluded. Extinction rates calculated as in Figure 9.3 (exclusion of the Somphocrinidae, whose species-level systematics are somewhat uncertain, only slightly reduces values for extinction rate in the Carnian). Subdivisions of Triassic stages are substages (names in Figure 9.2). Plotting of ranges in Triassic follows same approach as in Figure 9.2, with the exception of lineage 6 (a species from the Cassian Formation fairly certainly extending into the Julian; see Appendix 9.2; and Simms, in press).

for any subsequent stage. Of the 22 lineages known in the Cordevolian, 15 (lineages 1–12, 19, 33, 34) are represented in the Cassian Formation, and of these only three definitely survived into the Sevatian, when similar facies were developed. We can be sure, therefore, that at least the remaining 12 lineages genuinely became extinct before the Sevatian. Unlike the case of Cassian scallops we cannot be sure that these extinctions were not spread out over the whole Cordevolian–Sevatian interval. However, even if we take the longest estimate (15 million years; based on the Palmer, 1983 (DNAG) time-scale) for the period of time between the Cordevolian and Sevatian substages and calculate a corresponding extinction rate per million years, we obtain a value of 0.8 *for these lineages alone*. This is double the maximum value for Jurassic stages, which provide an indication of 'background' rates. We are thus fully justified in concluding that the Carnian peaks in extinction rates displayed in Figure 9.4 are a true indication of major extinctions before the end of the Triassic, although we have not yet confirmed that they actually took place in the Carnian.

Most extinctions of lineages occurring in the Cassian Formation were apparently as early as the time of deposition of the Raibl Formation (Julian), and include the disappearance of all the remaining representatives of the Encrinida. This order of relatively large, stenotopic forms could have been expected to be severely affected by environmental changes in the Julian. That two Cassian lineages (10, 19) of the three persisting beyond the Julian have a record in the latter substage is some evidence that environmental changes did actually cause the extinction (rather than emigration) of the encrinids at this time. An explanation for encrinid, and other, extinctions in terms of Julian environmental changes is supported by evidence from the Somphocrinidae. This family preferentially colonized 'Hallstatt Limestone' facies, which persisted locally throughout the Carnian, and somphocrinids continued in undiminished abundance through this interval. Although somphocrinids apparently suffered some extinctions in the Carnian (Figure 9.4), too much should not be made of this evidence, as the species-level taxonomy of the family is still rudimentary (see Appendix 9.2; and Simms, in press).

Somphocrinids are known, albeit more rarely, until the Sevatian. Their disappearance at this time may relate to a temporary loss of 'Hallstatt Limestone' facies in Europe. That they were unquestionably extinct by the Sinemurian is shown by their absence from 'Hallstatt Limestone' facies of this age (Adnet Limestone). The timing of extinction of the other four lineages unknown after the Sevatian is less easy to establish. Three (12, 35, 36) occur in the coral-rich Zlambach Beds, thus their absence in the Lower Jurassic of Europe might reflect emigration upon the development of unfavourable facies. Nevertheless, one of the lineages which survived from the Zlambach Beds into the Jurassic (19) is known in the very earliest Hettangian (Simms, in press), thus arguing for extinction rather than emigration of the other lineages. If one accepts what seems the most parsimonious solution, that those Sevatian lineages unknown in the Jurassic became extinct at the Triassic–Jurassic boundary, it should also be appreciated that the level of extinction was still lower than that which can be inferred in the 'Cordevolian–Julian' event. To those three lineages identified in Figure 9.4 which crossed the Triassic–Jurassic boundary, a further three can be added. Millericrinids and the genus *Paracomatula* are known in the Jurassic (albeit rather sparsely) and have been recorded in the Sevatian (Simms, in press). *Eocomatula*, also known in the Jurassic, is the plesiomorphic sister taxon of *Paracomatula* (see Simms, 1988b) and thus must also represent a lineage which crossed the Triassic–Jurassic boundary, although there are as yet no records of the genus in the Triassic. Taking this information with that in Figure 9.4, we obtain a value of 54% for extinctions across the Triassic–Jurassic boundary. This compares with 64% (14 of 22 lineages) for Cordevolian–Julian extinctions.

SUMMARY AND CONCLUSIONS

We have shown that there were major pre-Sevatian extinctions (much above 'background' rates) in both European scallops and crinoids. Very probably the majority of these took place in the mid-Carnian, contemporaneous with marked facies changes in Europe. The evidence from these two disparate groups, combined with the evidence of extinction in others (ammonoids, bryozoa, conodonts, reef-building organisms), suggests that a wide variety of marine invertebrates may have suffered heavy extinction in the Carnian. At least from the evidence of scallops and crinoids, extinctions in the Carnian appear to have been on a larger scale than any that took place at the very end of the Triassic. Naturally, detailed studies of other groups—of which brachiopods, gastropods and other (that is, non-pectinacean) bivalves would undoubtedly repay further investigation—are needed to assess whether this was generally the case.

We have shown that a high proportion of extinctions are explicable in terms of directly observable facies changes. Certainly, therefore, there is little reason to invoke the involvement of extraterrestrial agencies, an explanation in any case rendered less plausible by the occurrence of two episodes of extinction in the late Triassic. On the basis of information from carbon isotopes, clay mineralogy and the widespread distribution of sandstones and karstic phenomena, Simms and Ruffell (in press) have inferred a major increase in rainfall in the mid-Carnian. The question therefore arises as to whether this might have been the real cause of extinctions at this time, rather than facies changes (the influx of terrigenous clastics) which might have resulted from increased run-off. At least in the Alps this explanation can probably be ruled out, because the evidence points to tectonic, rather than climatic, control of facies changes (Bosellini *et al.*, 1978). As we have noted, there were nevertheless major extinctions associated with deposition of the Raibl Formation. The evidence generally of increased rainfall is, however, so strong, and the coincidence with extinctions so precise, that it is difficult to believe there was no causal connection. This hypothesis would be substantiated by demonstration of a correspondence in the areas affected by extinction (only Europe, as presently known) and the areas yielding evidence of increased rainfall (currently Europe, the Middle East and eastern North America). Although karstic features are developed at the Triassic–Jurassic boundary in the Alps (Hallam, 1981), these are on a smaller scale than Carnian karsts and provide little indication of climatic change. The regression, leading to reduced habitable area, which they imply, coupled with other marine environmental changes (see above), therefore seems to be the best explanation for marine invertebrate extinctions at this time (see also Hallam, 1981).

ACKNOWLEDGEMENTS

We thank A. Allasinaz, H. Hagdorn, A. Hallam, L. Krystyn and A. Ruffell for helpful comments and discussion; H. Remy (Bonn), G. Schairer (Munich), H. Summesberger (Vienna), M. Urlichs (Stuttgart), J. Wendt (Tübingen), R. Zardini (Cortina) and J. Zelenka (Budapest) kindly supervised our examinations of museum collections. Tom Easter (Goldsmiths' College) assisted with photographic work. The study reported herein was principally carried out during tenure of Fellowships from the Alexander von Humboldt-Stiftung (ALAJ) and NERC (MJS), which are gratefully acknowledged.

REFERENCES

Allasinaz, A., 1962, Il Trias in Lombardia (Studi geologici e palaeontologici). III: Studio paleontologico e biostratigrafico del Retico dei dintorni di Endine (Bergamo), *Rivista Italiana di Paleontologica e Stratigrafia*, **68** (3): 307–76.

Allasinaz, A., 1972, Revisione dei pettinidi triassici, *Rivista Italiana di Paleontologica e Stratigrafia*, **78** (2): 189–428.

Benton, M.J., 1986, More than one event in the late Triassic mass extinction, *Nature*, **321** (6073): 857–61.

Bizzarini, F., Laghi, G., Russo, F. and Urlichs, M., 1986, Preliminary biostratigraphic correlation between Ampezzo basin sections and the Cordevolian stratotype (late Triassic, Italian Dolomites), *Lavori, Società Veneziana di Scienze Naturali*, **11** (2): 151–8.

Bosellini, A., Dal Cin, R. and Gradenigo, A., 1978, Depositi litorali raibliani nella zona di Passo Falzarego (Dolomiti centrali), *Annali dell'Università di Ferrara, sezione 9, Scienze Geologiche e Paleontologiche*, **5** (13): 223–38.

Dubar, G., 1948, La faune domérienne du Jebel Bou-Dahar, *Notes et Mémoires du Service des Mines et de la Carte Géologique du Maroc*, **68**.

Fürsich, F.T. and Wendt, J., 1977, Biostratinomy and palaeoecology of the Cassian Formation (Triassic) of the Southern Alps, *Palaeogeography, Palaeoclimatology, Palaeoecology*, **22** (4): 257–323.

Hallam, A., 1981, The end-Triassic bivalve extinction event, *Palaeogeography, Palaeoclimatology, Palaeoceology*, **35** (1): 1–44.

Hallam, A., 1986, The Pliensbachian and Tithonian extinction events, *Nature*, **319** (6056): 765–8.

Hallam, A. and El Shaarawy, Z., 1982, Salinity reduction of the end-Triassic sea from the Alpine region into northwestern Europe, *Lethaia*, **15** (2): 169–78.

Harland, W.B., Cox, A.V., Llewellyn, P.G., Pickton, C.A.G., Smith, A.G. and Walters, R., 1982, *A geological time scale*, Cambridge University Press, Cambridge.

Hoffman, A., 1985, Patterns of family extinction depend on definition and geological timescale, *Nature*, **315** (6011): 659–62.

Johnson, A.L.A., 1984, The palaeobiology of the bivalve families Pectinidae and Propeamussiidae in the Jurassic of Europe, *Zitteliana*, **11**.

Johnson, A.L.A., 1985, The rate of evolutionary change in European Jurassic scallops, *Special Papers in Palaeontology*, **33**: 91–102.

Kristan-Tollmann, E., 1988, Unexpected communities among the crinoids within the Triassic Tethys and Panthalassa. In R.D. Burke, P.V. Mladenov, P. Lambert and R.L. Parsley (eds), *Echinoderm biology, Proceedings of the 6th International Echinoderm Conference, Victoria, British Columbia, 23–28 August, 1987*, Balkema, Rotterdam: 133–42.

Krumbeck, L., 1924, Die Brachiopoden, Lamellibranchiaten und Gastropoden der Trias von Timor. II: Paläontologischer Teil. In J. Wanner (ed.), *Paläontologie von Timor*, **13** (contribution 22), Schweizerbart, Stuttgart: 143–417.

Laws, R.A., 1982, Late Triassic depositional environments and molluscan associations from west-central Nevada, *Palaeogeography, Palaeoclimatology, Palaeoecology*, **37** (2–4): 131–48.

Moore, R.C. (ed.), 1969, *Treatise on invertebrate paleontology. Part N, volume 1; Mollusca, 6, Bivalvia, N*, Geological Society of America and University of Kansas Press, New York and Lawrence.

Morbey, S.J., 1975, The palynostratigraphy of the Rhaetian stage. Upper Triassic, in the Kendelbachgraben, Austria, *Palaeontographica*, **B 152** (1–3): 1–75.

Mostler, H., Schauring, B. and Urlichs, M., 1978, Zur Mega-, Mikrofauna und Mikroflora der Kössener Schichten (alpine Obertrias) vom Weissloferbach in Tirol unter besonderer Berücksichtigung der in der *suessi* und *marshi* Zone auftretenden Conodonten. In H. Zapfe (ed.), *Beiträge zur Biostratigraphie der Tethys-Trias, Schriftenreihe der Erdwissenschaftlichen Kommission, Österreichische Akademie der Wissenshaften*, **4**: 141–74.

Newton, C.R., Whalen, M.T., Thompson, J.B., Prins, N. and Delalla, D., 1987, Systematics and paleoecology of Norian (late Triassic) bivalves from a tropical island arc: Wallowa terrane, Oregon, *Memoir of the Paleontological Society*, **22**.

Olsen, P.E., Shubin, N.H. and Anders, M.H., 1987, New early Jurassic tetrapod assemblages constrain Triassic–Jurassic tetrapod extinction event, *Science*, **237** (4818): 1025–9.

Palmer, A.R., 1983, The decade of North American geology 1983 geologic time scale, *Geology*, **11** (9): 503–4.

Pearson, D.A.B., 1977, Rhaetian brachiopods of Europe, *Neue Denkschriften des Naturhistorischen Museums Wien*, **1**: 1–84.

Raup, D.M. and Sepkoski, J.J., Jr, 1984, Periodicity of extinctions in the geologic past, *Proceedings of the National Academy of Science U.S.A.*, **81** (3): 801–5.

Raup, D.M., Sepkoski, J.J., Jr, 1986, Periodic extinction of families and genera, *Science*, **231** (4740): 833–6.

Sepkoski, J.J., Jr, 1984, A kinetic model of Phanerozoic taxonomic diversity. III: Post-Paleozoic families and mass extinctions, *Paleobiology*, **10** (2): 246–67.

Signor, P.W., III and Lipps, J.H., 1982, Sampling bias, gradual extinction patterns, and catastrophes in the fossil record. In L.T. Silver and P.H. Schultz (eds), *Geological implications of impacts of large asteroids and comets on the Earth, Special Paper of the Geological Society of America*, **190**: 291–6.

Simms, M.J., 1988a, Patterns of evolution among Lower Jurassic crinoids. *Historical Biology*, **1** (1): 17–44.

Simms, M.J., 1988b, An intact comatulid crinoid from the Toarcian of southern Germany, *Stuttgarter Beiträge zur Naturkunde*, **B 140**.

Simms, M.J., 1989, British Lower Jurassic crinoids, *Palaeontographical Society Monographs*, London.

Simms, M.J., in press, Crinoids across the Triassic-Jurassic boundary *Cahiers Scientifiques de L'Institut Catholique de Lyon*.

Simms, M.J. and Ruffell, A., in press, Synchroneity of climatic change and extinctions in the late Triassic, *Geology*.

Tozer, E.T., 1979, Latest Triassic ammonoid faunas and biochronology, western Canada, *Geological Survey of Canada Paper*, **79-1B**: 127–35.

Urlichs, M., 1977, Zur Altersstellung der Pachycardientuffe und der Unteren Cassianer Schichten in den Dolomiten (Italien), *Mitteilungen der Bayerischen Staatssammlung für Paläontologie und historische Geologie*, **17**: 15–25.

Waller, T.R., 1978, Morphology, morphoclines and a new classification of the Pteriomorphia (Mollusca, Bivalvia), *Philosophical Transactions of the Royal Society*, London, **B 284** (1001): 345–65.

Zankl, H., 1971, Upper Triassic carbonate facies in the Northern Limestone Alps. In G. Müller (ed.), *Sedimentology of parts of central Europe, Guidebook, 8th International Sedimentology Congress, Heidelberg, 1971*, Kramer, Frankfurt/Main: 174–85.

Zapfe, H., 1974, Trias in Österreich. In H. Zapfe (ed.), *The stratigraphy of the Alpine-Mediterranean Triassic*, Schriftenreihe der Erdwissenschaftlichen Kommission, Osterreichische Akademie der Wissenschaften, **2**: 245–51.

Zardini, R., 1981, *Fossili Cassiani (Trias Medio-Superiore). Atlante dei bivalvi della Formazione di S. Cassiano raccolti nella regione dolomitica attorno a Cortina d'Ampezzo*, Edizioni Ghedina, Cortina d'Ampezzo.

APPENDIX 9.1: LATE TRIASSIC SCALLOP SPECIES

A full systematic treatment of European late Triassic scallop species is in preparation by ALAJ. The following diagnostic notes are provided to show the basis for the recognition of the species listed in the text. Indications are given where it is possible that species are not representative of the Pectinacea *sensu* Waller (1978). Generic and subgeneric taxonomy of species A–H and N follows Allasinaz (1972), although some doubt attaches to the affinities of at least species A and N. The supraspecific classification of Triassic scallops is currently under investigation by T.R. Waller and it can be expected that the higher-level systematics of the group will shortly be revised. In addition to the scallops listed in the text, *Entolium* was undoubtedly present in the late Triassic. However, it proved impossible to gain any clear idea of the number and range of species belonging to this unornamented genus from the rather poorly preserved material available. *Entolium* was therefore excluded from analysis. Also excluded was material from the Cassian Formation described by Zardini (1981) as *Radulonectites* sp. The forms concerned undoubtedly constitute an addition to the number of Cassian scallop species. However, there remains a possibility—not yet investigated—of an affinity with *Indopecten*, a form occurring in the Norian of the Near and Far East. Letters (A–N) below correspond to those identifying range bars and illustrations of species in Figure 9.2.

A. *'Leptochondria' subalternans* (d'Orbigny). Left valve (LV) exterior illustrated. LV moderately convex; ornamented with numerous narrow, somewhat nodulose, radial plicae and costae, of varying amplitude, increasing in number by intercalation. Right valve (RV) flat; ornamented with numerous fine radial costae; with deep byssal notch. Allasinaz (1972) referred d'Orbigny's species (and some synonymous species) to *Leptochondria* Bittner, but it is not clear whether these forms possess an external ligament area with the obtusely triangular resilifer characteristic of the genus (see Cox and Hertlein *in* Moore, 1969, pp. N338–9). Waller (1978) referred *Leptochondria* to the Buchiacea on the basis of a diminutive right anterior auricle. *'L.' subalternans* appears to have a right anterior auricle little different in size from typical pectinaceans (or aviculopectinaceans).

B. *Antijanira auristriata* (Münster). LV exterior illustrated. LV moderately convex; ornamented with numerous, generally smooth, radial plicae of varying width, but approximately equal amplitude, increasing in number by intercalation. Sulci usually equal in width to, or narrower than, adjacent plicae. RV almost flat; ornamented with a few low, narrow radial plicae, with deep byssal notch. Auricles of both valves bearing radial costae. According to Allasinaz (1972), this species possesses an 'aviculopectinid-type' ligament, with an elongate external ligament area containing a triangular resilifer (bounded by pseudoteeth). This feature, together with the shell structure described by Allasinaz (including crossed-lamellar material outside the pallial line) appears to be consistent with placement of *A. auristriata* in the Aviculopectinacea *sensu* Waller (1978).

C. *Chlamys (Granulochlamys) nodulifera* (Bittner). LV exterior illustrated. LV low convexity; ornamented with radial plicae bearing regularly spaced hood-shaped lamellae or nodules, themselves arranged in commarginal rows and sometimes linked by commarginal lamellae. RV unknown; probably bearing a deep byssal notch by analogy with *C. (G.) tubulifera*.

D. *Chlamys (Granulochlamys) tubulifera* (Münster). LV exterior illustrated. LV low convexity; ornamented with fine, somewhat divaricate, costae which expand to form cone-shaped lamellae, themselves arranged in commarginal rows and sometimes fused to form a continuous 'terrace'. RV similarly ornamented; almost flat; with deep byssal notch.

E. *Chlamys (Chlamys) interstriata* (Münster). ?RV exterior illustrated. LV moderate to high convexity; ornamented with smooth, rounded, radial plicae, equal in width to sulci. Usually 16–18 plicae, rarely added to during ontogeny. RV of similar convexity and ornamentation; with deep byssal notch.

F. *Chlamys (Praechlamys) ?stenodictya* (Salomon). RV exterior illustrated. LV low to moderate convexity; ornamented with numerous fine radial costae, increasing in number by intercalation and bearing somewhat irregularly spaced nodules and lamellae; interspaces sometimes with commarginal lamellae. RVs not known in Carnian, probably represented in Ladinian by RVs (including the types of Salomon's species) with similar ornament, of low convexity, bearing a deep byssal notch. Questionably conspecific forms from the Julian Raibl Formation (and equivalents) might represent early examples of *C. (C.) valoniensis*, or a further species.

G. *Filopecten filosus* (Hauer). LV interior illustrated. LV low to moderate convexity, RV low convexity. Exterior and interior of valves ornamented with fine divaricate costae and variably developed, low-amplitude radial costae. External ornament may be subdued by presence of a thin additional shell layer (Allasinaz, 1972), possibly also sometimes developed on internal surface. RV with small byssal notch.

H. *Entolioides deeckei* (Parona). ?RV interior illustrated. Both valves low convexity; LV usually slightly more convex than RV and ornamented with numerous low radial costae on the exterior; RV smooth. Both valves with two pairs of broad internal radial costae, not extending to margins, bearing distal tubercles; one pair near the junctions of the auricles and disc, the other somewhat more ventrally sited. Both valves relatively thick. RV with small byssal notch. The rather high resilifer, bordered by pseudoteeth, may indicate affinities with the Aviculopectinacea.

I. *Propeamussium schafhäutli* (Winkler). Composite mould (showing internal and external ornament of LV) illustrated. Both valves low convexity; LV slightly more convex than RV and ornamented with numerous fine radial costae; RV smooth. Both valves with 8–10 internal radial costae, not extending to margins. Depth of byssal notch uncertain, probably small.

J. *Chlamys (Chlamys) mayeri* (Winkler). LV exterior (and internal mould) illustrated. Both valves low convexity; LV slightly more convex than RV. LV ornamented with between 40 and 70 smooth, somewhat flattened, radial plicae, slightly wider than sulci, all present from early in ontogeny. RV apparently similarly ornamented; with deep byssal notch.

K. *Chlamys (Chlamys) coronata* (Schafhäutl). LV exterior illustrated. LV moderate convexity, RV low convexity. Both valves ornamented with rounded radial plicae (usually 23), not added to during ontogeny. Anterior and posterior plicae bearing lamellae. RV with deep byssal notch.

L. *Chlamys (Chlamys) valoniensis* (Defrance). ?LV exterior illustrated. LV moderately convex, RV almost flat. Both valves ornamented with numerous smooth radial plicae (more than 30 at height 20 mm), increasing in number by intercalation, slightly flattened on RV. RV with deep byssal notch. Possible records of *C. (C.) valoniensis* as far back as the Julian may be of *C. (C.) ?stenodictya*, or a further species. See also Johnson (1984).

M. *Chlamys (Chlamys) pollux* (d'Orbigny). LV exterior illustrated. Shape similar to *C. (C.) valoniensis*. Both valves ornamented with radial plicae (usually 24), some markedly higher on LV and bearing spines. Spines also on RV. See also Johnson (1984).

N. *'Chlamys' subcutiformis* (Kittl). LV exterior illustrated. LV low to moderate convexity, RV convexity similar or less. Both valves bearing regularly spaced, rounded commarginal swellings and commonly fine radial costae. Depth of byssal notch apparently small. The rather small LV anterior auricle in some forms of this species suggests buchiacean affinities, thus the generic assignment (after Allasinaz, 1972) must be regarded as provisional.

APPENDIX 9.2: CARNIAN–PLIENSBACHIAN CRINOID LINEAGES

The following list is of lineages represented at least at some juncture in the Carnian to Pliensbachian interval in Europe (including Turkey). Comatulidina and Milleri- crinida have been excluded for reasons of very sparse preservation. Also excluded are a few, generally rare, early Jurassic forms from continental Europe (such as *Amaltheocrinus, Capsicocrinus, Cotylederma, Gutticrinus, Quenstedticrinus, Shroshaecrinus*) which have not been studied in detail, and whose lineage ranges are uncertain (see Simms, 1989). The species constituting each lineage (where more than one) are either synonymous forms or species reckoned to be related by phyletic evolution. Further information, including authorship of species, is avail- able from the sources indicated below and, more generally, from Simms (1989). The numbers of lineages correspond to those in Figure 9.4.

1. *Traumatocrinus caudex* (Simms, in press, sp. 4).
2. *Chelocrinus cassianus* (Simms, in press, sp. 1).
3. *Traumatocrinus varians* (Simms, in press, sp. 2).
4. *Traumatocrinus granulosus* (Simms, in press, sp. 3).
5. *Encrinus* sp. nov. (Simms, in press, sp. 6).
6. *Ainigmacrinus calyconodalis* (Simms, in press, sp. 5).
7. *'Isocrinus' subcrenatus* (Simms, in press, sp. 11).
8. *'Isocrinus' fuchsi* (Simms, in press, sp. 12).
9. *Laevigatocrinus laevigatus* (Simms, in press, sp. 13).
10. *'Isocrinus' propinquus, 'I.' hercuniae* (Simms, in press, spp. 9, 10).
11. *Laevigatocrinus ?venustus, Singularocrinus singularis, 'Isocrinus' magnus* (Simms, in press, spp., 14, 17, 19). The latter two species, probable con- tinuants of *Laevigatocrinus*, are arbitrarily linked to *L. ?venustus* rather than *L. laevigatus*.
12. *Tulipacrinus tulipa* (Simms, in press, sp. 15).
13. *Hispidocrinus scalaris* (Simms, in press, sp. 23).
14. *Balanocrinus quiaiosensis* (Simms, in press, sp. 24); *B. subteroides* (Simms 1988a).
15. *Balanocrinus gracilis, B.* sp. nov. 1 (Simms, 1988a).
16. *Isocrinus basaltiformis, 'I.' nudus* (Simms, 1988a; 1989).
17. *Hispidocrinus schlumbergeri* (Simms, 1988a).
18. *Balanocrinus* sp. nov. 2 (Simms, 1988a).
19. *'Isocrinus' tyrolensis, 'I.' jaworskii, I. bavaricus, I. psilonoti, I. tuberculatus* (Simms, in press, spp. 8, 18, 20–22); *I. robustus, I. rollieri* (Simms 1988a; 1989).
20. *Holocrinus tenuispinosus* (Simms, in press, sp. 31).
21. *'Holocrinus' quinqueradiatus* (Simms, in press, sp. 32).
22. *Seirocrinus* sp. nov. (Simms, in press, sp. 36); *S. subangularis* (Simms, 1988a).

23. *Pentacrinites doreckae, P. fossilis* (Simms, in press, spp. 33, 34); *P. dichotomus* (Simms, 1988a).
24. *Ossicrinus reticulatus* (Simms, in press, sp. 46).
25. *Osteocrinus squamosus* (Simms, in press, sp. 53).
26. *Osteocrinus* aff. *acanthicus* (Simms, in press, sp. 48).
27. *Osteocrinus spinosus, O. rimosus, O. acus, O. saklibelensis* (Simms, in press, spp. 43–45, 48).
28. *Osteocrinus depressus, O. brevis* (Simms, in press, spp. 59, 60).
29. *Poculicrinus glaber, P. globosus* (Simms, in press, spp. 54, 55).
30. *Osteocrinus acanthicus, O. hessi, O. planus, O. sulcatus* (Simms, in press, spp. 47, 49–51).
31. *Osteocrinus longispinosus* (Simms, in press, sp. 52).
32. *Vasculicrinus inflatus, V. fastigatus* (Simms, in press, spp. 55, 57).
33. *Osteocrinus rectus rectus, O. rectus goestlingensis, O. virgatus* (Simms, in press, spp. 40–42; Kristan-Tollmann, 1988).
34. *Axicrinus alexandri* (Simms, in press, sp. 61).
35. *Lotocrinus reticulatus* (Simms, in press, sp. 62).
36. *Lanternocrinus lanterna* (Kristan-Tollmann, 1988).

Chapter 10

TERRESTRIAL ENVIRONMENTAL CHANGES AND EXTINCTION PATTERNS AT THE CRETACEOUS–TERTIARY BOUNDARY, NORTH AMERICA

Garland R. Upchurch, Jr.

INTRODUCTION

The terminal Cretaceous extinctions were one of the five largest extinction episodes in the history of marine invertebrates (Raup and Sepkoski, 1984) and terrestrial vertebrates (Benton, 1985). In the oceans, ecologically diverse groups, such as ammonites, calcareous nannoplankton, planktonic foraminifera, inoceramid and rudistid bivalves, and marine reptiles, either died out or were reduced to a fraction of their former diversity. On land, dinosaurs and pterosaurs are the most widely recognized victims of the terminal Cretaceous extinctions, with other clades of vertebrates such as marsupial mammals showing a major decline in diversity (Clemens, 1986). The nature and cause of the terminal Cretaceous extinctions have been the focus of an intense debate between gradualists, who typically postulate an acceleration of earthly background extinction processes operating over a time-period of thousands to millions of years, and catastrophists, who typically postulate one or more extraterrestrially driven environmental perturbations, each operating over a time-period of months to years. Common to all extinction scenarios is a fundamental alteration of the physical environment at the end of the Cretaceous, which most seriously affected clades adapted to warm climates and shallow marine habitats (see, for example, Kauffman, 1984).

Long-term global cooling and major marine regression are the most commonly invoked agents of extinction in gradualist scenarios (see, for example, Schopf, 1979; Clemens *et al.*, 1981; Hickey, 1981; Officer and Drake, 1983; 1985; Stanley, 1984; 1987), with acid rain forming a component of recent formulations of the 'volcanic winter' hypothesis (Officer *et al.*, 1987). Although a few authors have proposed a thermal maximum during the mid-Cretaceous and a steady decline in global temperatures to the early Tertiary (for example, Stanley, 1984), most restrict major temperature decline to the late Maastrichtian, which resulted from increased albedo due to vulcanism (Officer *et al.*, 1987) or world-wide marine regression (Kauffman, 1984). Major marine regression has been cited as an agent of extinction through a major reduction in habitat area for shallow marine organisms (Schopf, 1979; Kauffman, 1984).

In contrast, short-term global cooling, cessation of photosynthesis, and acid rain are the major agents of extinction in most catastrophic extinction scenarios (see, for example Alvarez *et al.*, 1980; Hsü, 1980; Smit and Hertogen, 1980; Prinn and Fegley, 1987). Most catastrophic extinction scenarios propose that a large asteroid or comet impacted the Earth at the end of the Cretaceous. This impact created a global dust cloud that blocked solar radiation for a period of months and created intense acid rain, which together caused mass-kill of the biota through defoliation of plants, freezing of living tissue, cessation of photosynthesis, collapse of food chains, and possibly cyanide poisoning. Some authors have also proposed extensive burning of terrestrial vegetation at the end of the Cretaceous (Wolbach *et al.*, 1985). Mass-kill, in turn, caused mass extinction. Some catastrophic extinction scenarios postulate long-term environmental alteration as the aftermath of a terminal Cretaceous catastrophe (for example, the 'Strangelove' ocean of Hsü and Mackenzie, 1985), which caused further extinction and inhibited ecosystem recovery. Some recent catastrophic extinction scenarios have developed a gradualistic component ('graded catastrophism'), due to evidence for stepwise mass extinction in the marine realm and clustered impacts (Hut *et al.*, 1987). Conversely, a certain degree of catastrophism is implicit in some scenarios of volcanically driven extinction, including recent formulations of 'volcanic winter' (Officer *et al.*, 1987).

This chapter will evaluate competing hypotheses on the cause of the terminal Cretaceous extinction by summarizing evidence for biologically significant environmental change during the late Cretaceous to Palaeocene. This evidence is based largely on data from terrestrial stratigraphic sequences in North America because: terrestrial organisms would be most sensitive to the atmospheric changes postulated by different extinction models; and terrestrial sequences outside of North America have received little study. Terrestrial plants are the primary focus of this chapter for three reasons. First, plants form the base of nearly all food chains and determine the three-dimensional structure of the terrestrial biotic environment. Second, land plants are sensitive indicators of environment and, thus, allow detailed reconstructions of vegetation and climate. Third, the fossil record of pollen, spores and dispersed plant cuticles shows a stratigraphic density rivaled only by the record of marine plankton, permitting inferences to be made on the rapidity of terminal Cretaceous extinction and environmental change in the terrestrial realm.

RECOGNITION OF GRADUAL AND CATASTROPHIC ENVIRONMENTAL CHANGE

Critical to an understanding of extinction dynamics and patterns of environmental change is the use of valid criteria for determining rates. To date, many studies of terminal Cretaceous extinction and environmental change have used invalid or undiagnostic criteria for inferring rates. One prime example is the use of low extinction over a time-period of millions of years to infer gradual, rather than catastrophic, extinction and environmental change (see, for example, Clemens *et al.*, 1981; Hutchinson and Archibald, 1986). Extinction magnitude simply indicates the severity of an event; a catastrophic event can cause little extinction if it is geographically restricted or its environmental effects permit the survival of most species. A second example is inferring gradual versus catastrophic extinction from the stratigraphic record without properly considering sampling interval and completeness of section. Catastrophic extinction will appear gradual if proper allowances are not made for sampling interval (cf. Alvarez, 1983). Gradual extinction will appear catastrophic if analysed on too coarse a time-scale or in stratigraphic sections containing hiatuses in crucial intervals. Thus, sampling effects need to be considered before inferring pattern of extinction; further, the ecological characteristics of survivors and victims are needed to evaluate stratigraphic evidence on extinction pattern.

A catastrophic extinction event should produce the following ecological signatures.

Evidence for mass-kill at an extinction boundary

A catastrophic event of mass extinction magnitude would cause the death of most living organisms. Mass-kill events are recognized either by taphonomic criteria (for example, upright, rooted tree trunks preserved in thick volcanic ash) or by floristic evidence for ecological succession. In the record of pollen, spores and leaves, mass-kill is recognized by the geographically widespread occurrence of early successional vegetation immediately above an event horizon.

Selection for dormancy in tropical and temperate ecosystems

Lineages with the ability to go dormant would preferentially survive a short-term deterioration in the environment. In vascular plants, both seeds and adult plants can undergo a period of dormancy. In most adult plants, dormancy is indicated by deciduous foliage and/or sharp growth rings in the wood. In algae, dormancy is typically achieved by the formation of resting spores. Selection for dormancy mechanisms would produce a poleward latitudinal gradient in extinction, with the highest extinction at low latitudes (Hickey, 1981). This is because organisms of progressively higher latitudes must cope with progressively greater seasonality of light and temperature. Extinction should be highest for evergreen vegetation with poorly developed dormancy mechanisms and lowest for deciduous vegetation. Clades with well-developed dormancy mechanisms should show less extinction than clades with poorly developed dormancy mechanisms.

Selection for freshwater taxa

Adaptation to life in the freshwater aquatic environment would improve the odds of surviving environmental perturbations. Aquatic environments are buffered against extremes in temperature. Further, primary productivity in freshwater aquatic ecosystems is based on algae that have well-developed dormancy mechanisms and short generation times, which ensures rapid recovery of food chains following short-term environmental perturbation. Terrestrial environments show greater potential for long-term disruption of food chains by environmental perturbations due to less thermal buffering, poorer development of dormancy mechanisms in many vegetational types and long generation time of woody plants relative to microscopic algae. In terrestrial environments, early successional vegetation can last from years to decades and typically has different physiognomy than late successional vegetation. Thus, microenvironment and three-dimensional habitat structure can be altered for a significant period of time in the terrestrial realm.

LATE CRETACEOUS TO PALEOCENE CLIMATE AND VEGETATION

Late Cretaceous

Vegetational changes during the Cretaceous resulted from both evolutionary changes in the diversity and dominance of major vascular plant clades and changes in the physical environment (Upchurch and Wolfe, 1987a; in press). In lowland regions, the most dramatic evolutionary change in the Cretaceous vegetation occurred during the Barremian to Cenomanian, coincident with the early diversification and rise to dominance of flowering plants (Crane, 1987; Upchurch and Wolfe, 1987a; in press). During the Barremian to Aptian, flowering plants were confined to stream-margin habitats at lower latitudes and were subordinate in relative abundance and diversity to ferns, conifers and cycadophytes. By the Middle Albian, flowering plants migrated to higher-middle latitudes and showed tendencies for local dominance, and their diversity was comparable to that of ferns and other groups of vascular plants. By the latest Albian to Middle Cenomanian, flowering plants were present in environments ranging from tropical to polar and formed a conspicuous, locally abundant element in megafloral assemblages from stream-margin and overbank facies. Nearly all major physiognomic types of foliage were present by the middle Cenomanian (Upchurch and Wolfe, 1987a), and latitudinal gradients indicate that many of the relations between climate and the physiognomy of flowering plants had evolved by this time as well (Wolfe and Upchurch, 1987b; Upchurch and Wolfe, in press).

The most detailed palaeobotanical reconstructions of late Cretaceous climates are based on megafloras from near-coastal regions of North America (megafloras from the North American continental interior are unknown). These floras indicate warm, equable conditions and significant poleward displacement of major isotherms, a result in general agreement with data from oxygen isotopes (see, for example, Savin, 1977). Based on data for North America, the boundary between megathermal and meso-thermal vegetation (a mean annual temperature of 20°C) occurred at 40–50°N palaeolatitude (pl), and the boundary between mesothermal and microthermal vegetation (a mean annual temperature of 13°C) occurred at 65–75°N pl (Wolfe and Upchurch, 1987b). This contrasts with the modern situation for near-coastal regions, where the megathermal–mesothermal boundary typically occurs south of 30°N and the mesothermal–micro-thermal boundary typically occurs south of 40°N (Wolfe, 1979). In lowland regions near coasts, the average mid-latitudinal temperature gradient was 0.3°C per 1° pl, a figure significantly lower than that of the modern-day (0.5°C per degree latitude). Most late Cretaceous megafloral assemblages from below 60°N pl represent broadleaved evergreen vegetation (Wolfe and Upchurch, 1987b), whose distribution is limited by a cold-month mean of 1°C and/or intense cold waves (Wolfe, 1979). Cold-sensitive life-forms such as rosette trees and shrubs (such as palms) show a pattern of distribution comparable to that of broadleaved evergreen vegetation (Upchurch and Wolfe, in press). Arctic megafloral assemblages represent broadleaved deciduous forest (Spicer, 1987; Wolfe and Upchurch, 1987b), whose distribution was controlled by winter darkness and possibly freezing temperatures.

Palaeobotanical temperature curves for the late Cretaceous contradict a postulated gradual decline in global temperature throughout the late Cretaceous (Savin, 1977, p. 331; Stanley, 1987, p. 162). For middle palaeolatitudes, data from eastern North America and the southern Western Interior indicate warm temperatures throughout the late Cretaceous, with a temperature increase from the Cenomanian to Santonian, a slight decline into the Campanian, a further decline into the early Maastrichtian, and a rebound to Santonian levels by the late Maastrichtian (Wolfe and Upchurch, 1987b). Within the late Maastrichtian, leaf megafloras from the southern Rocky Mountains show no evidence for a decrease in temperature approaching the Cretaceous–Tertiary boundary (cf. Wolfe and Upchurch, 1986, 1987a; see Wolfe and Upchurch, 1987b, p. 59, for an evaluation of apparently conflicting data from Asian megafloras). The record of dispersed leaf cuticles from the southern Rocky Mountains indicates persistence of a typical late Maastrichtian leaf flora up to the Cretaceous–Tertiary boundary clay (Wolfe and Upchurch, 1987a), implying no major cooling during the very latest Cretaceous. Thus, currently available palaeobotanical data for middle palaeolatitudes provide no clear evidence for a gradual pattern of temperature decline during the latest Cretaceous, such as that postulated by 'volcanic winter' (cf. Officer and Drake, 1983; Officer *et al.*, 1987) and other gradualist scenarios (for example, Stanley, 1984).

For high northern palaeolatitudes, data from the North Slope of Alaska show an increase in temperature from the Cenomanian to Coniacian (no Santonian assemblages are known) and a major decline during the Campanian–Maastrichtian (Parrish and Spicer, 1988a). The North slope data might indicate major Arctic cooling during the Campanian–Maastrichtian, in contrast to the situation for middle northern latitudes. However, the inferred temperature decline in the Arctic also might be explained by northward movement of Alaska (Smith *et al.*, 1981), combined with an early Maastrichtian age for the North Slope megafloras (Wolfe and Upchurch, 1987b). In either case, palaeobotanical data from middle and high palaeolatitudes of North America and from the Soviet Union (Krassilov, 1975; Vakhrameev, 1978) provide strong evidence for a temperature maximum near the middle part of the late Cretaceous, a conclusion reinforced by various marine temperature curves based on oxygen isotopes (summarized in Parrish and Spicer, 1988a, Figure 1). In addition, some oxygen isotope curves corroborate palaeobotanical evidence for warming during the Maastrichtian (see, for example, Yasamanov, 1981; Boersma, 1984; Barera *et al.*, 1987).

Palaeobotanical estimates of moisture indicate stratigraphic trends in precipitation during the late Cretaceous, but generally conditions were subhumid at lower-middle latitudes and wettest in the Arctic (Wolfe and Upchurch, 1987b). During the Cenomanian, leaf megafloras from the Atlantic and Gulf Coastal Plains show small leaf size and other features indicative of low moisture. More northerly assemblages show progressively larger leaf size, indicating a progressively higher ratio of precipitation to evaporation (cf. Wolfe and Upchurch, 1987b; Upchurch and Wolfe, in press). The weak development or absence of growth rings in woods from the Western Interior and the absence of extensive latewood for growth rings in woods from the North Slope indicate the absence of drought during the growing season (cf. Parrish and Spicer, 1988b; Upchurch and Wolfe, in press). Cenomanian leaf assemblages from low-middle palaeolatitudes in North America and Europe show wide variation in average leaf size, interpreted as evidence for strong fluctuations in average rainfall during the Cenomanian (Kvacek, 1983; Upchurch and Wolfe, 1987a; Wolfe and Upchurch, 1987b).

During the Campanian and Maastrichtian, leaf megafloras indicate generally subhumid conditions south of the late Cretaceous Arctic Circle, with a sharp increase in precipitation occurring at 46–48°N pl (Wolfe and Upchurch, 1987b). Vegetation south of 46–48°N pl shows leaf size significantly smaller than that of extant megathermal rainforests, indicating subhumid open-canopy vegetation (Wolfe and Upchurch, 1987b). Growth rings in woods are either absent or have little latewood, and angiosperm woods show high vulnerability to freezing or drought, which together indicate little seasonality to precipitation. North of 46–48°N pl, average leaf size is larger and similar to that of extant mesothermal broadleaved evergreen forests of mesic regions (Wolfe and Upchurch, 1987b); however, the low diversity of vines and species with drip tips indicates lower precipitation. Woods from north of 46–48°N pl invariably have growth rings, sometimes with extensive latewood, which indicates greater seasonality to precipitation than in regions to the south (Upchurch and Wolfe, in press). Late Maastrichtian palaeosols from north of 48°N pl (Hell Creek Formation of Montana) show features consistent with this interpretation (Fastovsky and McSweeney, 1987).

During the late Cretaceous, North America supported three areally extensive vegetational types (Fig. 10.1). Eastern North America and the southern Western Interior supported subhumid, megathermal, broadleaved evergreen woodland (Wolfe and Upchurch, 1987b). Conifers are thought to have formed an emergent stratum and angiosperms are thought to have formed a lower, open canopy beneath the conifer stratum. Plants with probable vine habit were of low diversity and never comprise more than 2–3% of the species in a megafossil assemblage. Close modern analogues include the vegetation of nutrient-poor sandy soils of the wet tropics, where plants are widely spaced and stature is limited by low nutrients and rapidly draining water (Upchurch and Wolfe, 1987a; Wolfe and Upchurch, 1987b). Angiosperms form 80% or more of the species in leaf megafossil assemblages from this vegetation, beginning in the Cenomanian (Upchurch and Wolfe, in press). Early Cretaceous holdover taxa such as ginkgophytes and cycadophytes are virtually absent from this region (Upchurch and Wolfe, in press).

The northern Western Interior and Pacific Coast supported subhumid, broad-leaved evergreen forest that existed under wetter, but more seasonal, moisture conditions than the vegetation of eastern North America and the southern Western Interior (Wolfe and Upchurch, 1987b; Upchurch and Wolfe, in press). Vines show higher relative diversity here than to the south, and some species have large leaves with drip tips, indicating the occurrence of closed-canopy conditions on at least some soil types (Wolfe and Upchurch, 1987b). Probable deciduous plants show a higher diversity in this forest than in vegetation to the south. Most of this forest grew under mesothermal conditions, but during warm intervals (late Maastrichtian) the southern margin of this forest existed under cool megathermal conditions. Angiosperm leaves generally comprise only two-thirds of the species in Cenomanian to Campanian megafloras from this vegetation, but almost always comprise over 80% of the species in Maas-

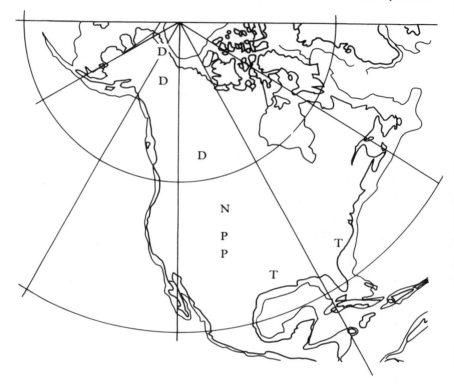

Figure 10.1. Generalized map of the late Cretaceous vegetation (from Upchurch and Wolfe, 1987a). T = tropical evergreen woodland; P = paratropical evergreen woodland and forest; N = notophyllous mesothermal evergreen forest and woodland; D = polar broad-leaved deciduous forest. Both tropical and paratropical vegetation are megathermal (mean annual temperature higher than 20°C).

trichtian megafloras (Upchurch and Wolfe, in press). Early Cretaceous holdover taxa such as ginkgophytes and cycadophytes show their highest relative diversity in this vegetation type and persist through the late Cretaceous, in contrast to the situation for other regions (Upchurch and Wolfe, in press). Stanley's (1987, p. 162) suggestion that this vegetational type became progressively more deciduous during the Maastrichtian seems unlikely, given evidence for a terrestrial temperature increase between the early and late Maastrichtian (Wolfe and Upchurch, 1987b).

The Arctic supported polar broad-leaved deciduous forest (Wolfe and Upchurch, 1987b). All woody angiosperms from this forest showed thin leaves and taphonomic evidence for deciduousness (Spicer, 1987), and the dominant conifers also possessed deciduous foliage, in contrast to the situation for most extant high-latitude conifers. Large leaf size, narrow bands of latewood in growth rings and abundant coals indicate that this forest grew under conditions of abundant moisture (Wolfe and Upchurch,

1987b; Parrish and Spicer, 1988b). The southern part of this forest probably existed under cool mesothermal conditions, while the northern part existed under microthermal temperatures (Wolfe and Upchurch, 1987b; Parrish and Spicer, 1988a). Ginkgophytes and cycadophytes were present in this vegetational type during the Cenomanian, but became extinct during subsequent stages of the late Cretaceous (Spicer, 1987), at least on the northern margin.

Vegetational and climatic change at the Cretaceous–Tertiary boundary

Vegetational change at the Cretaceous–Tertiary boundary resulted from widespread disruption of Cretaceous ecosystems and long-term ecosystem change that markedly altered the biotic environment (Upchurch and Wolfe, 1987a). Palynofloras and leaf floras from non-polar regions of North America clearly indicate a mass-kill event that was followed by a succession-like pattern of recovery. After initial recovery, the early Paleocene vegetation differed from the latest Cretaceous vegetation in two fundamental respects. First, multistratal structure was much more strongly developed than in the Cretaceous, especially south of 48°N pl. Second, latitudinal gradients in vegetation were markedly altered, such that broad-leaved deciduous forest directly abutted on paratropical rainforest.

The most widespread evidence for ecological disruption comes from marked changes in the relative abundance of key pollen and spore types immediately above the Cretaceous–Tertiary boundary clay in North America. In continuous stratigraphic sections ranging from New Mexico to southeastern Saskatchewan, the latest Cretaceous palynoflora is dominated by angiosperm pollen (typically 70–80% of the pollen and spores), and Cretaceous indicator species typically persist up to the K–T boundary clay. This clay is distinct from all other lithological units in having elevated levels of iridium, shock-metamorphosed minerals, and other indicators of a bolide impact (see Orth, Chapter 3, this volume). Immediately above the boundary clay, the relative abundance of angiosperm pollen sharply declines, and the relative abundance of fern spores rises from background levels of 10–40% to 70–100% (see, for example, Orth *et al.*, 1981; Tschudy and Tschudy, 1986; Nichols *et al.*, 1986; Fleming and Nichols, 1988). This 'fern spike' occurs in different lithologies, is only known from immediately above the Cretaceous–Tertiary boundary, and is dominated at each locality by only one of a few species (Fleming and Nichols, 1988). A high relative abundance of fern spores has been noted immediately above the K–T boundary in marine rocks from Japan (Saito *et al.*, 1986); however, whether this truly represents a 'fern spike' is uncertain due to high background levels of fern spores in the uppermost Maastrichtian part of the section. Within the basalmost 15 cm of Palaeocene, the relative abundance of angiosperm pollen rises sharply and the relative abundance of fern spores declines sharply to latest Cretaceous levels. The microstratigraphic floristic sequence of ferns–flowering plants is analogous to the early stages of modern-day primary succession following major volcanic devastation in wet tropical regions, including Krakatoa (Tschudy *et al.*, 1984; Tschudy and Tschudy, 1986) and El Chichón (Spicer *et al.*, 1985). An alternative explanation for the dominance of ferns in the basalmost centimetres of Paleocene is ecological change following acidification of the environment and a rise to dominance of mosses (paludification—Klinger, 1988), which was probably initiated by acid rain following an impact event (cf. Prinn and Fegley, 1987). However, no evidence currently exists for a major increase in moss abundance immediately above the K–T boundary clay, which, together with previous lines of evidence, implies that mass-kill was the major factor involved in the formation of the fern spike.

More northerly K–T boundary sections from Saskatchewan and Alberta also record a major and rapid change in palynomorph abundance reflective of vegetational change (Lerbeckmo *et al.*, 1987), but the extent to which this change resulted from mass-kill, rather than more gradual processes, is unclear. In these sections, the palynoflora is diverse and Cretaceous indicator species have their highest stratigraphic occurrence immediately below an iridium abundance anomaly. Immediately above this iridium anomaly, the palynoflora is depauperate and characterized by a few species of angiosperms, which can share dominance with ferns. Within 10–20 cm of section the palynoflora is dominated by ferns and conifers (mostly Taxodiaceae/Cupressaceae). This vegetational change is intimately associated with the initiation of areally widespread coal deposition, which implies that a rise in water tables and anoxic acidic conditions played a major role in vegetational change. In terms of a catastrophist model, the high relative abundance of flowering plants immediately following a bolide impact has two possible explanations. One is that northern vegetation showed less disruption at the end of the Cretaceous than southern vegetation due to the better development of dormancy mechanisms (polar broad-leaved deciduous forest is documented from as far south as central Alberta during the late Cretaceous). The other is that major ecological disruption occurred at the end of the Cretaceous, but flowering plants were common in the early stages of primary succession, much as they are today following ecological devastation in temperate regions such as Mount St. Helen (see, for example, Franklin *et al.*, 1985).

Detailed stratigraphic studies of plant megafossils and dispersed plant cuticles from the Raton Basin of New Mexico and Colorado corroborate palynological evidence for widespread ecological disruption at the end of the Cretaceous (Wolfe and Upchurch, 1987a). The latest Cretaceous megafossil and cuticle asemblages from this region show evidence for broad-leaved evergreen, open-canopy woodland. Immediately above the K–T boundary clay, in the palynological 'fern spike', megafloras are dominated by ferns, and dispersed cuticle assemblages show evidence for dominance by herbaceous plants (Wolfe and Upchurch, 1987a). Immediately above the palynological fern spike, in the basal 1–2 m of Paleocene, all megafloral assemblages and nearly all dispersed cuticle assemblages are dominated by one to three species. Some megafloral assemblages from carbonaceous shales are dominated by ferns, and most dispersed cuticle assemblages from coals are dominated by herb-type cuticle; however, palynofloras associated with these leaf assemblages are dominated by angiosperm pollen. This indicates that the palynological fern spike resulted from regional, rather than local, dominance of ferns. Other megafloral assemblages from the basal 1–2 m of Paleocene are dominated by one to three species of flowering plants, as are some dispersed cuticle assemblages from carbonaceous shales. Foliar physiognomy of the megafloral assemblages indicates early successional vegetation; however, the thickness of rock and the presence of several rooted horizons within one section indicates that this 'early successional' phase lasted longer than modern-day succession. Stratigraphically higher assemblages show physiognomic evidence for vegetation with later successional characteristics, which is corroborated by greater species diversity for individual assemblages. However, the species diversity of these early Paleocene leaf assemblages is generally lower than that for latest Cretaceous assemblages, and the rate of recovery in species diversity for megafossil assemblages was clearly lower than that for modern-day tropical succession (Wolfe and Upchurch, 1987a).

Early Paleocene climate and vegetation

Palaeobotanical estimates of mean annual temperature for the early Paleocene show a different latitudinal pattern than those for the late Cretaceous (Wolfe and Upchurch, 1986, 1987b; Upchurch and Wolfe, 1987a), which partly explains the disagreement over the pattern of temperature change at the end of the Cretaceous. In the Raton Basin and the Gulf Coastal Plain/Mississippi Embayment region, foliar physiognomy indicates no major temperature decline between the Maastrichtian and early Paleocene (Wolfe and Upchurch, 1986; 1987a). In the northern Rocky Mountains and Great Plains, however, foliar physiognomy would appear to indicate a major temperature decline. Here, mean annual temperatures are inferred to have declined as much as 10°C from the late Maastrichtian to early Paleocene, and mean cold-month temperatures are inferred to have been well below freezing during the early Paleocene (Hickey, 1980). This result is anomalous relative to the late Cretaceous, because inferred mean annual temperature shows a distinctly non-linear latitudinal decline along a major latitudinal gradient during the early Paleocene (Wolfe and Upchurch, 1986, Figure 3). If foliar physiognomic interpretations of vegetation and climate are taken at face value, megathermal climates with little or no freezing were present in the southern Western Interior, microthermal climates with much winter freezing were present in the northern

Western Interior, and mesothermal climates were restricted to the Denver Basin.

However, estimates of average temperatures for the early Paleocene of the northern Rocky Mountains and Great Plains cannot be taken at face value. First, the inferred latitudinal temperature gradient from 46° to 56°N pl for the early Paleocene is steeper than that for the North American continental interior today (about 0.8–0.9°C per degree latitude) (Wolfe and Upchurch 1986). Second, in North America today foliar physiognomy provides estimates of average temperature that are too cold relative to eastern Asia (Wolfe, 1979; cf. Dolph, 1985), probably due to the temperature extremes experienced by eastern North America relative to eastern Asia (Wolfe, 1979). Thus, winter temperature extremes can distort estimates of average temperatures based on foliar physiognomy towards values that are too cold. Third, the inferred cold average temperatures for the northern Western Interior are contradicted by the abundant occurrence of large ectothermic aquatic vertebrates such as crocodilians and champsosaurs in faunas from the Fort Union Formation (see, for example, Bartels, 1980; Hutchinson, 1982) and palms (*Sabalites*) in some of the leaf megafloras (Brown, 1962). Today, large aquatic ectotherms (Hutchinson, 1982) and palms (Box, 1981) cannot tolerate mean cold-month temperatures below freezing and appear to be absent from microthermal climates. Assuming that the temperature tolerances of large aquatic ectotherms and palms have not changed over the past 65 million years, foliar physiognomic estimates of mean annual temperature and mean cold-month temperature are probably too low.

This lack of congruence between temperature estimates has two possible explanations, both of which contradict simple models of global cooling at the end of the Cretaceous (see, for example, Stanley, 1984). One is that mesothermal evergreen taxa were severely affected by a mass-kill event at the Cretaceous–Tertiary boundary (Wolfe, 1987). According to this scenario, mesothermal evergreen lineages suffered high extinction due to the absence of refugia, unlike the situation for megathermal evergreen lineages. Because the megathermal–mesothermal boundary today is a floristic boundary for many evergreen taxa (Wolfe, 1979), evergreen lineages that survived the terminal Cretaceous event in megathermal regions could not have repopulated mesothermal regions, leaving deciduous lineages to dominate the regional vegetation. As a result, vegetation of the northern Western Interior and Great Plains was out of equilibrium with climate for a period of several million years (Wolfe, 1987). Leaf megafloras from the Powder River Basin of Wyoming show a gradual increase in the percentage of evergreen species through the Paleocene (Wolfe and Upchurch, 1987c), which would be expected if megathermal evergreen lineages were gradually developing the ability to tolerate mesothermal temperatures.

A second possible explanation is that average temperatures underwent little decline across the Cretaceous–Tertiary boundary, but that temperature extremes increased, producing infrequent, but biologically significant, freezing during the winter. This hypothesis is based on the modern situation for the southeastern United States, where foliar physiognomy provides estimates of mean annual temperature that are too cold. Today, lowland regions of the southeastern United States have mesothermal mean annual temperatures and cold-month means above 1°C (a combination of temperature parameters that permits the existence of broad-leaved evergreen vegetation in eastern Asia), but vegetation where the broad-leaved component is largely deciduous (see, for example, Braun, 1950, pp. 280–1; Wolfe, 1979). Here, intense cold waves of Arctic air limit the geographic distribution of broad-leaved evergreen forest, rather than a cold-month mean of 1°C as in eastern Asia (Wolfe, 1979). The combination of warm average temperatures and hard freezes during the winter creates the situation where the broad-leaved component of the vegetation is predominantly deciduous (especially the trees), but thermophiles such as large aquatic ectotherms (for example, *Alligator*) and shrubby palms (*Sabal*) are a component of the biota. This analogy is strengthened by the dominance of deciduous Taxodiaceae (bald cypress family) in both modern swamp environments from the southeastern United States (see, for example, Braun, 1950, p. 291–3) and early Paleocene swamps from the northern Rocky Mountains (Hickey, 1980).

Physiognomic analysis of leaf megafloras indicates a major increase in precipitation for North America during the early Paleocene, which was most pronounced in regions south of 48°N pl (Wolfe and Upchurch, 1986; 1987a; 1987b). Leaf size in an individual geographic region was nearly always larger in the early Paleocene than in the late Maastrichtian, and half or more of the evergreen species possessed drip tips. Megathermal regions supported closed-canopy, multistratal rainforest during the early Paleocene, rather than open-canopy woodland/forest. Plots of leaf size for the Western Interior and Gulf Coastal Plain show no detectable latitudinal gradient in moisture (in contrast to the Campanian–Maastrichtian), with leaf size similar to that of modern rainforests. Moisture, therefore, was not a limiting factor to plant growth during the early Paleocene.

North American vegetation during the early Paleocene (Fig. 10.2) showed distinctly different latitudinal organization than that of the late Cretaceous (Wolfe and Upchurch, 1986; Upchurch and Wolfe, 1987a). The Gulf Coastal Plain and southern Western Interior supported megathermal rainforest with strongly multistratal structure, rather than open-canopy woodland as in the Campanian–Maastrichtian. The northern Rocky Mountains supported broad-leaved deciduous forest with physiognomic characteristics indicative of abundant rainfall. This broad-leaved deciduous forest extended as far south as 50°N pl, into regions that were characterized by mesothermal evergreen woodland and forest during the late Cretaceous. Mesothermal evergreen forest was either absent during the early Paleocene or of restricted geographic extent. This radically different organization had major impact on the ecological evolution of

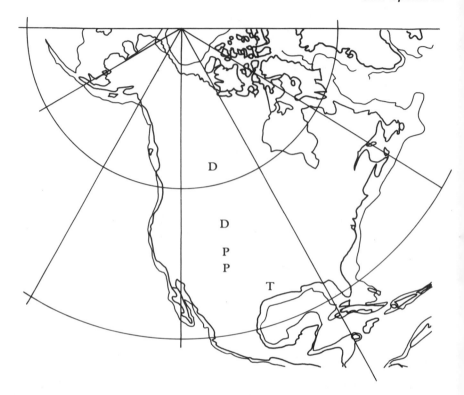

Figure 10.2. Map of the early Paleocene vegetation (from Upchurch and Wolfe, 1987a). T = tropical rainforest, P = paratropical rainforest, D = broad-leaved deciduous forest.

vascular plants (Upchurch and Wolfe, 1987a); its influence on the evolution of terrestrial vertebrates has yet to be assessed.

EXTINCTION PATTERNS AT THE CRETACEOUS–TERTIARY BOUNDARY

Early studies of plant extinction at the Cretaceous–Tertiary boundary (Hickey, 1981) postulated a pattern most compatible with a gradual, rather than catastrophic, mechanism. On a local level, extinction magnitude in the leaf megafossil record across the Cretaceous–Tertiary boundary was compared with that at the Paleocene–Eocene boundary, a time for which no global catastrophe had been proposed. On a global level, literature compilations implied that extinction magnitude in the pollen record was highest at higher-middle palaeolatitudes in the Northern Hemisphere and lower to the south, rather than highest in the tropics and lowest at the poles. However, extinction magnitude for pollen showed much scatter within an individual geographic region, and the majority of data points clustered at high-middle palaeolatitudes in the Northern Hemisphere. Subsequent investigations of terminal Cretaceous plant extinction in North America contradict Hickey's (1981) proposed pattern of plant extinction by showing rates and ecological patterns compatible with catastrophic change.

Detailed stratigraphic studies of terminal Cretaceous plant extinctions clearly indicate rapid extinction of some taxa at the end of the Cretaceous, as predicted by models of bolide impact. Palynological studies of K–T boundary sections ranging from northern New Mexico to central Alberta show persistence of Cretaceous indicator species up to within a few centimetres of the Cretaceous–Tertiary boundary clay (see, for example, Tschudy and Tschudy, 1986, and other previously cited palynological references). Some of the taxa that became extinct at the K–T boundary on at least a local level, such as *Proteacidites* and certain species of *Aquila-pollenites*, have obscure modern affinities, making an assessment of probable temperature tolerances difficult. Others, such as *Trisectoris* (Illiciales), *Gunnera* (Gunneraceae), and *Trichopeltinites* (an epiphyllous fungus), have affinities with extant taxa of mostly subtropical or tropical distribution, implying susceptibility to low temperatures. Many taxa of dispersed leaf cuticles from the Raton Basin of New Mexico and Colorado show evidence for similarly rapid extinction at the Cretaceous–Tertiary boundary (Wolfe and Upchurch, 1987a; Upchurch and Wolfe, 1987b), and most of these taxa belong to groups whose extant members show primarily tropical or subtropical distributions (Laurales, Illiciales). Thus, the modern affinities of taxa that suddenly become extinct at the K–T boundary, where known, are compatible with a low-temperature excursion due to bolide impact.

Latitudinal extinction gradients for terrestrial plants in North America differ between the pollen and leaf records, but provide no evidence for a pronounced extinction peak at higher-middle palaeolatitudes. Palynological studies of K–T boundary sections ranging from northern New Mexico to central Alberta show extinction of 20–30% at the level of species (Fleming, 1985; cf. Nichols *et al.*, 1986; Lerbeckmo *et al.*, 1987). This figure probably approximates extinction at the level of biological genus, because taxonomic analyses of latest Cretaceous pollen are based almost solely on light microscopy, and those features of pollen morphology visible under the light microscope better diagnose extant genera than extant species (see, for example, Leopold and McGinitie, 1972; Wolfe, 1973). Studies of the leaf megafossil record indicate species-level extinction as high as 75% in megathermal vegetation from the southern Rocky Mountains, with a northward decline to 25% in polar broad-leaved deciduous forest from central Alberta (Wolfe and Upchurch, 1986). This pattern of decreasing extinction with increasing latitude is precisely that predicted by catastrophic extinction theories. Much of the higher extinction observed in the leaf record may be due to better taxonomic resolution in the leaf megafossil record, and the tendency for species that became rare during the early Paleocene to be better represented in the palynological record. However, the absence of a latitudinal extinction gradient in the pollen record may be due to the fact that Laurales (the laurels and allies) show their highest diversity in the leaf record of the southern Rocky Mountains (Wolfe and Upchurch, 1986, Table 1; Upchurch and Wolfe, in press), but have pollen that rarely preserves in the geologic record (Muller, 1981).

Clades of both terrestrial and marine plants with well-developed dormancy mechanisms appear to have survived the terminal Cretaceous extinctions better than clades with poorly developed dormancy mechanisms (Kitchell et al., 1986; Wolfe and Upchurch, 1987a), implying that one or more events of short duration played an important role in extinction. In megafloras from the Western Interior, probable evergreen foliage shows higher extinction at the level of species than probable deciduous foliage (Wolfe and Upchurch, 1986; 1987a). This pattern also holds for marine phytoplankton. Diatoms, dinoflagellates and silicoflagellates, groups that characterize phytoplankton assemblages from higher palaeolatitudes and produce resting spores, show less extinction than coccoliths, a group that characterizes phytoplankton assemblages from lower palaeolatitudes and typically does not produce resting spores (cf. Thierstein, 1982, Table 1; Kitchell et al., 1986).

However, broad-scale stratigraphic analyses of land plants indicate that extinction by a catastrophic event at the end of the Cretaceous was not the only agent powering biotic change on land during the K–T boundary interval, just as in the record of marine invertebrates. Some clades of vascular plants became extinct in the Tertiary after declining in abundance at the K–T boundary (Tschudy and Tschudy, 1986; Upchurch and Wolfe, 1987a; 1987b), implying that long-term environmental change or subsequent short-term events played a role in their demise. Other clades of vascular plants with archaic morphologies (for example, Normapolles pollen) showed notable declines in diversity during the Maastrichtian (Tschudy, 1981), which indicates that biotic or climatic change during the latest Cretaceous also played a role in the modernization of the terrestrial flora. Whether or not this extinction correlates with events observed in the marine realm, and whether or not this extinction was caused by comet showers, vulcanism or other means, remains to be determined; it does not, however, appear to be caused by a gradual decline in temperature.

The degree to which terminal Cretaceous vertebrate extinctions can be explained by a catastrophic event is uncertain. This uncertainty results, in part, from: the small number of known dinosaur fossils (for example, Russell, 1984); debate over the extent to which dinosaurs could tolerate low temperatures and months of low food supply (see, for example, Brouwers et al., 1987); whether or not Paleocene occurrences of isolated dinosaur bones and teeth represent reworking (see, for example, Sloan et al. 1986; Fassett et al., 1987); and whether or not the absence of dinosaur bone within the uppermost 2 m of Cretaceous represents actual extinction

Table 10.1. Vertebrate extinction magnitude, freshwater aquatic food chains

TAXON	Number of Cretaceous representatives		Extinction (%)	
	Genera	Families	Genera	Families
Osteichthyans	13	10	38	10
Amphibia	13	10	33	13
Chelonia	16	4	11	0
Eosuchia	1	1	0	0
Crocodilia	4	1	25	0
Total	49	26	26	8

Source: Clemens (1986, Appendix 2). Question marks denote cases where there is uncertainty regarding taxonomy or extinction pattern (in Clemens, 1986).

(see, for example, Archibald and Clemens, 1982), a statistical artefact (Alvarez, 1983) or acid-dissolution of bone (Retallack and Leahy, 1986). Nevertheless, if dinosaurs actually did become extinct by the Paleocene, ecological characteristics of terrestrial vertebrate extinction in North America are consistent with extinction by one or more environmental catastrophes, even assuming that dinosaurs could tolerate several months of cold temperatures and low food supply. Using a time-averaged data set that assumes the absence of unreworked dinosaur bone in the Paleocene (Clemens, 1986, Appendix 2), major differences in extinction magnitude are apparent between clades that were ultimately dependent on the primary productivity of freshwater algae and clades that were ultimately

Table 10.2. Vertebrate extinction magnitude, terrestrial food chains

TAXON	Number of Cretaceous representatives		Extinction (%)	
	Genera	Families	Genera	Families
Eolacertilia	1	1	100	100
Lacertilia	15	7	27(?)	0
Serpentes	2	1	0(?)	0(?)
Ornithischia	14	8	100	100
Saurischia	8	5	100	100
Multituberculata	11	8	36	25
Marsupialis	4	3	75	66
Placentalia	9	4	11	25
Total	64	37	55	51

Source: as Table 10.1.

dependent on the primary productivity of terrestrial plants (Tables 10.1, 10.2). In particular, bony fishes, amphibians, turtles and crocodilians show far less extinction at the level of genus and family than lizards, snakes, dinosaurs and mammals. (Sharks, rays and pterosaurs were excluded from the analysis because of their probable dependence on marine productivity.) The strong difference in extinction magnitude between freshwater aquatic and terrestrial food chains is wholly consistent with extinction by one or more catastrophic events. This is because the time needed to re-establish three-dimensional ecosystem architecture and maximum levels of primary productivity/biomass after a major disruption is much faster in the aquatic realm, especially considering evidence for abundant precipitation following a terminal Cretaceous mass-kill event. However, the ultimate validity of this and other hypotheses of vertebrate extinction will only be determined by additional microstratigraphic, taphonomic and palaeo-environmental analyses of vertebrate remains, coupled with analyses of palaeobotany and geochemistry.

ACKNOWLEDGEMENTS

Research and writing of this chapter were supported by National Science Foundation Grant BSR-86-07298 to G. Upchurch and E. Kauffman and by a post-doctoral fellowship from the National Center for Atmospheric Research. The National Center for Atmospheric Research is supported by the National Science Foundation.

REFERENCES

Alvarez, L.W., 1983, Experimental evidence that an asteroid impact led to the extinction of many species 65 million years ago, *Proceedings of the National Academy of Sciences U.S.A.*, **80** (2): 627–42.

Alvarez, L.W., Alvarez, W., Asaro, F. and Michel, H.V., 1980, Extraterrestrial cause for the Cretaceous–Tertiary extinction, *Science*, **208** (4448): 1095–1108.

Archibald, J.D. and Clemens, W.A., 1982, Late Cretaceous extinctions, *American Scientist*, **70** (4): 377–85.

Barera, E., Huber, B.T., Savin, S.M. and Webb, P.N., 1987, Antarctic marine temperatures: late Campanian through early Paleocene, *Paleoceanography*, **2** (1): 21–47.

Bartels, W.S. 1980, Early Cenozoic reptiles and birds from the Bighorn Basin, Wyoming, *University of Michigan Papers on Paleontology*, **24**: 73–9.

Benton, M.J., 1985, Mass extinction among non-marine tetrapods, *Nature*, **316** (6031): 811–14.

Boersma, A., 1984, Campanian through Paleocene paleotemperature and carbon isotope sequence and the Cretaceous–Tertiary boundary in the Atlantic Ocean. In W.A. Berggren and J.A. Van Couvering, (eds), *Catastrophes in earth history: the new uniformitarianism*, Princeton University Press, Princeton, NJ: 247–78.

Box, E.O., 1981, *Macroclimate and plant forms: an introduction to predictive modeling in phytogeography*, W. Junk, Boston.

Braun, E.L., 1950, *Deciduous forests of eastern North America*, The Blakiston Company, Philadelphia.

Brouwers, E.M., Clemens, W.A., Spicer, R.A., Ager, T.A., Carter, L.D. and Sliter, W.V., 1987, Dinosaurs on the North Slope, Alaska: high latitude latest Cretaceous environments, *Science*, 237 (4822): 1608–10.

Brown, R.W., 1962, Paleocene flora of the Rocky Mountains and Great Plains, *U.S. Geological Survey Professional Paper*, 375.

Clemens, W.A., 1986, Evolution of the terrestrial vertebrate fauna during the Cretaceous–Tertiary transition. In D.K. Elliott (ed.), *Dynamics of extinction*, Wiley and Sons, New York: 63–85.

Clemens, W.A., Archibald, J.D. and Hickey, L.J., 1981, Out with a whimper not a bang, *Paleobiology*, 7 (3): 293–8.

Crane, P.R., 1987, Vegetational consequences of the angiosperm diversification. In E.M. Friis, W.G. Chaloner and P.R. Crane (eds), *The origins of angiosperms and their biological consequences*, Cambridge University Press, Cambridge: 107–44.

Dolph, G.E., 1985, Leaf margin variation in Indiana, *American Journal of Botany*, 72 (6): 891–2.

Fassett, J.E., Lucas, S.G. and O'Neill, F.M., 1987, Dinosaurs, pollen and spores, and the age of the Ojo Alamo Sandstone, San Juan Basin, New Mexico and Colorado. In J.E. Fassett and J.K. Rigby (eds), *The Cretaceous–Tertiary boundary in the San Juan and Raton Basins, New Mexico and Colorado*, Geological Society of America Special Paper, 209: 17–34.

Fastovsky, D.E. and McSweeney, K., 1987, Paleosols spanning the Cretaceous–Paleogene transition, eastern Montana and western North Dakota, *Bulletin of the Geological Society of America*, 99 (1): 66–77.

Fleming, R.F., 1985, Palynological observations of the Cretaceous–Tertiary boundary in the Raton Formation, New Mexico, *Palynology*, 9: 242.

Fleming, R.F. and Nichols, D.J., 1988, The ferns-spore abundance anomaly: a regional bioevent at the Cretaceous–Tertiary boundary, *Third International Conference on Global Bioevents: Abrupt Changes in the Global Biota, Abstracts*: 16.

Franklin, J.F., MacMahon, J.A., Swanson, F.J. and Sedell, J.R., 1985, Ecosystem responses to the eruption of Mount St Helens, *National Geographic Research*, 1: 198–216.

Hickey, L.J., 1980, Paleocene stratigraphy and flora of the Clark's Fork Basin, *University of Michigan Papers on Paleontology*, 24: 33–49.

Hickey, L.J., 1981, Land plant evidence compatible with gradual, not catastrophic, change at the end of the Cretaceous, *Nature*, 292 (5823): 529–31.

Hsü, K.J., 1980, Terrestrial catastrophe caused by cometary impact at the end of the Cretaceous, *Nature*, 285 (5762): 201–3.

Hsü, K.J. and McKenzie, J.A., 1985, A 'Strangelove' ocean in the earliest Tertiary. In E.T. Sundquist and W.S. Broeker (eds), *The carbon cycle and atmospheric CO_2: natural variations Archean to Present*, Geophysical Monograph, 32: 487–92.

Hut, P., Alvarez, W., Elder, W.P., Hansen, T., Kauffman, E.G., Keller, G., Shoemaker, E.M. and Weissman, P.R., 1987, Comet showers as a cause of mass extinctions, *Nature*, 329 (6135): 118–26.

Hutchinson, J.H., 1982, Turtle, crocodilian, and champsosaur diversity changes in the Cenozoic of the north-central region of western United States, *Palaeogeography, Palaeoclimatology, Palaeoecology*, 37 (2–4): 149–64.

Hutchinson, J.H. and Archibald, J.D., 1986, Diversity of turtles across the Cretaceous–Tertiary boundary in northeastern Montana, *Palaeogeography, Palaeoclimatology, Palaeoecology*, 55 (1): 1–22.

Kauffman, E.G., 1984, The fabric of Cretaceous marine extinctions. In W.A. Berggren and J.A. Van Couvering (eds), *Catastrophes and earth history: the new uniformitarianism*, Princeton University Press, Princeton, NJ: 151–246.

Kitchell, J.A., Clark, D.L. and Gombos, A.M., Jr, 1986, Biological selectivity of extinction: a link between background and mass extinction, *Palaios*, 1 (5): 504–11.

Klinger, L.F., 1988, Widespread habitat change through paludification as an interactive mechanism in mass extinction events. In *Global Catastrophes in Earth History: An Interdisciplinary Conference on Impacts, Vulcanism, and Mass Mortality, Snowbird, Utah, October 20–23, 1988, Abstracts*: 92.

Krassilov, V.A., 1975, Climatic changes in Eastern Asia as indicated by fossil floras. II: Late Cretaceous and Danian, *Palaeogeography, Palaeoclimatology, Palaeoecology*, 17 (2): 157–72.

Kvacek, Z., 1983, Cuticular studies in angiosperms of the Bohemian Cenomanian, *Acta Palaeontologica Polonica*, 28 (1–2): 159–70.

Leopold, E.B. and McGinitie, H.D., 1972, Development and affinities of Tertiary floras of the Rocky Mountains. In A. Graham (ed.), *Floristics and paleofloristics of Asia and eastern North America*, Elsevier, Amsterdam: 147–200.

Lerbeckmo, J.F., Sweet, A.R. and St Louis, R.M., 1987, The relationship between the iridium anomaly and palynological floral events at three Cretaceous–Tertiary boundary localities in western Canada, *Bulletin of the Geological Society of America*, 99 (3): 325–30.

Muller, J., 1981, Fossil pollen records of extant angiosperms, *The Botanical Review*, 47 (1): 1–142.

Nichols, D.J., Jarzen, D.M., Orth, C.J. and Oliver, P.Q., 1986, Palynological and iridium anomalies at the Cretaceous–Tertiary boundary, south-central Saskatchewan, *Science*, 231 (4739): 714–17.

Officer, C.B. and Drake, C.L., 1983, The Cretaceous–Tertiary transition, *Science*, 219 (4591): 1383–90.

Officer, C.B. and Drake, C.L., 1985, Terminal cretaceous environmental events, *Science*, 227 (4691): 1161–7.

Officer, C.B., Hallam, A., Drake, C.L. and Devine, J.D., 1987, Late Cretaceous and paroxysmal Cretaceous/Tertiary extinctions, *Nature*, 326 (6109): 143–7.

Orth, C.J., Gilmore, J.S., Knight, J.D., Pillmore, C.L., Tschudy, R.H. and Fassett, J.E., 1981, An iridium abundance anomaly at the palynological Cretaceous–Tertiary boundary in northern New Mexico, *Science*, 214 (4527): 1341–3.

Parrish, J.T. and Spicer, R.A., 1988a, Late Cretaceous terrestrial vegetation: a near-polar temperature curve, *Geology*, 16 (1): 22–5.

Parrish, J.T. and Spicer, R.A., 1988b, Middle Cretaceous wood from the Nanushuk Group, central North Slope, Alaska, *Palaeontology*, 31 (1): 19–34.

Prinn, R.G. and Fegley, B., Jr, 1987, Bolide impacts, acid rain, and biospheric traumas at the Cretaceous–Tertiary boundary, *Earth and Planetary Science Letters*, 83 (1): 1–15.

Raup, D.M. and Sepkoski, J.J., Jr, 1984, Periodicity of extinctions in the geologic past, *Proceedings of the National Academy of Sciences U.S.A.*, **81** (3): 801–5.

Retallack, G. and Leahy, G.D., 1986, Letter to Science, *Science*, **234** (4781): 1170–1.

Russell, D.A., 1984, Terminal Cretaceous extinctions of large reptiles. In W.A. Berggren and J.A. Van Couvering (eds), *Catastrophes and earth history: the new uniformitarianism*, Princeton University Press, Princeton, NJ: 373–84.

Saito, T., Yamanoi, T. and Kaiho, K., 1986, End-Cretaceous devastation of terrestrial flora in the boreal Far East, *Nature*, **323** (6085): 253–5.

Savin, S.M., 1977, The history of the Earth's surface temperature over the past 100 million years, *Annual Review of Earth and Planetary Sciences*, **5**: 319–56.

Schopf, T.J.M., 1979, The role of biogeographic provinces in regulating marine faunal diversity through geologic time. In J. Gray and A.J. Boucot (eds), *Historical biogeography, plate tectonics, and the changing environment*, Oregon State University Press, Corvallis: 449–57.

Sloan, R.E., Rigby, J.K., Jr, Van Valen, L.M. and Gabriel, D., 1986, Gradual dinosaur extinction and simultaneous ungulate radiation in the Hell Creek Formation, *Science*, **232** (4750): 629–33.

Smit, J. and Hertogen, J., 1980, An extraterrestrial event at the Cretaceous–Tertiary boundary, *Nature*, **285** (5762): 198–200.

Smith, A.G., Hurley, A.M. and Briden, J.C., 1981, *Phanerozoic palaeocontinental world maps*, Cambridge University Press, Cambridge.

Spicer, R.A., 1987, The significance of the Cretaceous flora of northern Alaska for the reconstruction of the climate of the Cretaceous, *Geologisches Jahrbuch*, **96**: 265–92.

Spicer, R.A., Burnham, R.J., Grant, P. and Glicken, H., 1985, *Pityrogramma calomelanos*, the primary, post-eruption colonizer of Volcan Chichonal, Chiapas, Mexico, *American Fern Journal*, **75** (1): 1–5.

Stanley, S.M., 1984, Temperature and biotic crises in the marine realm, *Geology*, **12** (4): 205–8.

Stanley, S.M., 1987, *Extinction*, Scientific American Books, New York.

Thierstein, H.R., 1982, Terminal Cretaceous plankton extinctions: a critical assessment. In L.T. Silver and P.H. Schultz (eds), *Geological implications of impacts of large asteroids and comets on the Earth, Special Paper of the Geological Society of America*, **190**: 385–99.

Tschudy, R.H., 1981, Geographic distribution and dispersal of Normapolles genera in North America, *Review of Paleobotany and Palynology*, **35** (4): 283–314.

Tschudy, R.H., Orth, C.J., Gilmore, J.S. and Knight, J.D., 1984, Disruption of the terrestrial plant ecosystem at the Cretaceous–Tertiary boundary, *Science*, **225** (4666): 1030–2.

Tschudy, R.H. and Tschudy, B.D., 1986, Extinction and survival of plant life following the Cretaceous/Tertiary boundary event, Western Interior, North America, *Geology*, **14** (8): 667–70.

Upchurch, G.R., Jr and Wolfe, J.A., 1987a, Mid-Cretaceous to early Tertiary vegetation and climate· evidence from fossil leaves and wood. In E.M. Friis, W.G. Chaloner and P.K. Crane (eds), *The origins of angiosperms and their biological consequences*, Cambridge University Press, Cambridge: 75–105.

Upchurch, G.R., Jr and Wolfe, J.A., 1987b, Plant extinction patterns at the Cretaceous–Tertiary boundary, Raton and Denver basins, *Geological Society of America Abstracts with Programs*, **19** (7): 874.

Upchurch, G.R., Jr and Wolfe, J.A., in press, Cretaceous vegetation of the Western Interior and adjacent regions of North America, *Special Paper of the Geological Association of Canada.*

Vakhrameev, V.A., 1978, Climates of the Northern Hemisphere in the Cretaceous in light of paleobotanical data, *Paleontologicheski Zhurnal,* **1978** (2): 3–17.

Wolbach, W.S., Lewis, R.S. and Anders, E., 1985, Cretaceous extinctions; evidence for wildfires and search for meteoric material, *Science,* **230** (4722): 167–70.

Wolfe, J.A., 1973, Fossil forms of Amentiferae, *Brittonia,* **25** (4): 334–55.

Wolfe, J.A., 1979, Temperature parameters of humid to mesic forests of Eastern Asia and relation to forests of other regions of the Northern Hemisphere and Australasia, *U.S. Geological Survey Professional Paper,* **1106.**

Wolfe, J.A., 1987, Late Cretaceous–Cenozoic history of deciduousness and the terminal Cretaceous event, *Paleobiology,* **13** (2): 215–26.

Wolfe, J.A. and Upchurch, G.R., Jr, 1986, Vegetation, climatic and floral changes at the Cretaceous–Tertiary boundary, *Nature,* **324** (6093): 148–52.

Wolfe, J.A. and Upchurch, G.R., Jr, 1987a, Leaf assemblages across the Cretaceous–Tertiary boundary in the Raton Basin, New Mexico and Colorado, *Proceedings of the National Academy of Sciences U.S.A.,* **84** (15): 5096–100.

Wolfe, J.A. and Upchurch, G.R., Jr, 1987b, North American nonmarine climates and vegetation during the late Cretaceous, *Palaeogeography, Palaeoclimatology, Palaeoecology,* **61** (1–2): 33–77.

Wolfe, J.A. and Upchurch, G.R., Jr, 1987c, Maastrichtian–Paleocene vegetation and climate of the Powder River Basin, Wyoming and Montana, *Geological Society of America Abstracts with Programs,* **19** (7): 896.

Yasamanov, N.A. 1981, Paleothermometry of Jurassic, Cretaceous, and Paleogene periods of some regions of the USSR, *International Geology Review,* **23** (5): 700–5.

Chapter 11

STEPWISE EXTINCTIONS AND CLIMATIC DECLINE DURING THE LATER EOCENE AND OLIGOCENE

Donald R. Prothero

INTRODUCTION

Although the Cretaceous–Tertiary extinction has garnered the most attention, and the Permian–Triassic extinction was probably the most severe, the Eocene–Oligocene extinctions are also of great geologic interest. Not only were they among the most important extinction events in the Phanerozoic, but they have an additional advantage in that they were relatively recent and, therefore, have an excellent geologic record. Numerous outstanding marine sections, especially from deep-sea cores, have recovered the details of this interval in many places in the world ocean. There are also many land sections containing plant and animal fossils that were deposited during this time. Most of these sections have tight stratigraphic control, not only from biostratigraphy, but also from magnetostratigraphy, isotope stratigraphy and seismic stratigraphy. Unlike earlier extinctions, the victims were mostly members of extant higher groups, so it is much easier to infer their palaeoecology and to reconstruct the palaeoclimatic changes that must have occurred.

The Eocene–Oligocene extinctions have figured prominently in the discussions of extinction periodicity, first proposed by Fischer and Arthur (1977) and Raup and Sepkoski (1984). Some authors (for example, Alvarez *et al.*, 1982; Asaro *et al.*, 1982; Ganapathy, 1982) have used the Eocene–Oligocene extinctions as evidence for extraterrestrial causes. Unfortunately, all of these studies have been built around the assumption that there is a single extinction event. This misconception was fostered by the overemphasis of the Terminal Eocene Event (TEE), which was the subject of IGCP Project 174 (Pomerol and Premoli-Silva, 1986). It is now becoming clear that the Eocene–Oligocene extinctions were a complex series of stepwise events, spanning over 10 million years, and not a single abrupt event like some of those discussed in this volume. Since this is the case, simplistic models with a single, abrupt cause are no longer relevant.

Complicating the discussion has been a dispute over the time-scale. The controversy has been unusually acute for events which have many data sources and are relatively recent (Snelling, 1985). The differences between versions of the time-scale are not trivial. Recent estimates of the age of the Eocene–Oligocene boundary, for example, have ranged between 32 and 38 million years. Most of the discrepancies appear to be due to the heavy use (Curry and Odin, 1982) of glauconite dates, which are notoriously unreliable. In this chapter, I follow the time-scale of Berggren *et al.* (1985), which has been substantiated by almost all the new dates published subsequently (Montanari *et al.*, 1985; Berggren, 1986; Aubry *et al.*, 1988).

One minor point about this time-scale is frequently confused. In the formal sense, only the Priabonian is 'late' Eocene; the preceding Lutetian and Bartonian are technically 'middle' Eocene. Since most of the events discussed in this chapter take place in the later third of the Eocene, most authors have referred to them as 'late Eocene'. To avoid confusion, I refer to events happening in the Bartonian or Priabonian as 'later' Eocene, and restrict 'late Eocene' to the Priabonian. Likewise, since the Oligocene is divided into only two stages, the Rupelian and Chattian, there is no 'middle' Oligocene. I use 'mid-Oligocene' in the informal sense to denote events near the Rupelian–Chattian boundary.

THE EVIDENCE

Data from the later Eocene and Oligocene have been emerging from many sources since the mid- to late 1970s. Most of the best evidence has come from the study of pelagic sediments and their contained microfossils. These allow the high resolution necessary, and also have the potential for indicating palaeoceanographic changes from fluctuations in the biota and the stable isotopes. Some of these sections also have magnetostratigraphic control and one (Gubbio) produces high-temperature radiometric dates (Montanari *et al.*, 1985). From these sections the most complete picture of this time interval is preserved.

Shallow marine sections, with their benthic invertebrates, are less complete and generally have lower resolution. Nevertheless, some information of good quality is beginning to emerge from the study of their contained faunas. Terrestrial sections are typically the rarest and least complete. Nevertheless, they have many advantages which make them important to the overall story. They often have excellent records of the changes in land faunas and floras. They also have the potential for magnetostratigraphy, and, in many cases, already have excellent high-temperature radiometric dates associated with the fossils and magnetics. In some intervals, such as much of the Oligocene, the marine record is notoriously poor, but there is an excellent land section which currently gives the most detailed record of the mid-Oligocene extinction (Prothero, 1985a, 1985b).

In this section, I will review the evidence from these different sources of data, beginning with the pelagic record. Only after the empirical evidence has been reviewed will the discussion turn to possible causes. Too many previous studies have gone the other way, predicting an extinction at a certain time and then finding an event to fit it, while ignoring the rest of the data. Naturally, the conclusions of such selective interpretations of the fossil record are suspect.

The pelagic record

The most complete record with highest resolution of all the events is offered by the deep sea. Since pelagic sediments (both biogenic and non-biogenic) settle out of the surface waters in an almost steady 'rain', they produce a remarkably complete record, with fewer unconformities than are found in the shallow marine or terrestrial environments. The deep sea is also unusual in that sedimentation and plankton distribution is controlled by large-scale water masses, which are orders of magnitude larger than the depositional environments and biogeographic regions that control shallow marine and terrestrial sediments and organisms. Thus, there are relatively few problems with local lithofacies or biofacies, and most of the differences are due to major changes in oceanographic circulation or world climate.

Keller (1983a, 1983b) has reviewed the evidence from planktonic fora-miniferans from a number of deep-sea sites around the world. She recognizes five distinct events in the later Eocene and Oligocene. They are (in order from oldest to youngest):

The Lutetian–Bartonian event (middle Eocene, boundary between plank tonic foram zones P12 and P13, top magnetic Chron C18R, about 43 million years BP). This was a relatively minor event, marked by the extinc-tion or decline of a few key foraminiferan species. It apparently represents the beginning of cooling after the predominantly warm early and middle Eocene. It is marked by hiatuses in several deep-sea sections, indicating changes in oceanic circulation causing carbonate dissolution and/or bottom-water erosion. It is also marked by a lesser, but still significant, sea-level drop, labelled Tejas A3.4 by Haq *et al.* (1987). There are currently no detailed oxygen- or carbon-isotope data for this interval, but it does not stand out in the relatively low-resolution oxygen-isotope data of Shackle-ton and Kennett (1975) and Savin *et al.* (1975), or even the more recent data of Miller *et al.* (1987).

The Bartonian–Priabonian event (middle–late Eocene boundary, P14–P15 boundary, Chron C17R, about 41 million years BP). This was a major

extinction event, marked by the disappearance of spinose tropical foraminiferans and the migration towards the equator of mid-latitude faunas
with the cooler water masses. This is also the most severe extinction event
in the calcareous nannoplankton, wiping out almost 50% of the species
(Aubry, 1983). There are major hiatuses in most deep-sea sections, and a
major sea-level drop at the top of the Tejas A3 sequence of Haq *et al.*
(1987). There is also a significant cooling event apparent in the oxygen
isotopes (Miller and Curry, 1982). No event is apparent in the carbon
isotopes, although the data still have very low resolution.

The late Priabonian event (late Eocene, P15–P16 boundary, Chron C15R,
about 38 million years BP). This is a relatively minor event, marked by the
dominance of cooler water foraminiferans in high latitudes. There are
widespread deep-sea hiatuses, although sea level was already low through
the Priabonian (Haq *et al.*, 1987). The oxygen isotopes again indicate
further cooling, and there is a major change in the carbon isotopes (Miller,
Curry, and Ostermann, 1985). This is the only event associated with
microtektites (Hut *et al.*, 1987).

The Terminal Eocene Event (Eocene–Oligocene boundary, P17–P18
boundary, mid-Chron C13R, about 36.5 million years BP). This event has
been extensively discussed in Pomerol and Premoli-Silva (1986), and has
received much attention in recent years. Ironically, when it was first
christened by Wolfe (1978), he was actually referring to the mid-Oligocene
event (Prothero, 1985b). However, there are relatively few extinctions in
the foraminiferans, but simply an increase in dominance of cold-water
forms (Corliss, 1981; Corliss *et al.*, 1984; Keller, 1983a, 1983b). The
extinction in the calcareous nannoplankton is also minor compared to the
Bartonian event (Aubry, 1983). The extinctions of ostracodes (Benson,
1975) and benthic foraminiferans, however, were quite severe. There is
also a major shift in the oxygen isotopes (Keigwin, 1980; Miller and Curry,
1982), indicating global cooling of about 2–3°C, and another decline in the
carbon isotopes (Miller, Curry, and Ostermann, 1985). There are no
significant hiatuses in low- or middle-latitude sequences, but possibly some
hiatuses at high latitudes. The sea level dropped slightly, although it was
already at its Priabonian low (Haq *et al.*, 1987).

The mid-Oligocene event (late Rupelian, top P19, mid-Chron C11R, about
32.5 million years BP). This event was first described by Wolfe (1971) as the
'Oligocene deterioration' and then later mislabelled the 'Terminal Eocene
Event' (Wolfe, 1978). In the pelagic record, it is marked by the extinction
of many cool-water foraminiferans that dominated the early Oligocene,
although the surviving forms are also associated with cold water. There is
also a significant cooling event reflected in the nannoplankton (Haq *et al.*,
1977). There are major oxygen- and carbon-isotope excursions at this
event (Keigwin and Keller, 1984), although they are not as severe as the

isotopic events at the Eocene–Oligocene boundary. There are major marine hiatuses in the late Rupelian, reflected mostly by the cutting of submarine canyons (Miller, Mountain, and Tucholke, 1985; Miller *et al.*, 1987). The most impressive record in the mid-Oligocene, however, is the largest offlap event in the entire Vail onlap-offlap curve. As currently calibrated by Haq *et al.* (1987), the Tejas A4.5 sequence begins about 32 million years ago and reaches its maximum offlap at 30 million years BP. Whether this is entirely due to eustatic sea-level change is still controversial, but the evidence of Miller, Mountain and Tucholke (1985) indicates that it must have been one of the biggest sea-level drops in Tertiary history.

Between these major events, most of the climatic indicators show evidence of warming and some restoration of pre-event conditions. In many cases, there is a renewed radiation of planktonic organisms to fill the gaps left by the extinctions. However, these radiations do not completely fill the niches, since there is an overall decline in diversity through the interval. Likewise, each post-event warming episode does not completely return to the prior temperature, so there is net cooling from the early Eocene to the mid-Oligocene.

The shallow marine record

Unlike the pelagic environment, the shallow marine environment is at or near erosional base level, so it is prone to erosional unconformities during lowstands in sea-level. As a consequence, the shallow marine record of the late Palaeogene is much less complete than the deep-sea record. In addition, the organisms have not been studied in nearly the detail that is seen for planktonic microfossils of the pelagic realm. Thus, the data for the shallow marine environment are much coarser in resolution, and, in many cases, have not been synthesized in a fashion useful to this discussion.

One of the few studies which has attempted to relate shallow marine organisms to the complex pattern seen in pelagic organisms was by Hansen (1987; also see Hut *et al.*, 1987). Hansen recognized several major extinctions in Gulf Coast molluscs through the later Eocene. After a middle Eocene peak in diversity, there were three episodes of extinction before the Oligocene. The first occurs at the Bartonian–Priabonian boundary, when 89% of the gastropod species and 84% of the bivalves became extinct. The second occurred in the late Priabonian, when 72% of the gastropods and 63% of the bivalves (those that survived or evolved since the last extinction) became extinct. The third occurred at the Eocene–Oligocene boundary, when 97% of the gastropods and 89% of the bivalves became extinct. There is no evidence of the Lutetian–Bartonian event, and the Gulf Coast record has a large Oligocene unconformity which wiped out any record of the mid-Oligocene.

Other studies have documented a similar pattern, although not in sufficient detail to resolve the five separate extinction events. Hickman (1980) showed that warm-adapted species of molluscs declined and then disappeared along the Pacific Coast during the late Eocene. Zinsmeister (1982) showed that there was heavy extinction of molluscs in the Antarctic region during the Eocene–Oligocene transition. McKinney (1986) reported a major decline in echinoid diversity through the late Eocene, although the resolution of his data is not yet sufficient to determine if they match the pattern of events seen in the pelagic record.

Some work has been done on the isotopic composition of mollusc shells (Cavelier *et al.*, 1981). For example, Burchardt (1978) studied oxygen isotopes from mollusc shells from the North Sea. He found a steady decline in temperature from the Lutetian to the early Oligocene, with the most dramatic drop at the Eocene–Oligocene boundary. However, these data do not have the resolution to determine if there are other sudden isotopic shifts in the later Eocene. Nor were there sufficient data to determine if there is a discrete mid-Oligocene event, although the oxygen-isotope curve reached its 'coolest' point in the mid-Oligocene.

In addition to biological and isotopic evidence, there is now more direct evidence of climate in shallow marine sediments. Recent drilling in the Weddell Sea of Antarctica (Legs 113 and 114 of the Deep Sea Drilling Project) recovered evidence of glacial ice in Antarctica in the early Oligocene (Kennett and Barker, 1987; Wise *et al.*, 1987). There is also good evidence of another ice advance in the mid-Oligocene. As we shall see below, the isotopic evidence has demanded Oligocene glaciation for some time and the latest results seem to support this.

The terrestrial record

Of the pelagic, shallow marine, and terrestrial records, the last is the most incomplete. Most terrestrial sediments are deposited far above base level, so they are rarely preserved and are prone to unconformities. Nevertheless, there are a number of remarkably complete sections that preserve parts of the later Eocene and Oligocene record, although there is no single place that records the entire interval, as do the best pelagic sections. The record of land plants and animals is so large on many continents that a clear signal emerges, even if the dating and resolution are not yet at the precision possible for the deep sea.

Several studies have attempted to assess climatic changes through changes in land plants. The most complete results are those of Wolfe (1971; 1978; Wolfe and Hopkins, 1967) for North America. Although there is not much resolution in the data, the floral evidence clearly shows a peak in warming around 43 million years BP (Lutetian), maximum cooling around 41 million years BP (the Bartonian–Priabonian event), warming again to a peak around 34 million years BP (early Oligocene), and then the most severe cooling event at 32 million years BP. This last decline is clearly the mid-Oligocene event, labelled the 'Oligocene deterioration' by Wolfe (1971). These data do not have sufficient resolution to determine if the other three events might be present as well. Surprisingly, there is no clear floral evidence for the Eocene–Oligocene extinction (despite the mislabelling by Wolfe, 1978, Figure 1). The Bartonian–Priabonian and mid-Oligocene extinctions were each marked by decline in inferred mean annual temperatures of as much as 10°C, a remarkable cooling.

Although they have not been quantified in the same way as Wolfe's data, Collinson *et al.* (1981) and Collinson and Hooker (1987) found a similar floral transition in the Palaeogene of England. Middle Eocene floras are predominantly tropical, but, by the late Eocene, the tropical taxa have become extinct and are replaced by taxodiaceous swamps and reed marshes, with patches of woodland or forest. These reed marshes become dominant at the end of the Eocene and through the early Oligocene. This evidence from the megaflora is supported by evidence from pollen (Boulter and Hubbard, 1982; Hubbard and Boulter, 1983). Chateauneuf (1980) studied floras of the Paris Basin. Diversity of tropical forms reached a high in the late Lutetian and early Bartonian, and then declined through the later Eocene. There were two sharp extinction events, one in the late Priabonian, and one that appears to be the Terminal Eocene Event. There are no data for the mid-Oligocene. Other floral evidence from Europe is reviewed by Cavelier *et al.* (1981).

Floral evidence from other parts of the world is in close agreement with the data from North America and Europe, even if the resolution is not very precise. Kemp (1978) summarized the floral data for Australia, Antarctica, and New Zealand throughout the Tertiary. Like other parts of the world, the middle Eocene was characterized by tropical rainforests. By the late Eocene, this vegetation had declined considerably and Oligocene vegetation is characterized by a low diversity of cool temperate forms.

Finally, in sections which do not preserve plant fossils directly, there are other methods of obtaining palaeobotanical evidence. Retallack (1983) examined palaeosols from the late Eocene of Oregon and the Oligocene of the Big Badlands of South Dakota. Late Eocene floras are again semi-tropical, but, at the Terminal Eocene Event, humid-climate floras are replaced by those of subhumid climates. Another palaeosol change occurs at the mid-Oligocene event, where palaeosols formed under subhumid climates are replaced by those formed under subarid conditions. These changes are so abrupt that they are responsible for major changes in sedimentological character of the lithologic units.

The land vertebrate record has been studied by a number of researchers. Hutchinson (1982) tabulated the generic diversity of turtles, crocodilians and champsosaurs through the Tertiary. Although he used resolution on the scale of the North American land mammal 'ages' (typically 2–4 million years in length), some clear trends emerge. There is a striking decline in total diversity from the middle Eocene to the late Eocene, composed mostly of the decline in crocodilians, champsosaurs and aquatic turtles.

There is a small increase in terrestrial turtles (mostly tortoises). Both of these trends indicate cooling and drying conditions. At the mid-Oligocene event (Chadronian–Orellan boundary), there is almost total extinction of aquatic turtles and crocodilians, and another increase in terrestrial turtles, which make up almost all the herpetofauna at this time. Indeed, land tortoises are such common fossils in the Orellan deposits of the Big Badlands that they were once known as the 'Turtle-Oreodon beds'. Rage (1986) noted similar changes in European herpetofaunas in the Eocene and Oligocene.

The best evidence of terrestrial change, however, comes from the excellent record of fossil mammals. This record is just now being calibrated against the magnetic polarity time-scale, making it possible to correlate it precisely with the marine record for the first time. For example, Prothero (1985b) calibrated North American Oligocene mammal-bearing sections against the magnetic polarity time-scale, and tabulated familial and generic diversity in million-year increments. He found a peak of both extinctions and originations in the late Duchesnean (late Priabonian), a peak of originations in the earliest Chadronian (Eocene–Oligocene boundary), and a peak of extinctions (but not originations) at the Chadronian–Orellan boundary (mid Oligocene event). Thus, the major turnover was concentrated, not at the Eocene–Oligocene boundary, but in the late Priabonian, some 2–3 million years before. The biostratigraphic and magnetostratigraphic data for the middle Eocene are not yet published, but preliminary results by Prothero (in prep.) and by Stucky (1989) suggest that there were extinction events at the Uintan–Duchesnean boundary that correspond to the Bartonian–Priabonian event or possibly the Lutetian–Bartonian event.

The qualitative aspects of this change are also interesting (Prothero, 1985b). A number of archaic groups (many of which were clearly browsers or arboreal forms) that characterized the middle Eocene began a steady decline through the later Eocene, and were extinct by the end of the Eocene in North America. These include mixodectids, microsyopids, taeniodonts, achaenodonts, uintatheres, nyctitheriids, anaptomorphine primates, dermopterans, sciuravid rodents, dichobunid artiodactyls, limnocyonid and 'miacid' carnivores, mesonychids, hyopsodonts, isectolophid tapiroids and the ceratomorph *Hyrachyus*. These were replaced by the 'White River Chronofauna' (Emry, 1981), which was dominated by a number of families better suited for more open terrain and less forested habitat. Some were even grazers. Many of the key taxa are members of extant families which originated at this time, including the rabbits, dogs, camels, pocket gophers, squirrels, rhinos and shrews. This 'White River Chronofauna' maintained its stability throughout most of the Oligocene and was replaced in the early Miocene. The one exception to this stability was the wave of extinctions at the Chadronian–Orellan boundary (mid-Oligocene event), which wiped out the last vestige of archaic forms characteristic of the Eocene. These included the titanotheres, multituberculates, pantolestids, oromerycids, epoicotheres, and paramyid and cylindrodont rodents. They are replaced by leptauchenine oreodonts and eumyine cricetid rodents, both groups with much more abrasive-resistant dentitions.

Similar patterns have been observed in European mammals, although these faunas have not yet been calibrated by magnetostratigraphy. At one time, the discussion centred around 'la Grande Coupure' of Stehlin (1909), the dramatic faunal turnover event that completely changed European mammalian faunas between the Eocene and Oligocene. Recent studies by Brunet (1977), Legendre (1987) and Hartenberger (1986; 1987) have revealed a more complex picture. There is high turnover throughout the late Eocene, so that the 'Grande Coupure' may represent a composite of both the late Priabonian and Terminal Eocene Events. There is also another peak of turnover that may correspond to the Bartonian–Priabonian event as well (Hartenberger, 1986, Figure 4). Like North American faunas, European faunas show a peak of extinctions, but not originations, at the mid-Oligocene event (Sannoisian–Stampian event). There is apparently no evidence for a Lutetian–Bartonian event in European land mammals, however.

In the middle Eocene, the European fauna was dominated by primates, multituberculates, 'insectivores', creodonts, archaic ungulates ('condylarths'), tillodonts, pantodonts, and archaic perissodactyls (such as palaeotheres and lophiodonts) and artiodactyls (mostly 'dichobunids', xiphodonts, choeropotamids, cebochoerids, mixtotheriids, dacrytheriids, anoplotheriids, amphimerycids, and cainotheriids). As a consequence of the 'Grande Coupure', the fauna was radically rearranged (Hartenberger, 1986; 1987). By the early Oligocene, European faunas are dominated by rodents (mostly theridomyids), advanced carnivores (mostly amphicyonids, mustelids, viverrids, procyonids, ursids and nimravids), advanced artiodactyls (particularly anthracotheres, leptomerycids, entelodonts, and tayassuids) and advanced perissodactyls (particularly rhinocerotoids and chalicotheres). Rabbits also make their first appearance and become an important element of the fauna. Archaic groups, particularly the creodonts, archaic ungulates, multituberculates, tillodonts and pantodonts, were extinct, and once-dominant arboreal groups, such as primates, were locally extinct.

Collinson and Hooker (1987) reviewed the changes in land mammals in the London Clay through the Palaeogene. In the Bartonian, there was a great reduction in small mammals, particularly insectivores, and an increase in large ground mammals and browsing herbivores. In the Priabonian, there was a drastic reduction in arboreal mammals, as well as small mammals and insectivores, and a great increase in large ground mammals and browsing herbivores. The rodent fauna changed from frugivorous pseudosciurids to browsing theridomyids. There was a major extinction of soft-browsing perissodactyls and replacement with coarse-browsing forms. Finally, at the Eocene–Oligocene boundary, the arboreal types disappeared, and large mammals dominated. At this point, granivorous rodents made their first appearance. All of these changes are consistent with the vegetational changes seen by Collinson and Hooker (1987), where tropical forests of the Lutetian are replaced by taxodiaceous swamps and fresh-water reed marshes in the late Eocene.

Similar changes have been observed for the Eocene and Oligocene faunas of Asia and South America, although the details of the turnover have not been worked out as carefully as they have for North America and Europe (Savage and Russell, 1983). Without question, the Eocene–Oligocene transition is the most important turnover event in the history of Cenozoic mammals (Lillegraven, 1972). It marked the end of the forest-dwelling, browsing fauna that dominated the Palaeocene and Eocene, and the beginning of the 'modern' fauna, which is much more adapted to open country (particularly savannahs) and grazing.

CAUSES OF THE EOCENE/OLIGOCENE EXTINCTIONS

Now that we have reviewed the empirical data for the patterns of extinction, it is appropriate to speculate on the causes. It is clear that there are at least five steps of extinction (43, 41, 38, 36.5 and 32.5 million years BP) spaced out over 10 million years. Thus, models which attribute these extinctions to a single catastrophic event (such as the asteroid model of Alvarez *et al.*, 1982; Asaro *et al.*, 1982; Ganapathy, 1982) cannot be correct. We must search for causative agents which are capable of inducing climatic stress over a 10-million-year interval. Such agents can either be terrestrial or extraterrestrial. I will review the terrestrial evidence first.

In recent years, the evidence for major global climatic changes during the Eocene–Oligocene transition has become overwhelming. Much of this evidence is reviewed above. It is clear that the Earth underwent a stepwise cooling of over 10°C during this interval (Savin *et al.*, 1975), with pulses of cooling in the Bartonian, the Terminal Eocene Event and the mid-Oligocene event. This global cooling, and the associated changes in weather patterns and moisture, were undoubtedly responsible for the largest part of the extinctions, especially in the terrestrial and shallow marine realms (Stanley, 1984; 1987). The real question centres around the causes of this global cooling.

Abundant evidence suggests that the major cause was oceanographic changes. The oxygen-isotopic evidence is now interpreted by many authors (among them Matthew and Poore, 1980; Miller and Fairbanks, 1983; 1985; Keigwin and Keller, 1984) as evidence for Oligocene Antarctic glaciation. The recently reported data from the Weddell Sea of Antarctica (Kennett and Barker, 1987; Wise *et al.*, 1987) confirm the presence of early Oligocene sea-ice, although the full development of the Antarctic ice-cap may not have occurred until the Miocene. Along with growth of the Antarctic ice-cap was the development of cold, deep bottom-water masses (the 'psychrosphere'), which shows up not only in the isotopic data, but also in dissolution and erosion of the deep-sea record. According to Benson (1975), Kennett (1977), Keigwin (1980), and Keller (1983a, 1983b) the psychrosphere first developed at the Terminal Eocene Event. These oceanographic changes are attributed to rearrangements of oceanic circulation caused by plate tectonics. The obvious candidate (Kennett, 1977) is the separation of Australia from Antarctica, which allowed water to circulate between the two continents. Once this circulation developed, it triggered the beginning of circum-Antarctic circulation. Today, this current traps cold water in a continuous cycle around Antarctica, refrigerating the South Pole and generating cold deep bottom water. It is also responsible for isolating the polar waters and generating a larger gradient in temperatures between pole and equator. By contrast, during the Eocene, polar waters in the southern Indian Ocean or southern Pacific exchanged with warmer, lower-latitude waters, allowing mixing and a less extreme temperature gradient between pole and equator.

The key question is how this model explains the five-step decline in world climate. Kennett (1977) and Murphy and Kennett (1986) cited evidence to show that shallow-water circulation between Australia and Antarctica began about 38 million years BP, in the latest Eocene. Benson *et al.* (1986) suggested that this psychrospheric circulation developed even earlier in the late Eocene, which might explain the major Bartonian–Priabonian event. The Eocene–Oligocene boundary is the final threshold of this transition. Much new evidence is needed to determine the timing of these events and see if they are really as discrete as other data suggest. Kennett (1977) and Murphy and Kennett (1986) cited evidence to show that deep-water circulation between Australia and Antartica developed in the middle Oligocene. There is evidence for a major mid-Oligocene glaciation (Miller, Mountain, and Tucholke, 1985), which might explain the mid-Oligocene extinctions and the record drop in sea level.

Although these Southern Hemisphere events undoubtedly had the major effect on ice formation and global cooling, the Northern Hemisphere may have amplified the effect. The opening of the Greenland–Norway passage, allowing exchange between the Arctic and Atlantic, is believed to have occurred near the Eocene–Oligocene boundary (Talwani and Eldholm, 1977; Berggren, 1982). More recent data (Miller and Fairbanks, 1983), however, show evidence for the development of the North Atlantic Deep Water (NADW) by the earliest Oligocene.

These palaeoceanographic events were undoubtedly the major forcing factors of global climate and thus probably the primary causes of the Eocene–Oligocene extinctions. In recent years, however, a number of non-oceanographic models have been proposed. As we have seen, the prolonged and stepwise nature of the extinctions and climatic change rule out single catastrophic events, such as asteroids. Two other possibilities remain that could fit the evidence: vulcanism and comet showers.

Kennett *et al.* (1985) point out that there is a pulse of vulcanism in the southwest Pacific just before the Terminal Eocene Event. It is well known that volcanic aerosols, particularly sulphuric acid, cause climatic change. The 'volcano weather' caused by the eruption of Krakatoa is a well-known example. Kennett *et al.* (1985) suggest that this southwest Pacific vulcanism amplified the climatic changes already associated with the development of Antarctic circulation and glaciation. Volcanic ashes are abundant through the late Priabonian and early Oligocene, so they may have influenced two of the five extinction pulses.

Rampino and Stothers (1988) pointed out that many of the great mass extinctions in the last 250 million years were associated with eruptions of flood basalts. Officer and Drake (1985) made a strong case for the K–T extinctions being caused by the eruption of the Deccan traps. There are similar traps which erupted in Ethiopia during the early phases of the opening of the East African Rift. According to Rampino and Stothers, the start of the main phase of eruption began around 35 ± 2 million years BP, and continued through the Oligocene. This suggests that 'volcano weather' might be a possible factor in the Oligocene, but does nothing to explain the long Eocene decline. In addition, Rampino and Stothers attribute the supposed periodicity of these flood basalts to impact cratering by comets. This seems highly implausible, since flood basalts are generated from deep in the lower crust and mantle where no comet impact could penetrate. In addition, each episode of flood basalt eruption is associated with rifting and has known plate tectonic causes. If the periodicity is real, it is more likely that there is some sort of periodicity in mantle overturn (Sheridan, 1987).

The most glamorous models for mass extinctions are extraterrestrial in nature. Despite the great attention they receive in the popular press, they are inadequate to explain the Eocene–Oligocene climatic and faunal changes. The single catastrophic asteroid model fails completely for reasons discussed above. The iridium anomaly reported by Ganapathy (1982) and Alvarez *et al.* (1982) apparently has no relation to any of the extinction events discussed above. In addition, Kyte and Wasson (1986) find no such iridium anomaly anywhere in the Eocene or Oligocene. Consequently, a series of comet showers spread out over millions of years has been offered as an alternative which fits the prolonged, stepwise nature of the extinctions (Hut *et al.*, 1987). As Hut *et al.* (1987) point out, however, only one of the five Eocene–Oligocene extinctions seems to be associated with microtektites, and it is one of the least severe (the late Priabonian event). Most of the microtektite horizons correspond to dissolution horizons, or do not have any correspondence to known episodes of extinction (Keller *et al.*, 1983; Hut *et al.*, 1987, Figure 4).

Despite this tenuous evidence, Hut *et al.* (1987) persisted in attributing at least some of the Eocene–Oligocene extinctions to extraterrestrial causes. It seems obvious from the discussion above, however, that the earth's climate and biota were much more strongly affected by major palaeoceanographic changes. These must have been the primary forcing agent, no matter how much material was or was not coming in from outer space. At best, it appears that extraterrestrial materials influenced only the weakest and least dramatic of the extinction events. All of the rest are clearly a result of terrestrial causes.

SUMMARY

Extinctions in the late Eocene and Oligocene took place in five steps spaced out over 10 million years (43, 41, 38, 36.5, and 32.5 million years BP). These extinctions can be seen in pelagic, shallow marine, and terrestrial animals and plants to varying degrees. The Eocene–Oligocene extinctions are associated with many indicators of climatic change, including severe cooling, glaciation and changes in oceanographic circulation, which were undoubtedly the proximal causes of extinction. This climatic change is primarily the result of the development of the circum-Antarctic current due to the separation of Antarctica from Australia. Other factors, such as vulcanism, or comet showers, may have had a minor reinforcing effect on one or two of these extinction events, but clearly did not cause most of the climatic change that was primarily responsible for the extinctions.

ACKNOWLEDGEMENTS

I thank S.K. Donovan for the invitation to participate in this volume. R.H. Benson, W.A. Berggren, G. Keller, J.P. Kennett and K.G. Miller graciously read the manuscript, and made many helpful comments. Research for this chapter was partially supported by a grant from the Donors of the Petroleum Research Fund of the American Chemical Society, and by NSF Grant EAR87-08221.

NOTE ADDED IN PROOF

The numerical calibration of the late Paleogene timescale has been changing almost weekly since this chapter was last revised. New argon–argon dates on a number of marine and terrestrial sections suggest that the Eocene–Oligocene boundary is between 33.5 and 34 million years in age, contrary to the age of 36.5 used by Berggren *et al.* (1985) and Haq *et al.* (1987). These changes mean that the entire timescale will need to be revised in the near future, but at the moment it is difficult to determine what numerical ages will be assigned to the events described in this paper. If my estimates are correct, then the Lutetian–Bartonian event would occur at about 40 million years BP, the Bartonian–Priabonian event at about 38, the late Priabonian event at about 36, the Eocene–Oligocene boundary at about 33.7, and the mid-Oligocene event at about 30.5. Regardless of how the numerical age assignments of these events are revised, the data and discussions in this paper are based on the relative stratigraphic sequence, so the general conclusions about marine events are unaffected by these new dates.

However, these new argon–argon dates on terrestrial sequences radically change the correlation of North American land mammal "ages" with the marine chronologies. Calibration of the terrestrial magnetostratigraphy (Prothero, 1985a, 1985b) with the new dates gives the following results: Chadronian—from C16N to top C13R (33–37 million years ago); Orellan —from top C13R to mid-C12R (31.5–33 million years); Whitneyan—from mid-C12R to base C11N (29–31.5 million years); Arikareean—from 21–29 million years. More importantly, these correlations imply that the Bartonian/Priabonian event is the Chadronian/Duchesnean transition, the Chadronian is entirely late Eocene, and the Chadronian/Orellan boundary (labeled the mid-Oligocene event in this paper) is actually the Terminal Eocene Event. The mid-Oligocene event probably falls in the Whitneyan.

REFERENCES

Alvarez, W., Asaro, F., Michel, H.V. and Alvarez, L.W., 1982, Iridium anomaly approximately synchronous with terminal Eocene extinctions, *Science*, **216** (4548): 886–8.

Asaro, F., Alvarez, L.W., Alvarez, W. and Michel, H.V., 1982, Geochemical anomalies near the Eocene/Oligocene and Permian/Triassic boundaries. In L.T. Silver and P.H. Schultz (eds), *Geological implications of impacts of large asteroids and comets on the Earth, Special Paper of the Geological Society of America*, **190**: 517–28.

Aubry, M.-P., 1983, Late Eocene to early Oligocene calcareous nannoplankton biostratigraphy and biogeography, *Bulletin of the American Association of Petroleum Geologists*, **67** (3): 415.

Aubry, M.-P., Berggren, W.A., Kent, D.V., Flynn, J.J., Klitgord, K.D., Obradovich, J.D. and Prothero, D.R., 1989, Paleogene geochronology; an integrated approach, *Paleoceanography*, **3** (6): 707–42.

Benson, R.H., 1975, The origin of the psychrosphere as recorded in changes of deep-sea ostracod assemblages, *Lethaia*, **8** (1): 69–83.

Benson, R.H., Chapman, R.E. and Deck, L.T., 1986, Paleoceanographic events and deep-sea ostracods. In K.J. Hsü *et al.* (eds), *South Atlantic Paleoceanography*, Cambridge University Press, Cambridge: 325–50.

Berggren, W.A., 1982, Role of ocean gateways in climate change. In W.H. Berger and J.C. Crowell (eds), *Climate in Earth history*, National Academic Press, Washington, DC: 118–25.

Berggren, W.A., 1986, Geochronology of the Eocene–Oligocene boundary. In C. Pomerol and I. Premoli-Silva (eds), *Terminal Eocene events*, Elsevier, Amsterdam: 349–56.

Berggren, W.A., Kent, D.V. and Flynn, J.J., 1985, Paleogene geochronology and chronostratigraphy. In N.J. Snelling (ed.), *The chronology of the geological record, Memoir of the Geological Society of London*, **10**: 141–95.

Boulter, M.C. and Hubbard, R.N.L.B., 1982, Objective palaeoecological and biostratigraphic interpretation of Tertiary palynological data by multivariate statistical analysis, *Palynology*, **6** (1): 55–68.

Brunet, M., 1977, Les mammifères et le problème de la limite Eocène-Oligocène en Europe, *Mémoires Spéciaux Geobios*, **1**: 11–27.

Burchardt, B., 1978, Oxygen isotope palaeotemperature from the Tertiary Period in the North Sea area, *Nature*, **275** (5676): 121–3.

Cavelier, C., Chateauneuf, J.-J., Pomerol, C., Rabussier, D., Renard, M. and Vergnaud-Grazzini, C., 1981, The geological events at the Eocene–Oligocene boundary, *Palaeogeography, Palaeoclimatology, Palaeoecology*, **36** (3–4): 223–48.

Chateauneuf, J.-J., 1980, Palynostratigraphie et paléoclimatologie de l'Eocène supérieur et de l'Oligocène du Bassin du Paris (France), *Mémoires de la Bureau de Recherches Géologiques et Minieres*, **116**: 1–357.

Collinson, M.E., Fowler, K. and Boulter, M.C., 1981, Floristic changes indicate a cooling climate in the Eocene of southern England, *Nature*, **291** (5813): 315–17.

Collinson, M.E. and Hooker, J.J., 1987, Vegetational and mammalian faunal changes in the early Tertiary of southern England. In E.M. Friis, W.G. Chaloner and P.R. Crane (eds), *The origin of angiosperms and their biological consequences*, Cambridge University Press, Cambridge: 259–304.

Corliss, B., 1981, Deep-sea benthonic foraminiferal faunal turnover near the Eocene/Oligocene boundary, *Marine Micropaleontology*, **6** (4): 357–84.

Corliss, B., Aubry, M.-P., Berggren, W.A., Fenner, J.M., Keigwin, L.D., Jr and Keller, G., 1984, The Eocene/Oligocene boundary event in the deep sea, *Science*, **226** (4676): 806–10.

Curry, D. and Odin, G.S., 1982, Dating of the Palaeogene. In G.S. Odin (ed.), *Numerical dating in stratigraphy*, Wiley and Sons, New York: 607–30.

Emry, R.J., 1981, Additions to the mammalian fauna of the type Duchesnean, with comments on the status of the Duchesnean 'Age', *Journal of Paleontology*, **55** (3): 563–70.

Fischer, A.G. and Arthur, M.A., 1977, Secular variations in the pelagic realm, *Special Publication of the Society of Economic Paleontologists and Mineralogists*, **25**: 19–50.

Ganapathy, R., 1982, Evidence for a major meteorite impact on the Earth 34 million years ago: implications for Eocene extinctions, *Science*, **216** (4549): 885–6.

Hansen, T.A., 1987, Extinction of late Eocene to Oligocene molluscs: relationship to shelf area, temperature changes, and impact events, *Palaios*, **2** (1): 69–75.

Haq, B.U., Hardenbol, J. and Vail, P.R., 1987, Chronology of fluctuating sea levels since the Triassic, *Science*, **235** (4793): 1156–67.

Haq, B.U., Premoli-Silva, I. and Lohmann, G.P., 1977, Calcareous plankton paleobiogeographic evidence for major climatic fluctuations in the early Cenozoic Atlantic Ocean, *Journal of Geophysical Research*, **82** (27): 3861–76.

Hartenberger, J.-L., 1986, Crises biologiques en milieu continental au cours du Paléogène: exemple des mammifères d'Europe, *Bulletin des Centres de Recherches Exploration-Production Elf-Aquitaine*, **10** (2): 489–500.

Hartenberger, J.-L., 1987, Modalités des extinctions et apparations chez les mammifères du Paléogène d'Europe, *Mémoires de la Société Géologique de France*, new series, **150**: 133–43.

Hickman, C.S., 1980, Paleogene marine gastropods of the Keasey Formation of Oregon, *Bulletins of American Paleontology*, **78** (310): 1–112.

Hubbard, R.N.L.B. and Boulter, M.C., 1983, Reconstruction of Palaeogene climate from palynological evidence, *Nature*, **301** (5896): 147–50.

Hut, P., Alvarez, W., Elder, W.P., Hansen, T.A., Kauffman, E.G., Keller, G., Shoemaker, E.G. and Weissman, P.R., 1987, Comet showers as a cause of mass extinctions, *Nature*, **329** (6135): 118–25.

Hutchinson, J.H., 1982, Turtle, crocodilian, and champsosaur diversity changes in the Cenozoic of the north-central region of the western United States, *Palaeogeography, Palaeoclimatology, Palaeoecology*, **37** (2–4): 149–64.

Keigwin, L.D., Jr, 1980, Palaeoceanographic change in the Pacific at the Eocene–Oligocene boundary, *Nature*, **287** (5784): 722–5.

Keigwin, L.D., Jr and Keller, G., 1984, Middle Oligocene cooling from equatorial Pacific DSDP Site 77B, *Geology*, **12** (1): 16–19.

Keller, G., 1983a, Paleoclimatic analysis of middle Eocene through Oligocene planktic foraminiferal faunas, *Palaeogeography, Palaeoclimatology, Palaeoecology*, **43** (1): 73–94.

Keller, G., 1983b, Biochronology and paleoclimatic implications of middle Eocene to Oligocene planktic foraminiferal faunas, *Marine Micropaleontology*, **7** (1982/1983): 463–86.

Keller, G., d'Hondt, S. and Vallier, T.L., 1983, Multiple microtektite horizons in Upper Eocene marine sediments: no evidence for mass extinctions, *Science*, **221** (4606): 150–3.

Kemp, E.M., 1978, Tertiary climatic evolution and vegetation history in the southeast Indian Ocean region, *Palaeogeography, Palaeoclimatology, Palaeoecology*, **24** (3): 169–208.

Kennett, J.P., 1977, Cenozoic evolution of Antarctic glaciation, the Circum-Antarctic Ocean, and their impact on global paleoceanography, *Journal of Geophysical Research*, **82** (27): 3843–59.

Kennett, J.P. and Barker, P., 1987, Cenozoic paleoclimatic and paleoceanographic history of Antartica: overview from ODP Leg 113, *Geological Society of America Abstracts with Programs*, **19** (7): 725.

Kennett, J.P. *et al.*, 1985, Palaeotectonic implications of increased late Eocene–early Oligocene volcanism from south Pacific DSDP sites, *Nature*, **316** (6028): 507–11.

Kyte, F.T. and Wasson, J.T., 1986, Accretion rate of extraterrestrial matter: iridium deposited 33 to 67 million years ago, *Science*, **232** (4755): 1225–9.

Legendre, S. 1987, Concordance entre paléontologie continentale et les événements paléocéanographiques: example des faunes de mammifères du Paléogène du Quercy, *Centre du Recherche de l'Academie des Sciences de Paris*, **304**: 45–9.

Lillegraven, J.A., 1972, Ordinal and familial diversity of Cenozoic mammals, *Taxon*, **21** (2–3): 261–74.

McKinney, M.L., 1986, Cenozoic echinoid diversity and mass extinction patterns closely tied to temperature, *Abstracts 4th North American Paleontological Convention*, A32.

Matthews, R.K. and Poore, R.Z., 1980, Tertiary $\delta^{18}O$ record and glacio-eustatic sea-level fluctuations, *Geology*, **8** (10): 501–4.

Miller, K.G. and Curry, W.B., 1982, Eocene to Oligocene benthic foraminiferal isotopic record in the Bay of Biscay, *Nature*, **296** (5855): 347–50.

Miller, K.G., Curry, W.B. and Ostermann, D.R., 1985, Late Paleogene benthic foraminiferal paleoceanography of the Goban Spur region, DSDP Leg 80, *Initial Reports of the Deep Sea Drilling Project*, **80**: 505–38.

Miller, K.G. and Fairbanks, R.G., 1983, Evidence for Oligocene–Middle Miocene abyssal circulation changes in the western north Atlantic, *Nature*, **306** (5940): 250–3.

Miller, K.G. and Fairbanks, R.G., 1985, Oligocene–Miocene global carbon and abyssal circulation changes, *Geophysical Monograph Series*, **32**: 469–86.

Miller, K.G., Fairbanks, R.G. and Mountain, G.S., 1987, Tertiary oxygen isotope synthesis, sea level history, and continental margin erosion, *Paleoceanography*, **2** (1): 1–19.

Miller, K.G., Mountain, G.S. and Tucholke, B.E., 1985, Oligocene glacio-eustasy and erosion on the margins of the North Atlantic, *Geology*, **13** (1): 10–13.

Montanari, A., Drake, R., Bice, D.M., Alvarez, W., Curtis, G.H., Turrin, B.D. and DePaolo, D.J., 1985, Radiometric time scale for the upper Eocene and Oligocene based on K/Ar and Rb/Sr dating of volcanic biotites from the pelagic sequence of Gubbio, Italy, *Geology*, **13** (9): 596–9.

Murphy, M.G. and Kennett, J.P., 1986, Development of latitudinal thermal gradients during the Oligocene: oxygen isotope evidence from the southwest Pacific, *Initial Reports of the Deep Sea Drilling Project*, **90**: 1347–60.

Officer, C.B. and Drake, C.L., 1985, Terminal Cretaceous environmental events, *Science*, **227** (4691): 1161–7.

Pomerol, C. and Premoli-Silva, I., (eds), 1986, *Terminal Eocene events*, Elsevier, Amsterdam.

Prothero, D.R., 1985a, Mid-Oligocene extinction events in North American land mammals, *Science*, **229** (4713): 550–1.

Prothero, D.R., 1985b, North American mammalian diversity and Eocene–Oligocene extinctions, *Paleobiology*, **11** (4): 389–405.

Rage, J.C., 1986, The amphibians and reptiles at the Eocene–Oligocene transition in western Europe: an outline of the faunal alterations. In C. Pomerol and I. Premoli-Silva (eds), *Terminal Eocene events*, Elsevier, Amsterdam: 309–10.

Rampino, M.R. and Stothers, R.B., 1988, Flood basalt volcanism during the past 250 million years, *Science*, **241** (4866): 663–8.

Raup, D.M. and Sepkoski, J.J., Jr, 1984, Periodicity of extinctions in the geologic past, *Proceedings of the National Academy of Science U.S.A.*, **81** (3): 801–5.

Retallack, G.J., 1983, Late Eocene and Oligocene paleosols from Badlands National Park, South Dakota, *Special Paper of the Geological Society of America*, **193**: 1–82.

Savage, D.E. and Russell, D.E., 1983, *Mammalian paleofaunas of the world*, Addison-Wesley, Reading, MA.

Savin, S.M., Douglas, R.G. and Stehli, F.G., 1975, Tertiary marine paleotemperatures, *Bulletin of the Geological Society of America*, **86** (11): 1499–1510.

Shackleton, N.J. and Kennett, J.P., 1975, Paleotemperature history of the Cenozoic and the initiation of Antarctic glaciation: oxygen and carbon isotopic analyses of DSDP Sites 277, 279, and 281, *Initial Reports of the Deep Sea Drilling Project*, **29**: 743–55.

Sheridan, R.E., 1987, Pulsation tectonics and the control of continental breakup. *Tectonophysics*, **143** (1): 59–73.

Snelling, N.J., 1985, An interim time scale. In N.J. Snelling (ed.), *The chronology of the geological record*, Memoir of the Geological Society of London, **10**: 261–5.

Stanley, S.M., 1984, Marine mass extinction: a dominant role for temperature. In M.H. Nitecki (ed.), *Extinctions*, University of Chicago Press, Chicago: 69–117.

Stanley, S.M., 1987, *Extinction*, Scientific American Books, New York.

Stehlin, H.G., 1909, Remarques sur les faunules de mammifères des couches éocènes et oligocènes du Bassin de Paris, *Bulletin de la Société Géologique de France*, series 4, **9**: 488–520.

Stucky, R.K., 1989, Evolution of land mammal diversity in North America during the Cenozoic. In H.H. Grenoways (ed.), *Current mammalogy 2*, Plenum Press, New York.

Talwani, M. and Eldholm, O., 1977, Evolution of the Norwegian–Greenland Sea, *Bulletin of the Geological Society of America*, **88** (7): 969–99.

Wise, S.W., *et al.*, 1987, Early Oligocene ice on the Antarctic continent, *Geological Society of America Abstracts with Programs*, **19** (7): 893.

Wolfe, J.A., 1971, Tertiary climatic fluctuations and methods of analysis of Tertiary floras, *Palaeogeography, Palaeoclimatology, Palaeoecology*, **9** (1): 27–57.

Wolfe, J.A., 1978, A paleobotanical interpretation of Tertiary climates in the Northern Hemisphere, *American Scientist*, **66** (6): 694–703.

Wolfe, J.A. and Hopkins, D.M., 1967, Climatic changes recorded by Tertiary land floras in northwestern North America. In K. Hatai (ed.), *Tertiary correlations and climatic changes in the Pacific*, Sasaki Publishers, Sendai, Japan: 67–76.

Zinsmeister, W.J., 1982, Late Cretaceous–Early Tertiary molluscan biogeography of the southern Circum-Pacific, *Journal of Paleontology*, **56** (1): 84–102.

Chapter 12

THE LATE PLEISTOCENE EVENT AS A PARADIGM FOR WIDESPREAD MAMMAL EXTINCTION

Anthony D. Barnosky

INTRODUCTION

Extinction is a fact of life. For living species, the question generally is not if extinction will occur, rather, the question is when. For many extinct species, we know when extinction took place, at least in general terms, and we are left with a different paradox: why did (and does) extinction occur? One chapter in Earth history, the Quaternary, casts a uniquely focused light on this question. Including the Pleistocene (around 1.8 million to 10 000 years BP) and Recent, or Holocene (10 000 years BP to the present), epochs, this snapshot of time reveals a major extinction event that took place at the terminal Pleistocene, within the time radiocarbon dating is effective (since 50 000 years BP). Moreover, the extinction affected primarily mammal species whose ecological requirements can be inferred with a clarity unattainable in earlier geologic epochs. Hence, both the speed of the extinction and likely reasons—climate change and the entry of human predators on the scene—are known in some detail (Martin and Klein, 1984; Barnosky, 1986; Owen-Smith, 1987).

Because of the well-dated time-scale and the possible role of humans, Pleistocene extinctions are commonly regarded as somehow atypical compared to previous extinction events. Here I take a different point of view. What can the finer focus of the terminal Pleistocene event, particularly in North America, tell us about the rules that govern extinction? I will attempt to answer this question by first discussing why the Pleistocene event should not be considered unique, then by reviewing what the enhanced detail of the Pleistocene demonstrates about causes of widespread mammal extinction. Finally, I will use the Pleistocene example to derive some 'rules of widespread extinction' that are testable with data from the Tertiary mammal record.

MAGNITUDE OF PLEISTOCENE EXTINCTION

The terminal Pleistocene extinction affected mainly large mammals in North America where 43 genera died. Approximately 91% of these were large-bodied (weighing over 5 kg including 73% of the megafauna (weighing over 44 kg). Also affected was South America (at least 46 extinct genera, including approximately 80% of the megafauna) and Australia (21 extinct genera, including 86% of the megafauna) (Webb, 1984; Martin, 1984a; Murray, 1984). Europe, Africa, and probably Asia endured only a few true extinctions (but many extirpations). Besides mammals, at least 19 genera of birds also vanished in North America (Steadman and Martin, 1984). In addition, a myriad of taxa survived. Nevertheless, many of these survivors, both animals and plants, responded to the extinction-causing event by markedly changing their geographic range (Guilday, 1984) and reassembling into new communities (Graham and Lundelius, 1984; Guthrie, 1984a; Owen-Smith, 1987; Webb *et al.*, 1987). In these ways terminal Pleistocene extinction affected much of the world's terrestrial biota. Marine extinctions, however, were insignificant (Stanley, 1984).

Thus, it seems clear that in general magnitude, the terminal Pleistocene extinction was not as severe as, say, the end Cretaceous extinction, when entire classes of vertebrates and many marine invertebrates breathed their last (Martin, 1984b; Hallam, 1987). Also, it follows that the terminal Pleistocene event was different in kind or magnitude from the events that caused numerous extinction episodes in the marine realm.

MAMMAL EXTINCTION

A cause for confusion, however, is how the Pleistocene event relates to other episodes of mammal extinction. Faunal turnover, that is, the sum of extinction plus origination (by immigration and evolution) has resulted in at least 18 temporally superposed, distinct mammal faunas in North America since the beginning of the Paleocene. These faunas form the basis for dividing the Cenozoic into the North American Land Mammal Ages (NALMAs), which themselves can be further subdivided on the basis of faunal turnover (Woodburne, 1987).

Within these faunal turnovers are periods of extinction when at least 30 genera of North American mammals disappeared from earth within a time-span of at most 3 million years. (The actual time spanned by extinction might be shorter, but in most cases cannot be precisely resolved.) Such extinction episodes I arbitrarily call 'widespread mammal extinction'. They affect entire continents, if not the entire world. Besides the Pleistocene extinction, widespread mammal extinction took place at least during the late Uintan (late Eocene, 42–44 million years BP), possibly Barstovian (middle Miocene, 11.5–16.5 million years BP), Clarendonian (late-middle Miocene, 9–11 million years BP), late Hemphillian (late Miocene, 4.5–6 million years BP), and late Blancan (late Pliocene, 1.8–3 million years BP) NALMAs (Webb, 1984; Stucky, 1989). Does the Pleistocene extinction differ from these enough to require a special explanation? Or does it express general principles that explain how and why widespread mammalian extinction takes place?

Rate of faunal turnover

One feature that makes the Pleistocene episode look special at face value is the extinction rate. Extinction rates, calculated as genera becoming extinct per million years, show that the Pleistocene rate of extinction approached 200, whereas Tertiary rates never exceeded 50 (Gingerich, 1984; 1987). But this is a misleading way to calculate rates. First, the shorter the time interval over which the rate is averaged, the faster the rate appears (Gingerich, 1987, p. 1059). Pleistocene rates are generally averaged over 1 million years or less, whereas other rates are averaged over more than 1 million years. Second, generic diversity increases through geologic time, possibly exponentially (Gingerich, 1987, p. 1055). Whether this represents a real increase in the diversity of life or simply reflects better representation of geologically young strata is unclear, but in either case the effect on calculating turnover rates is the same. Pleistocene rates are artificially inflated simply because there are more Pleistocene genera known. To avoid this sample bias, Gingerich (1984; 1987) calculated extinction rates as genera extinct per total genera available for extinction in a given time interval (Figure 12.1(a); redrawn from Gingerich, 1984, Figure 10.2(b)). By this method, Pleistocene extinctions are not any more pronounced than those that occurred at the following times: early Eocene, middle Eocene, late Eocene, early Oligocene, middle Oligocene, late Oligocene, early Miocene, 'Pliocene', and late Pliocene (*sensu* Gingerich, 1984; 1987; whose data are from Romer, 1966; his early 'Pliocene' is late Miocene in modern usage). In fact Pleistocene extinctions are about average for those that occur throughout the Cenozoic.

Extinction rates cannot be considered independent of origination rates. When rates are calculated as genera extinct (or originating) per total genera for given time intervals, Gingerich's analyses suggest that origination rates peaked at the same time as extinction rates peaked for all Tertiary extinction episodes (Figure 12.1(a)). In the middle Palaeocene, origination rate far exceeded extinction rate, but there was a progressive decline in the difference between the two until the middle Oligocene, when the two rates nearly met. Thereafter in the Tertiary, origination peaks continued to correspond with extinction peaks, but were only slightly

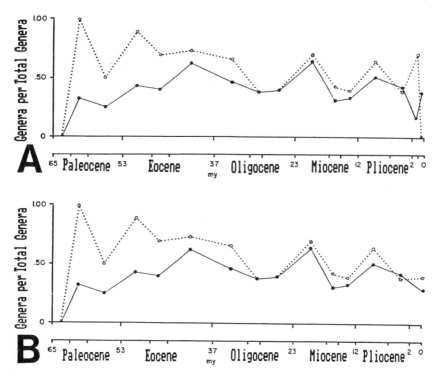

Figure 12.1. Origination (dotted line) and extinction (solid line) rates for genera of rodents, artiodactyls and fissiped carnivores. (a) Pleistocene considered as two 1-million-year intervals. (b) Pleistocene considered as one 2-million-year interval to make it more comparable to pre-Pleistocene intervals. Data are from Gingerich (1984, p. 214), who compiled it from Romer (1966); hence the time-scale is out of date. For example, Romer's 'early Pliocene' is now considered late Miocene. However, as Gingerich (1984; 1987) noted, this makes little difference for recognizing general trends.

higher. At the beginning of the Pleistocene, however, extinction rates dropped to an all-time low for the Cenozoic, at the same time as origination rates increased dramatically.

The origination rate in the Pleistocene was fed largely by immigration. Immigration rates for the Blancan, Irvingtonian and Rancholabrean NALMAs exceeded the average rate for any earlier Neogene immigration episode (Webb, 1984; Webb and Barnosky, 1989). Rapid evolution also took place in some groups, for example, the arvicoline rodents (Repenning, 1987).

Viewed as genera extinct or originating per total genera, the Pleistocene faunal turnover was special not for its extinction rate, which did not even exceed average values, but for its high origination rates, which increased faunal diversity to exceptionally high levels (Gingerich, 1984; 1987). So-called disharmonious communities, which mix extant taxa (both animals and plants) that today are allopatric, bolster the view of at least locally increased biotic diversity during the Pleistocene (Graham and Lundelius, 1984; Graham and Mead, 1987; Webb *et al.*, 1987). If the exponential increase in number of genera noted by Gingerich (1987) is not a sampling artefact, global diversity may have increased as well. The landscape may have been able to support increased diversity because of the 'fragmentation and diversification of habitats accompanying successive Pleistocene glaciations', which increased spatial heterogeneity (Gingerich, 1984, p. 16).

Was the lag between Pleistocene originations and extinctions special at all? Figure 12.1(a) averages all of the Tertiary extinction and origination peaks over time-intervals of 2 million years in the best case (middle Oligocene) and 8 million years in the worst case (middle Eocene). Most of the Tertiary time intervals range from 3 to 6 million years in duration (data from Gingerich, 1984, Table 10.3). In contrast, the Pleistocene peaks and valleys in Figure 12.1(a) are based on 1-million-year time intervals. When the two Pleistocene intervals are lumped together to make them more comparable to the Tertiary intervals, the lag between the immigration peak and extinction peak disappears (Fig. 12.1(b)). The lag-time between Pleistocene origination and extinction events in Figure 12.1(a) simply reflects the better time resolution for Pleistocene events.

This recognition allows us to look at the Pleistocene origination–extinction lag in a different way. Perhaps the lag is not unique to the Pleistocene, but a trade mark of all widespread extinctions made evident through the enhanced resolving power of the Pleistocene microscope. Stucky (1989) presented an analysis that provides support for this viewpoint. He plotted raw numbers of North American genera originating and disappearing through the Cenozoic at intervals ranging from less than 1 million years in best cases to about 5 million years in worst cases, with most intervals being between 0.75 million and 2 million years. Thereby he recognized four pre-Pleistocene extinction peaks when generic extinctions per interval exceeded 30: late Uintan (late Eocene), Barstovian (middle Miocene), late Hemphillian (late Miocene), and late Blancan (late Pliocene). Peaks of origination preceded peaks of extinction by at least 1 million years for the three Neogene extinctions, but the two peaks appear contemporaneous for the late Eocene event. The Eocene peaks merit more detailed examination because they are based on a time interval of 2 million years; even the Pleistocene peaks appear contemporaneous at this interval.

Body size

The late Pleistocene extinction has been considered notable for the decimation mainly of the large-mammal fauna (Martin, 1984a; 1984b). By Webb's (1984) reckoning, 43 genera disappeared in North America. Thirty-nine of these, or 91%, were large mammals characterized by body-

Table 12.1. *Late Neogene extinction episodes. Data are from Webb (1984).*

Episode	Time (millions of years BP)	Total genera extinct	Percentage of large-body genera (> 5 kg)
Terminal Pleistocene	0.01 – 0.4	43	91
Late Irvingtonian	0.4 – 1.0	9	56
Early Irvingtonian	1.0 – 1.8	10	50
Late Blancan	1.8 – 3.0	35	57
Early Blancan	3.0 – 4.5	9	22
Late Hemphillian	4.5 – 6.0	62	56
Early Hemphillian	6.0 – 9.0	27	78
Clarendonian	9.0 –11.0	39	62

weight exceeding 5 kg. Thirty-three, or 77%, had body weights exceeding 44 kg (Martin, 1984a). Is such a bias towards large animals common in terrestrial extinctions? Webb (1984, p. 192) provided data to test this question for eight mammalian extinction episodes (including the terminal Pleistocene one) in the late Neogene (Table 12.1). Chi-squared comparisons verify that only the late Irvingtonian and early Hemphillian extinctions did not statistically differ from the terminal Pleistocene one in showing a less extreme bias towards taxa with large bodies. However, a plot of number of extinct genera versus percentage of extinct genera with large bodies for each extinction episode reveals an interesting relationship (Fig. 12.2(a)). The more genera that became extinct, the higher the proportion of large-bodied forms that contributed to the extinction count. The terminal Pleistocene event agrees well with this generalization; removal of the Pleistocene event from the plot actually decreases the correlation coefficient (Figure 12.2(b)).

The positive correlation between body size and severity of the cause of extinction (measured by the number of genera that died) probably results from an observation articulated by Guilday (1984, p. 256):

Large mammals are inherently more vulnerable to environmental changes simply because they are large and require a greater expanse of primary habitat to sustain themselves because of greater individual demands for food or space to play out their reproductive and defensive strategies, cover, flight, herding, etc. Greater demands are placed upon the habitat by an elephant than by a small rodent . . .

In addition, large mammals are generally characterized by small populations and small numbers of species within a genus, whereas most small mammals are characterized by large populations and several species per genus. It is statistically easier to destroy the few individuals that compose an elephant population than it is to destroy the many that constitute a

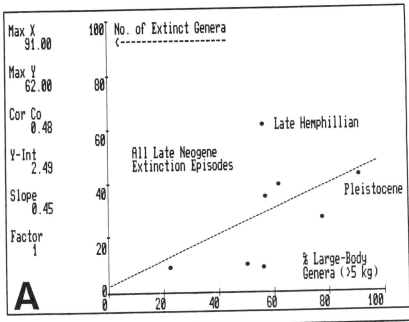

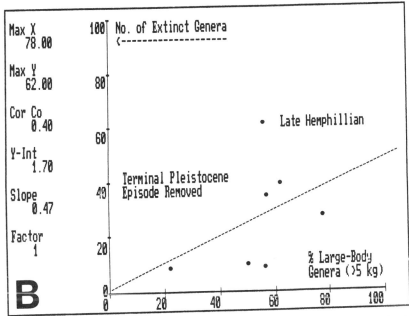

Figure 12.2. Simple linear regression showing the relationship between body size and total number of genera affected by extinction episodes. Data are from Webb (1984) and are presented in Table 12.1. (a) Plot for all late Neogene extinction episodes. (b) Plot for all late Neogene extinction episodes excluding the terminal Pleistocene event. Note that excluding the Pleistocene data has little effect on the correlation coefficient or slope.

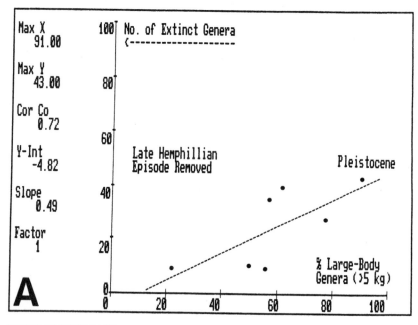

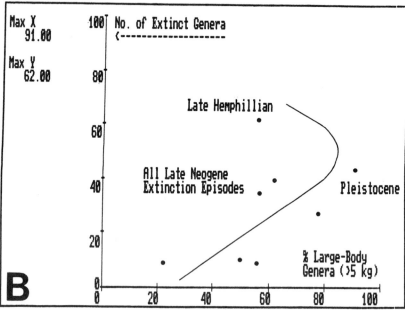

Figure 12.3. (a) as Figure 12.2(a), except the late Hemphillian data point is excluded from the analysis; note that this increases the correlation coefficient considerably, but has little effect on the slope. (b) Hypothesized relationship between body size and total number of genera affected by extinction episodes. The relationship may be linear up to a critical value of about 45 extinct genera, at which point the fauna will be so depleted of large-bodied genera that the proportion of small-bodied genera will begin to rise. Data are from Webb (1984) and are presented in Table 12.1.

mouse population. Similarly, it is easier to wipe out a genus made up of few species than one that includes many species.

If any Neogene extinction needs special explanation in terms of body size, it is the late Hemphillian one. Removal of the late Hemphillian extinction from Figure 12.2(a) increases the correlation coefficient, as illustrated in Figure 12.3(a), which suggests that the late Hemphillian event resulted in 'too many' small mammals becoming extinct in comparison to the other late Neogene events. This probably reflects the fact that large-mammal genera are much less numerous than small-mammal genera, so theoretically there should come a point at which extinction affects so many taxa that most large mammals will be gone from the pool of genera on which extinction can act. When the number of extinct genera reaches this critical value, the proportion of small-bodied forms that become extinct necessarily begins to rise. Thus, the late Hemphillian event may not be simply an 'outlier' on a linear regression; instead, it may give a clue that the critical value at which the relationship becomes non-linear is near 45 extinct genera (Fig. 12.3(b)).

Timing of extinction

That most Pleistocene extinctions, particularly of large mammals, were concentrated towards the end of the epoch is shown by calculating extinction rates for time intervals within the Pleistocene (Gingerich, 1984; Martin, 1984b). This terminal Pleistocene extinction is the best-dated extinction in the geologic record because it falls within the effective range of radiocarbon dating, which resolves time to plus-or-minus hundreds of years rather than the plus-or-minus thousands, hundreds of thousands or more years typical of other dating techniques. The radiocarbon record leaves little doubt that the terminal Pleistocene extinction took at longest 10 000 years. Mead and Meltzer (1984, p. 447) documented the youngest reliable radiocarbon dates for 23 extinct large-mammal taxa. All of the reliable dates fall between 9000 and 18 000 years BP. Expressed more graphically, 18 000 years ago North America was filled with strange beasts; 9000 years ago they were gone. The extinction was instantaneous on a geological time-scale (but possibly very slow in ecological time). Does such a geologically rapid pulse of dying typify most widespread mammalian extinctions?

As Webb (1984, p. 190) has noted, our arguments for answering 'yes' or 'no' are built on equally shaky ground. With increasing age of the sedimentary record, the ability to differentiate short intervals of time decreases, both because the sedimentary record becomes less complete and because high-resolution radiometric dating techniques are lacking. Either we assume the extinctions of each taxon were distributed through the whole stratigraphic interval under consideration (for example, through a whole land-mammal age), or we assume that all the extinctions were concentrated in a certain shorter time within the interval, commonly at its stratigraphic top. The only assumption that has been supported objectively —by the Pleistocene record—is the latter one. Therefore, the most reasonable working hypothesis is 'Most widespread extinctions were rapid, like the terminal Pleistocene event', rather than 'The Pleistocene extinction was unique in being rapid'.

Cause of extinction

A purported cause must fulfil three criteria before we can reasonably infer that it triggered an extinction (Grayson, 1984, p. 816). First, the suspected cause and the extinction must demonstrably coincide in time. Second, the magnitude of the cause should match the magnitude of the extinction. And third, a mechanism must be provided to show how the purported cause would result in the extinction.

For the terminal Pleistocene extinction, all three of these criteria have been more fully attained than for any other extinction event (Martin and Klein, 1984). In North America the timing of extinction coincided with two major perturbations to existing biotic systems: climate change and the arrival of humans. Convincing mechanisms have been advanced to argue that either event by itself could have decimated the Pleistocene fauna (Mossiman and Martin, 1975; Martin, 1984a; 1984b; Graham and Lundelius, 1984; Guthrie, 1984a; Barnosky, 1986; Owen-Smith, 1987). Indeed, current debates focus not on if Pleistocene extinctions resulted from these causes, but on which of the two was most important.

Climate change

On both local (King and Saunders, 1984; Barnosky, 1986) and regional scales, climate change can be shown to correlate with terminal Pleistocene extinctions in numerous parts of the world. Perhaps nowhere is the regional relationship better documented than in the mid-Appalachian Mountain area (Guilday, 1984) and northcentral Great Plains of the United States (Wendlund *et al.*, 1987).

In the mid-Appalachian region, which includes much of Pennsylvania, West Virginia, western Virginia, western Maryland, and parts of Ohio, Kentucky, and Tennessee between latitudes 36° and 45°N, a regional change in climate is indicated by palynological data (Gaudreau and Webb, 1985; Watts, 1979, 1983; Delcourt and Delcourt, 1981; 1985). From about 18 000 years BP to 10 500 years BP, the northernmost vegetation of this area resembled a periglacial tundra, and the remainder a parkland composed of *Picea, Pinus banksiana, Abies, Betula* and an understorey of woody shrubs, grasses, sedges and herbs. Although the vegetational composition was without modern analogue, in general it indicates boreal conditions. By 10 000 years BP regional vegetation began to shift from open coniferous forest to the present closed-canopy deciduous forest, indicating postglacial warming. It was during this same time that the fauna changed drastically:

Eighteen large and one small species of mammal [became] biologically extinct . . . three large and ten small species [were extirpated]. Of those mammal species that still occur in the area, four are rare and local boreal relicts; nine others have become less common or have undergone some measure of ecological readjustment expressed in range reductions. Nine mammals have undergone a Holocene size reduction, while four have increased in size within the area during the last 11,000 years, paralleling, for the most part, modern latitudinal size clines (Guilday, 1984, p. 254).

Similar kinds of faunal changes also took place in the northcentral Great Plains. In Iowa between around 18 000 and 10 500 years BP the fauna was composed of about 70% boreal species and 20% steppe or deciduous species. Near 10 500 years BP these percentages switched to less than 30% boreal species and more than 50% steppe or deciduous species, and remained basically that way until the present (Wendlund *et al.*, 1987, p. 461). In Illinois and Missouri, the change was less drastic, but still evident, from about 15% boreal species between 16 000 and 13 000 years BP to less than 5% boreal species by 11 000 years BP. Extinct taxa are commonly found in the faunas of boreal aspect, but not in the more temperate ones.

Evidence that these biotic responses were linked to climate change is provided by comparing them to climate model simulations. Webb *et al.* (1987) showed how fossil-pollen percentages relate to climate variables by using response surfaces, and noted a general agreement between palaeo-climate inferences derived from fossil-pollen data and climate model simulations. For fossil vertebrates in the northcentral Great Plains, the terminal Pleistocene event corresponded both in time and kind with predictions arising from climate-model experiments (Graham and Mead, 1987; Wright, 1987; Kutzbach and Wright, 1985). Model results suggest that between 18 000 and 12 000 years BP, the westerly jet stream split around the Laurentide ice-sheet, deflecting storm tracks to the south of their present position (Kutzbach, 1987). A glacial anticyclone centred over the ice-sheet and caused surface easterlies to blow south of the ice. As air descended down the ice it was adiabatically warmed. In the northcentral United States, adiabatic warming and the constraints imposed by the Milankovitch cycles would result in less extreme differences between summer and winter during late-glacial time as compared to the present, even though mean annual temperature would have been cooler. In contrast, by about 9000 years BP, further retreat of the ice-sheet and changes in isolation associated with the Milankovitch cycles increased seasonality, so that summers were warmer and winters were cooler than today.

The biological consequences that one might expect of such a change in seasonality parallel closely what is observed in the fossil record. With more seasonal climates, communities should reassemble as animals limited by winter temperatures withdrew to the south, and those limited by hot summer temperatures withdrew to the north. This pattern is evident in the destruction of the late Pleistocene disharmonious mammal assemblages, which were characterized by sympatry of taxa that are allopatric today

(Graham and Lundelius, 1984; Graham and Mead, 1987). Changes in vegetational patterns—from no-analogue assemblages to modern compositions (Webb *et al.*, 1987)—accompanied destruction of the disharmonious faunas. In general, the large herbivores that survived were ruminants, whose digestive systems are adapted to feeding in habitats with low vegetational diversity, and to digesting toxins in their multiple stomachs: bison, deer, moose, sheep and the like. They were able to follow their preferred food plants, even though the plant communities were restructured. In contrast, herbivores whose monogastric digestive systems required them to eat a wide variety of vegetation and did not tolerate toxins became extinct or were extirpated over wide regions: mammoths, mastodonts, horses, camels, sloths and peccaries (Guthrie, 1984a). Such taxa required vegetational mosaics, where many different kinds of food plants can be found within the distance they normally forage. Extinction was also the fate of herbivores whose physiology was tightly linked with the 'old' seasonality pattern, for example cervids with extremely large antlers such as *Cervalces* (Guthrie, 1984b). The disappearance of large herbivores led to the extinction of large carnivores and scavengers dependent on them. Disappearance of certain taxa, particularly proboscideans, may have disrupted intricate grazing successions, further altering community compositions and thus contributing to extirpation and extinction (Graham and Lundelius, 1984; Owen-Smith, 1987).

In short, communities that had evolved to suit a particular set of climatic conditions fell apart when the conditions changed. This model of extinction has been dubbed 'coevolutionary disequilibrium' by Graham and Lundelius (1984), who discuss it in detail.

Overkill

In North America, an important element in the restructuring of terminal Pleistocene communities was the addition of a new predator: *Homo sapiens*. Humans with an efficient hunting technology arrived in North America around 12 000 years ago, as indicated by the earliest dates on sites with Clovis-style artefacts (Haynes, 1984). It is probably no coincidence that this date corresponds so closely with the glacial–interglacial transition, as dispersal through Beringia would have required an ice-free corridor between the Cordilleran and Laurentide ice-sheets. Hunting by humans is thought to have contributed to Pleistocene extinction in North America because human arrival correlates closely in time with extinction (Martin, 1984a); predominantly large herbivores—animals most likely to be hunted—succumbed (Martin, 1984a; 1984b); Shasta ground sloths (*Nothrotheriops shastensis*) became extinct, even though their supposed habitat remained (Martin, 1984b); artefacts have been found associated with five of the 37 genera of large vertebrates that disappeared (mammoths, mastodonts, horses, camels and giant tortoises) (Webb and Barnosky, 1989); and world-wide extinction of the megafauna may be diachronous, with extinction time roughly correlating with the first appearance of humans (Martin,

1984a). The case for diachronous extinction rests mainly on the observations that in Africa and Europe, where megafauna had coevolved with humans, late Pleistocene extinctions were minimal, and that human predation was responsible for the decimation of the New Zealand/Madagascar fauna between around 0.5 and 1 Ka. Despite the reasoning of Martin (1984a, pp. 375, 395), it remains to be demonstrated by radiocarbon dates that extinction in South America and Australia coincided with the arrival of humans (Grayson, 1984, p. 808). Nevertheless, the North American record, coupled with Diamond's (1984) observations of the naivety of large-game species unfamiliar with human predators, supports the view that human activity played a role in terminal Pleistocene extinctions.

The components of extinction

It seems clear that both climate change and humans contributed to late Pleistocene extinctions in North America. Whether 'one set up the punch which the other delivered . . . or vice versa' may be unknowable (Guilday, 1984, p. 257), but the data do suggest that both physical and biological stresses must act together to cause extinction. The physical stress in the case of the late Pleistocene was a rapid change in climate. The biological stress was a spatial reorganization of the biota as species responded individually to changing climate and changing biogeographic ranges of competitors and predators. The addition of human predators was but one facet of this community reorganization, and in this respect can be accommodated within Graham and Lundelius's (1984) model of coevolutionary disequilbrium.

How strong must the physical and biological punches be, and how close together must they hit, to cause extinction? Is one more important than the other? A clue is provided by comparing the timing of immigration events, glacial–interglacial transitions and extinction episodes of the last million years. Three immigration events, one each at approximately 850 000, 400 000 and 150 000 years BP, are recognizable mainly by the record of arvicoline rodents (Repenning, 1987). The first of these three events, at 850 000 years BP, falls near the boundary of oxygen-isotope warm stage 23 and cold stage 22 (Repenning, 1984). The second event, at 400 000 years BP, correlates roughly with a glacial–interglacial transition of oxygen-isotope stage 12 to stage 11 (Repenning, 1984), but may be as much as 40 000 years later (Bradley, 1985, p. 187) or earlier (Repenning, 1984, p. 106) than the actual transition. The third event, at 150 000 years BP, took place in the middle of a glaciation signalled by oxygen-isotope stage 6. None of these events coincides with a rise in extinction rate (Gingerich, 1984, p. 220). The two earlier events thus illustrate that immigration plus a major climate change does not always equal extinction, so it is no surprise that the third event shows that immigration by itself does not cause widespread extinction. Moreover, the absence of widespread extinction at any of the pre-Wisconsinan glacial–interglacial transitions suggests that pronounced climate change alone is an insufficient stimulus.

The extinction equation thus seems to require not only change in the physical environment *plus* appearance of new players on the biological scene, but that at least one of these components be of a certain critical magnitude. Recognizing the critical magnitude depends on identifying how the terminal Pleistocene changes differed from other glacial–interglacial changeovers.

The glacial–interglacial cycles are driven by orbitally induced Milankovitch cycles: the eccentricity of the Earth's orbit around the Sun (periodicity of 100 000 years), the tilt of the earth's axis relative to the ecliptic (periodicity of 41 000 years), and the season of perihelion, which is the Earth's closest approach to the Sun (periodicity of 23 000 years). The three different periodicities are superimposed upon one another such that there is constant change through time in the relative relationships of Earth's distance from the Sun, its tilt, and timing of perihelion. Their relationship affects the latitudinal and seasonal distribution of solar radiation, which in turn modulates atmospheric and oceanic circulation, ice-sheet growth and levels of CO_2. Bartlein and Prentice (1989) speculated that species evolve to withstand the environmental perturbations caused by frequently recurring combinations of the three cycles, for example, combinations that in the late Pleistocene forced alternation between glacial and interglacial conditions about every 75 000–100 000 years (Hays *et al.*, 1976). Hence, climate changes of this magnitude should not result in extinction, which is in agreement with the pre-Wisconsinan Pleistocene record. However, infrequent combinations of the cycles, perhaps superimposed on climate change on different time-scales such as that triggered by epeirogeny (Raymo *et al.*, 1988), might result in widespread extinction. The transition from the end of the Pleistocene into the present interglacial exemplifies an infrequent combination of Milankovitch forcings that caused particularly extreme climate change. Rather than being typical of glacial–interglacial transitions, the climatic differences between the Wisconsinan and Holocene represent nearly the maximum change that would be expected in insolation and ice-volume (Bartlein and Prentice, 1989). The resulting hypothesis is that, in order to contribute to a widespread extinction, the magnitude of climatic change does not have to increase very much over what is normally produced by Milankovitch-based orbital variations on about the 100 000-year time-scale, but that it does have to increase a little. If infrequently occurring combinations of the three Milankovitch cycles are an important trigger of extinction, such combinations should be found to have a periodicity longer than the average life-span of mammalian species, which is a little over 1 million years (Stanley, 1978, p. 30).

In North America, we can also implicate arrival of a single species—humans—in contributing to extinction. Numerous Pleistocene immigration events prior to the Wisconsinan are unassociated with extinction. The biological input to extinction therefore depends as much (or more) on the kind of taxa inserted into the fauna as on the sheer numbers of new taxa. Immigrants such as efficient predators, large herbivores and pathogens, which interact with a large proportion of the native fauna, are more likely to contribute to extinction than are those that simply compete with single or small groups of species, for example, most small herbivores. However, with increasing numbers of immigrants, the chance that one of the immigrants will be a significant taxon increases.

A PARADIGM FOR WIDESPREAD MAMMAL EXTINCTION

If the Pleistocene extinction was typical of widespread mammal extinctions, its enhanced detail and time-resolution may spotlight some rules that glimmer so faintly as to be overlooked in the more distant geologic past. These rules, derived from the Pleistocene, offer a general paradigm for how the process of widespread mammal extinction might work, but a paradigm is useful only in so far as it can be tested. Tests that arise from each of the rules are therefore articulated as well.

Rule 1: *There is a time-lag between a period of high origination rates (immigration plus evolution) and high extinction rates.* The origination event must precede the extinction event in order to inflate faunal diversity nearly to oversaturation, and stimulate production of complexly coevolved communities.

Test: When calculated over refined time-intervals, pre-Pleistocene peaks in origination rates should precede peaks in extinction rates. Generally, intervals of around 1 million years should be fine enough, because that interval makes the Pleistocene lag evident.

Rule 2. If Rule 1 is fulfilled, *widespread extinction will result when climate changes by a critical amount* and *one or more critical taxa arrive in the fauna.* By itself, neither climate change nor the new taxa will act fast enough to cull the fauna on a continental or world-wide scale. If only climate change or arrival of new taxa takes place, extinctions may occur, but they should be fewer in number than what characterizes widespread extinction (at least 30 genera on one continent). In the case of only climate change, there should be enough time and spatial opportunity for most faunal elements to migrate with suitable habitats. In the face of only new taxa, the biotic system should accommodate the intruders either by niche partitioning or extinction of only direct competitors.

Test: Within a short stratigraphic interval, many mammal taxa should disappear, there should be changes in depositional environment that would be consistent with the purported climate change and critical taxa should first appear (see Rule 4 for more discussion on critical taxa).

Rule 3. *The magnitude of climate change must increase over that typically caused by Milankovitch orbital variations on the 100000-year time-scale.* However, the increase may be small and result from infrequently occurring combinations of the three cycles, as well as from superimposition of

Milankovitch cycles on longer-term causes of climatic change (Bartlein and Prentice, 1989).

Test: As detailed climate models become available for pre-Pleistocene segments of the geologic time-scale, the predicted timing and results of 'unusual' Milankovitch combinations can be compared with timing of widespread mammal extinction. Likewise, the timing of non-Milankovitch-based climate change can be examined. The model predictions and the extinctions should coincide in time.

Rule 4. *The new taxa that contribute to extinctions must be of a kind or number that interacts with a wide variety of the native fauna.* The greater the magnitude of an immigration event or evolutionary radiation, the greater the probability that this will happen. But even arrival of a single taxon is sufficient if it is the right taxon.

Test: Taxa whose first appearance seems to coincide with an extinction event should include those that would influence a wide spectrum of the palaeocommunity, for example, more efficient predators, or large herbivores that substantially impact vegetational composition and distribution (like proboscideans or large herds of bison).

Rule 5. *The main mechanism of widespread extinction is coevolutionary disequilbrium.* Each species responds to climate change and newly-arrived taxa individually, leading to the dissolution of previously tightly coevolved communities. Some taxa become extinct because they were affected directly by a new climate, predator or competitor; others because they were dependent on linked components within no-longer existing communities.

Test: Some species that survive the extinction should show a substantially altered geographic range. Some that were sympatric before the event should become allopatric after it, or *vice versa*. The species most affected by extinction should be those most dependent on complex communities, such as herbivores that might fit into grazing successions or require a wide variety of plants in a small area.

Rule 6. *Widespread extinction is concentrated in a short time interval, compared to intervals between widespread extinction episodes.* This follows from Rules 2–5, which when summed mean that widespread extinction takes place when the amplitude of environmental oscillations (climatic plus biological) becomes higher than the amplitudes a biotic system has evolved to withstand. At this critical threshold, the rules of survival change fast; species either learn them quickly, or they are out of the game.

Test: Most disappearances of taxa should occur in the upper part of a stratigraphic unit that records an extinction event, rather than being more or less evenly distributed throughout the vertical extent of the unit.

Rule 7. *The proportion of large-bodied genera that become extinct is a function primarily of magnitude of extinction.* The proportion increases linearly in a predictable way when plotted against total number of genera wiped out by an extinction event, up to about 45 extinct genera. Above 45 extinct genera, the proportion of small-bodied genera should begin to increase.

Test: For a given extinction episode, the proportion of extinct large-bodied genera to total extinct genera should not differ significantly from the relationships indicated in Figures 12.2(a) and 12.3(b).

The proof, of course, lies in putting the rules to the tests. Should the rules hold fast against the weight of pre-Pleistocene evidence, we gain confidence that nature has a motor of extinction, just as she has a motor of evolution. Should the tests result in demoting the 'rules' to 'special explanations', we learn that different extinctions need not be bound by a common process; each may be just a chance combination of unlucky events. The road between these extremes will be marked by other insights into extinction, waiting to be discovered. Whatever the outcome of testing the paradigm, we stand to gain in our quest to understand the process of extinction.

ACKNOWLEDGEMENTS

Discussions with C. Barnosky, L. Krishtalka, K. Pfaff and R. Stucky, as well as their comments on the manuscript, were helpful in formulating my ideas. This chapter is an outgrowth of research supported by NSF Grant EAR-8615373 and Graham Netting Research Grants of the Carnegie Museum of Natural History.

REFERENCES

Barnosky, A.D., 1986, 'Big game' extinction caused by climatic change: Irish elk (*Megaloceros giganteus*) in Ireland, *Quaternary Research*, **25** (1): 128–35.
Bartlein, P.J. and Prentice, I.C., 1989, Orbital variations, climate, and palaeoecology, *Trends in Ecology and Evolution*.
Bradley, R.S., 1985, *Quaternary paleoclimatology*, Allen and Unwin, Winchester, MA.
Delcourt, H.R. and Delcourt, P.A., 1985, Quaternary palynology and vegetational history of the southeastern United States. In V.M. Bryant, Jr and R.G. Holloway (eds), *Pollen records of late Quaternary North American sediments*, American Association of Stratigraphic Palynologists Foundation, Dallas, TX: 1–38.
Delcourt, P.A. and Delcourt, H.R., 1981, Vegetation maps for eastern North America: 40,000 yrs BP to the present. In R.C. Romans (ed.), *Geobotany II*, Plenum Press, New York: 123–65.
Diamond, J.M., 1984, Historic extinction: a Rosetta Stone for understanding prehistoric extinctions. In P.S. Martin and R.G. Klein (eds), *Quaternary Extinctions*, University of Arizona Press, Tucson: 824–66.
Gaudreau, D.C. and Webb, T., III, 1985, Late-Quaternary pollen stratigraphy and isochrone maps for the northeastern United States. In V.M. Bryant, Jr and R.G. Holloway (eds), *Pollen records of late Quaternary North American sediments*, American Association of Stratigraphic Palynologists Foundation, Dallas, TX: 245–80.

Gingerich, P.D., 1984, Pleistocene extinctions in the context of origination-extinction equilibria. In P.S. Martin and R.G. Klein (eds), *Quaternary extinctions*, University of Arizona Press, Tucson: 211–22.

Gingerich, P.D., 1987, Evolution and the fossil record: patterns, rates, and processes, *Canadian Journal of Zoology*, **65** (5): 1053–60.

Graham, R.W. and Lundelius, E.L., Jr, 1984, Coevolutionary disequilibrium and Pleistocene extinctions. In P.S. Martin and R.G. Klein (eds), *Quaternary extinctions*, University of Arizona Press, Tucson: 223–49.

Graham, R.W. and Mead, J.I., 1987, Environmental fluctuations and evolution of mammalian faunas during the last deglaciation in North America. In W.F. Ruddiman and H.E. Wright, Jr (eds), *North America and adjacent oceans during the last deglaciation: the geology of North America K-3*, Geological Society of America, Boulder, CO: 371–402.

Grayson, D.K., 1984, Explaining Pleistocene extinctions: thoughts on the structure of a debate. In P.S. Martin and R.G. Klein (eds), *Quaternary extinctions*, University of Arizona Press, Tucson: 807–23.

Guilday, J.E., 1984, Pleistocene extinction and environmental change, case study of the Appalachians. In P.S. Martin and R.G. Klein (eds), *Quaternary extinctions*, University of Arizona Press, Tucson: 250–8.

Guthrie, R.D., 1984a, Mosaics, allelochemics, and nutrients: an ecological theory of late Pleistocene megafaunal extinctions. In P.S. Martin and R.G. Klein (eds), *Quaternary extinctions*, University of Arizona Press, Tucson: 259–98.

Guthrie, R.D. 1984b, Alaskan megabucks, megabulls, and megarams: the issue of Pleistocene gigantism. In H.H. Genoways and M.R. Dawson (eds), *Contributions in Quaternary vertebrate paleontology: a volume in memorial to John E. Guilday, Carnegie Museum of Natural History Special Publication*, **8**: 482–510.

Hallam, A., 1987, End-Cretaceous mass extinction event: argument for terrestrial causation, *Science*, **238** (4831): 1237–42.

Haynes, C.V., 1984, Stratigraphy and late Pleistocene extinction in the United States. In P.S. Martin and R.G. Klein, (eds), *Quaternary extinctions*, University of Arizona Press, Tucson: 345–54.

Hays, J.D., Imbrie, J. and Shackleton, N.J., 1976, Variation in the Earth's orbit: pacemaker of the ice ages, *Science*, **194** (4270): 1121–32.

King, J.E. and Saunders, J.J., 1984, Environmental insularity and the extinction of the American mastodon. In P.S. Martin and R.G. Klein (eds), *Quaternary extinctions*, University of Arizona Press, Tucson: 315–44.

Kutzbach, J.E., 1987, Model simulations of the climatic patterns during the deglaciation of North America. In W.F. Ruddiman and H.E. Wright, Jr (eds), *North America and adjacent oceans during the last deglaciation: the geology of North America K-3*, Geological Society of America, Boulder, CO: 425–46.

Kutzbach, J.E. and Wright, H.E., Jr, 1985, Simulation of the climate of 18,000 yr BP; results for the North Atlantic/European sector and comparison with the geologic record of North America, *Quaternary Science Reviews*, **4** (2): 147–87.

Martin, P.S., 1984a, Prehistoric overkill: the global model. In P.S. Martin and R.G. Klein (eds), *Quaternary extinctions*, University of Arizona Press, Tucson: 354–403.

Martin, P.S., 1984b, Catastrophic extinctions and late Pleistocene blitzkreig: two radiocarbon tests. In M.H. Nitecki (ed.), *Extinctions*, University of Chicago Press, Chicago: 153–90.

Martin, P.S. and Klein, R.G. (eds), 1984, *Quaternary extinctions*, University of Arizona Press, Tucson.

Mead, J.I. and Meltzer, D.J., 1984, North American late Quaternary extinctions and the radiocarbon record. In P.S. Martin and R.G. Klein (eds), *Quaternary extinctions*, University of Arizona Press, Tucson: 440–50.

Mossiman, J.E. and Martin, P.S., 1975, Simulating overkill by Paleoindians, *American Scientist*, **63** (3): 304–13.

Murray, P., 1984, Extinctions downunder: a bestiary of extinct Australian late Pleistocene monotremes and marsupials. In P.S. Martin and R.G. Klein (eds), *Quaternary extinctions*, University of Arizona Press, Tucson: 600–28.

Owen-Smith, N., 1987, Pleistocene extinctions: the pivotal role of megaherbivores, *Paleobiology*, **13** (3): 351–62.

Raymo, M.E., Ruddiman, W.F. and Froelich, P.N., 1988, Influence of late Cenozoic mountain building on ocean geochemical cycles, *Geology*, **16** (7): 649–53.

Repenning, C.A., 1984, Quaternary rodent biochronology and its correlation with climatic and magnetic stratigraphies. In W.C. Mahaney (ed.), *Correlation of Quaternary chronologies*, GeoBooks, Norwich: 105–18.

Repenning, C.A., 1987, Biochronology of the microtine rodents of the United States. In M.O. Woodburne (ed.), *Cenozoic mammals of North America, geochronology and biostratigraphy*, University of California Press, Berkeley: 236–78.

Romer, A.S., 1966, *Vertebrate paleontology*, University of Chicago Press, Chicago.

Stanley, S.M., 1978, Chronospecies longevities, the origin of genera, and the punctuational model of evolution, *Paleobiology*, **4** (1): 26–40.

Stanley, S.M., 1984, Marine mass extinctions; a dominant role for temperatures. In M.H. Nitecki (ed.), *Extinctions*, University of Chicago Press, Chicago: 69–118.

Steadman, D.W. and Martin, P.S., 1984, Extinction of birds in the late Pleistocene of North America. In P.S. Martin and R.G. Klein (eds) *Quaternary extinctions*, University of Arizona Press, Tucson; 466–80.

Stucky, R.K., 1989, Evolution of land mammal diversity in North America during the Cenozoic. In H.H. Genoways (ed.), *Current mammalogy 2*, Plenum Press, New York.

Watts, W.A., 1979, Late Quaternary vegetation of central Appalachia and the New Jersey coastal plain, *Ecological Monographs*, **49** (4): 427–69.

Watts, W.A., 1983, Vegetational history of the eastern United States 25,000 to 10,000 years ago. In H.E. Wright, Jr and S.C. Porter (eds), *Late Quaternary environments of the United States, volume 1, the late Pleistocene*, University of Minnesota Press, Minneapolis: 294–310.

Webb, S.D., 1984, Ten million years of mammal extinctions in North America. In P.S. Martin and R.G. Klein (eds), *Quaternary extinctions*, University of Arizona Press, Tucson: 189–210.

Webb, S.D. and Barnosky, A.D., 1989, Faunal dynamics of Pleistocene mammals, *Annual Review of Earth and Planetary Sciences*, **17**.

Webb, T., III, Bartlein, P.J. and Kutzbach, J.E., 1987, Climatic change in eastern North America during the past 18,000 years; comparisons of pollen data with model results. In W.F. Ruddiman and H.E. Wright, (eds), *North America and adjacent oceans during the last deglaciation: the geology of North America K-3*, Geological Society of America, Boulder, CO: 447–62.

Wendlund, W.M., Benn, A. and Semken, H.A., Jr, 1987, Evaluation of climatic changes on the North American Great Plains determined from faunal evidence. In R.W. Graham, H.A. Semken, Jr and M.A. Graham (eds), *Late Quaternary mammalian biogeography and environments of the Great Plains and Prairies, Illinois State Museum Scientific Papers*, **22**: 460–73.

Woodburne, M.O. (ed.), 1987, *Cenozoic mammals of North America, geochronology and biostratigraphy*, University of California Press, Berkeley.

Wright, H.E., Jr, 1987, Synthesis; the land south of the ice sheets. In W.F. Ruddiman and H.E. Wright, Jr (eds), *North America and adjacent oceans during the last deglaciation: the geology of North America K-3*, Geological Society of America, Boulder, CO: 479–88.

SYSTEMATIC INDEX

(Taxa mentioned in figure captions and tables are included).

255

SUBJECT INDEX